D0958130

PERSPECTIVES IN ECOLOGICAL THEORY

PERSPECTIVES IN *Ecological Theory*

Edited by

JONATHAN ROUGHGARDEN, ROBERT M. MAY

and

SIMON A. LEVIN

PRINCETON UNIVERSITY PRESS

Princeton, New Jersey

 Published by Princeton University Press, 41 William Street, Princeton, New Jersey 08540
In the United Kingdom: Princeton University Press, Guildford, Surrey

Library of Congress Cataloging-in-Publication Data
Perspectives in ecological theory / edited by Jonathan Roughgarden, Robert M. May, and Simon A. Levin.
p. cm.
Bibliography: p.
Includes indexes.
ISBN 0-691-08507-2 (alk. paper) ISBN 0-691-08508-0 (pbk.)
1. Ecology. I. Roughgarden, Jonathan. II. May, Robert M. (Robert McCredie), 1936– . III. Levin, Simon A.
QH541.P425 1989
574.5—dc19 88-25378

This book has been composed in Linotron 202 Granjon

Clothbound editions of Princeton University Press books are printed on acid-free paper, and binding materials are chosen for strength and durability. Paperbacks, although satisfactory for personal collections, are not usually suitable for library rebinding

Printed in the United States of America by Princeton University Press, Princeton, New Jersey

Contents

PERSPECTIVES IN ECOLOGICAL THEORY

Introduction

JONATHAN ROUGHGARDEN,
ROBERT M. MAY,
AND SIMON A. LEVIN

Darwin's views of the role of theory in evolutionary biology will warm any theorist's heart: "let theory guide your observations"; "all observation must be for or against some view if it is to be of any service!" (Gruber and Barrett 1974). In common with physical scientists, Darwin saw theoretical ideas as providing the plans that give shape and coherence to facts and observations that otherwise would lie around like a pile of bricks at a building site.

Ecological theory can, of course, take many forms, many of which are not at all mathematical. Darwin was the most influential ecological and evolutionary theorist, despite his self-acknowledged mathematical incompetence (DeBeer 1964, p. 179, writes that "Darwin himself often regretted that he had never proceeded far enough to understand 'something of the great leading principles of mathematics', for, as he sadly admitted, 'men thus endowed seem to have an extra sense' "). Nevertheless, much of ecological theory today is expressed in largely mathematical terms, if only because (in Joel Cohen's phrase) "mathematics is a way of making commonsense precise." Mathematical formulations force us to make the assumptions clear and unambiguous, and the conclusions are correspondingly unambiguous.

Whether or not in mathematical form, theoretical ideas serve many purposes. They suggest observational protocols and manipulative experiments both in the field and in the laboratory. They provide frameworks around which curricula can be organized, so that the body of facts can be given coherence rather than presented as a jumble. Most important, perhaps, theory can imagine and explore a wider range of worlds than the unique one we inhabit, and by so doing can lead to fresh perceptions and new questions about why our actual world came to be as it is (see, for example, Jacob 1982).

Ecological theory today spans a large range of topics, from the physiology and behavior of individuals or groups of organisms, through population dynamics and community ecology, to the ecology of ecosystems and the geochemical cycles of the entire biosphere. Ecological theory also embraces large parts of evolutionary biology, including paleontology and systematics, and of the earth sciences, especially oceanography and tectonics. Most of this work is, however, scattered among the various subdisciplines; very properly, the theorists are most interested in communicating with the appropriate empiricists, be they physiologists, biogeographers, oceanographers, or whatever.

This book attempts to draw together these contributions into a comprehensive statement of what ecological theory has accomplished over the broad range of its inclusive niche, and of what ecological theory can contribute to defining and to solving the fundamental problems currently facing the discipline of ecology. It aims to outline current accomplishments and current questions, across the panorama of ecological theory. More interestingly, however, the book seeks to air thoughts about likely future directions and about areas in which the dialogue between theorists and empiricists may be enlivened in specific ways.

The book had its origins in a meeting held at Asilomar in 1987, which brought together a group of some forty theoretical and empirical ecologists, representative of the broad sweep of subjects just referred to (see back of book for a complete list of participants). Valuable contributions were also made by representatives of several federal agencies and foundations. Eight general topics were identified, and these topics were discussed against the background of precirculated papers. Hapless chairmen—some of whom did not know what they were getting themselves into—then wrote chapters for the book, based on the discussion. Some of these chapters summarize directions for research and other points of interest that emerged in the discussion, while others take these kinds of ideas as background for their authors' more personal speculations about current and likely future directions; we did not think it a good idea to fit these diverse contributions into some Procrustean conformity.

ORGANIZATION OF THE BOOK

The book is divided into eight sections. The first section deals with the interface between individual organisms and the dynamics of populations. Too often, population biologists must be satisfied with phenomenological descriptions of population processes, without benefit of the understanding that comes from deriving these coefficients from the underlying physiology and behavior of individuals. Such an interplay of individual and population properties allows an explicit consideration of coevolutionary processes, which depend upon variation among individuals within populations and the expressed phenotypic properties of individuals. Conversely, organismal biologists and behavioral ecologists tend to focus on how physiology and behavior are shaped by evolution, and usually give little attention to the consequences at the population level, or to the importance of population interactions in guiding evolution.

In this section Gross gives a theoretician's perspective on plant physiology, particularly as it relates to photosynthesis, and pursues some of the consequences for the population biology of plants. Pulliam gives an overview of work on models that aim to understand the foraging behavior of animals, and shows how these models guide, and are guided by, field and laboratory experiments. Koehl summarizes the discussion, indicating likely growth points in this general area.

The second section deals with the dynamics of populations and with interactions among species. The past decade or so has witnessed the development and experimental testing of "second generation" models of intermediate complexity, which aim to go beyond the very broad generalities of earlier models, to take account of the particular features of specific categories of systems. Pacala reviews mathematical models for the dynamics of plant populations, and discusses their relation to existing and proposed future experiments; he also discusses the relation between these models and the more familiar ones that have been constructed for animal populations. Kareiva surveys models for the dynamics of herbivorous insect populations, emphasizing the role played by dispersal and by the patchy structure of most habitats; he also shows how these models may be combined with manipulative experiments in the field. Tilman's chapter summarizes the current status of theories and experiments concerning the dynamics of interacting populations, and gives suggestions about possible lines of future research.

The third section focuses on the interface between ecological and evolutionary theory. Travis and Mueller give an overview of recent work on ecological genetics, which is leading to a clearer understanding of the adaptive significance of genetic variation within populations. Theoretical advances here include a better understanding of frequency-dependent and density-dependent natural selection (and extensions to coevolution between interacting populations) and of age-specific aspects of selection and applications to the theory of life histories, as well as developments in the theory of quantitative inheritance (and concomitant measurement of the heritability of quantitative traits). Lifting his sights to the broad sweep of the paleontological record, Stanley discusses the significance of differential rates of extinction and their possible explanation in ecological terms; he also speculates on the role of "species selection" as an evolutionary factor. Feldman provides a commentary on a range of topics concerned with the interplay between ecology and evolution, and with mathematical models that combine population ecology with population genetics.

The fourth section wrestles with a difficult set of questions concerning how theoretical and experimental studies of ecosystems are to be structured, when ultimately one is dealing with systems and processes that involve many widely different spatial and temporal scales. How, for instance, does one put bounds on a study of a food web: when studying an intertidal system, does one include the owls that may eat the odd rodent that rarely consumes an item from the seashore? What is the role of "tourist" species (in Southwood's phrase), which occasionally wander into a food web but are not a permanent part of it? How are ecosystem structure and function influenced by the rare but important events that may occur every century or so? When and how is it permissible to deal with species aggregates ("mollusks," "phytoplankton"), rather than individual species? Is a functional classification of species, or, for example, one organized according to body size more appropriate than a taxonomic classification?

In this section O'Neill presents ideas about the possible use of hierarchy theory to organize studies of ecosystems, and to understand the relations among the different kinds of information that are present at different levels in a complex system. Powell shows how different physical and biological processes operate on different spatial and temporal scales in aquatic systems, with special emphasis on plankton patchiness and its connection with hydrodynamic processes. Steele, whose book on the structure of marine ecosystems pioneered explicit consideration of the influence of scale upon ecosystem dynamics, gives a summary of suggestions for further research on the general theme of scaling in space and time.

The fifth section deals with theoretical work on communities and food webs as such. Roughgarden shows how earlier work on conditions for coexistence among competing species and on prey-predator interactions (tracing back to MacArthur in the 1960s and beyond) is being succeeded by more complex studies of multilevel trophic webs and the assembly of communities of interacting species. He illustrates his discussion by reference to the buildup of lizard communities in the Lesser Antilles, a group of islands that also seems to reveal a "disassembly" of larger communities as a result of plate tectonic processes. Cohen summarizes some interesting patterns that have emerged from the analysis of systematic comilations of data about terrestrial and marine food webs, and shows how many of these patterns follow—in a statistically well-defined way—from a few simple assumptions or "rules" (which themselves, however, remain unexplained at this time). Cody's chapter summarizes the discussion that took place on this general theme, to provide an overview of the relations between theoretical and empirical ecology at the community level; his chapter also identifies areas of community ecology that seem in need of fresh ideas.

The sixth section deals with ecosystem structure and function. Levin reviews fundamental and applied aspects of mathematical models addressed to questions of productivity, nutrient cycling, and patterns of succession, with emphasis on how various kinds of environmental stress can affect such processes. Approaching some of the issues of Section IV from a different perspective, Levin argues that our increasing appreciation of local and global problems in the environment makes it necessary to couple biological and physical factors (including the effects of climate change), which inevitably involves the recognition that components of biotic and abiotic systems can operate on vastly different scales of space and time. Horn, Shugart, and Urban review models for forest ecosystems. Schlesinger's overview records suggestions for other ways in which theory may combine with existing and new technology in pursuit of a better understanding of how ecosystems work.

The final two sections focus on applications of ecological theory to resource management, conservation biology, and the control of pests and diseases. The seventh section begins with Clark's review of how population biology and economics interact in the management of what are usually called "renewable resources." Among other things, Clark shows how, even if there is a "sole owner" (thus avoiding the "tragedy of the commons"), narrowly economic considerations can lead to the overexploitation, or even extinction, of a biological resource. This tends to happen if the maximum sustainable biological yield, expressed as a percentage of the "standing crop," is below the inflation-corrected bank interest rate; in these circumstances, it makes economic sense to liquidate the capital stock and reinvest elsewhere. Such perceptions help to explain many bioeconomic happenings, including the overexploitation of whales. The work makes it clear that many controversies, as for instance between loggers and conservationists, are essentially differences of opinion about the rate at which the future should be discounted; bioeconomic theory thus can clarify the essential nature of many quarrels about political and social alternatives. Gilpin and Pimm discuss the many different kinds of contributions that ecological theory is making to pressing problems in conservation biology, ranging from the dynamics, genetics, and behavior of small populations of endangered organisms in zoos or in reserves, to insights that may guide difficult choices in acquiring land for preservation. Ehrlich's essay surveys these and other ways in which ecological theory has been useful in resource management and in conservation biology, and suggests lines of further application.

In the eighth section, the paper by Hassell and May reviews the interplay between theoretical and empirical studies pertaining to the population biology of host-parasite and insect prey-predator associations (with "parasite" defined broadly to include viruses, bacteria, protozoans, and helminths). The paper surveys basic ideas about host-microparasite, host-macroprasite, and host-parasitoid systems, along with experimental tests of these ideas. Hassell and May summarize the main conclusions, and give a list of important unresolved questions. This chapter aims to emphasize similarities and differences between the population biology of interactions between hosts and viral, bacterial, protozoan, helminth, parasitoid, and other insect predators, which explains why there is one long chapter rather than separate ones on pathogens and on insect predators. Building on this chapter and on the group discussion, Anderson surveys current accomplishments, and likely future directions, in the use of mathematical models to provide insights about the control of crop pests and insect vectors of disease, or in the design of programs of immunization or chemotherapy against pathogens and parasites.

THE ROLE OF THEORY IN ECOLOGY

As the pages in this volume indicate, theory is the pathway to extrapolation, generalization, and understanding. It provides an antidote to the helpless feeling engendered by the view that nature is so complicated, and evolutionary processes so contingent on accident and history, that all we can ever hope to achieve is detailed understanding of specific situations—Geertz (1973) "thick description" of particular times and places—rather than any general rules or patterns.

Whether we are dealing with the physiology or behavior of individuals, the dynamics or genetics of populations, the structure of communities, or broad paleoecological or biogeographic patterns, theory can suggest possibilities for experiments or observational protocols, can prompt tentative and testable generalizations, can serve as a crude guide to action in circumstances where action cannot wait on the certitude of detailed site-specific studies, as well as provide detailed models that summarize experimental findings in appropriate cases. Knowing which kind of model seems appropriate to which situation is the art of good theory. For example, in some applications, one may prefer a simple model tempered with judgment and experience. Yet a very detailed model seems needed in climate analysis, where a mere degree can trigger an ice age, and where accurate and detailed models are possible. Still, in models with strong nonlinearities, it is increasingly recognized that very detailed models usually provide only an illusion of precision, are not robust in their predictions, and are not reliable tools for management (Ludwig and Walters 1985).

Some contemporary discussion about the role of theory in ecology reflects a more general discussion about how science should be done. In Darwin's day, the prevailing scientific orthodoxy was the Baconian method, and in his books Darwin accordingly portrays himself as first marshalling the facts, and then seeing what conclusions emerge, "patiently accumulat[ing] and reflect[ing] on all sorts of facts which could possibly have any bearing. . . . After five years' work I allowed myself to speculate on the subject, and drew up some short notes . . ." (Darwin 1859). His notebooks, however, tell a very different tale, much more familiar to a practicing scientist. As Gruber and Barrett (1974, p. 122) put it, "The pandemonium of Darwin's notebooks and his actual way of working, in which many different processes tumble over each other in untidy sequences—theorizing, experimenting, casual observing, cagey questioning, reading, etc.—would never have passed muster in a methodological court of inquiry." Today, the early writings of Popper are much in vogue, and some would interpret these as denying legitimacy to any research program not cast in terms of falsifiable hypotheses. It should not be forgotten, however, that philosophies of science are themselves theories rather than revealed truths, and the hypothesis that any one of them provides the only correct way to do science seems to us to be abundantly falsified by the actions of practicing scientists. Lehman (1986) writes: "Those who embrace constraints crafted by others in the form of Popperian or hypothetico-deductive straight jackets may have divined a means to restrict their imagination, but there is no evidence in my view that those constraints encourage breakthroughs in biological sciences." The recent book by Gleick (1987) on *Chaos: Making A New Science,* with its account of the contingent, hurley-burley way in which new ideas and new experiments—or, perhaps more important, old ideas and old experiments seen in a new light—were prompted by the vagaries of accidental meetings and conversations, seems to us much closer to how science actually advances than are tidy schemes in which scientific discovery is reduced to a kind of painting-by-numbers. In the words of Feyerabend (1975), "Science is an essentially anarchistic enterprise: theoretical anarchism is more humanitarian and more likely to encourage progress than its law-and-order alternatives."

Just as there is a plurality of ways to do science, there is a plurality of interests and aptitudes among the people doing ecological research today. Somehow theory and empirical activities need to be combined, either by collaborations, or by individuals that take their hand to both. A big need is to develop theories for systems and questions that have been previously neglected. It is especially important not to fill a theoretical vacuum by importing a model developed for one system to a strange and clearly inappropriate context. Instead, new theory will need to be devised and analyzed.

Our aim in this book has been to capture the excitement of theoretical ecology by presenting some of the past and present accomplishments of ecological theory, on a canvas of wider sweep than is seen in volumes focused on just one aspect of the subject. More than this, the book offers speculations and suggestions about possible future directions of research. Sometimes the focus is on fundamental aspects—for example, deriving population parameters from the physiology or behavior of individuals, or understanding the ecological significance of genetic variation, and so forth; at other times the focus is on applications of theory—for example, to resource management, conservation biology, or the control of pests and pathogens.

In all this, our overarching aim is to share a representative collection of empirical and theoretical ecologists' views of where ecological theory is now, and where it may be going next.

We thank the National Science Foundation for support in making this project possible (BSR-8712008), Judy Thompson of Hopkins Marine Station for help in coordinating travel and facilities, and Judith May and the staff of Princeton University Press for their skill in publishing this volume.

REFERENCES

DeBeer, G. 1964. *Charles Darwin: A Scientific Biography.* Doubleday, New York.

Feyerabend, P. 1975. *Against Method.* Verso, London.

Darwin, C. 1859. *On The Origin of Species by Means of Natural Selection, or the Preservation of Favoured Races in the Struggle for Life.* John Murray, London.

Geertz, C. 1973. *The Interpretation of Cultures.* Basic Books, New York.

Gleick, J. 1987. *Chaos: Making a New Science.* Viking, New York.

Gruber, H. E., and P. H. Barrett. 1974. *Darwin on Man: A Psychological Study of Scientific Creativity, Together with Darwin's Early and Unpublished Notebooks.* Dutton, New York.

Jacob, F. 1982. *The Possible and the Actual.* University of Washington Press, Seattle.

Lehman, J. 1986. The goal of understanding in limnology. *Limnol. Oceanogr.* 31:1160–66.

Ludwig, D., and C. J. Walters. 1985. Are age-structured models appropriate for catch-effort data? *Can. J. Fish. Aquat. Sci.* 42:1066–72.

FROM INDIVIDUALS TO POPULATIONS I

Chapter 1

Plant Physiological Ecology: A Theoretician's Perspective

LOUIS J. GROSS

The central issues of plant physiological ecology concern the effects of environment on individual plant growth, survival, and reproduction. In this regard, physiology is viewed as the mechanism through which the joint effects of heredity and environment are coupled to determine the growth form and reproductive success of an individual (Kramer 1948). My goal here is to provide a very brief review of the major questions that the field addresses, with emphasis on the use of theory; give a few examples of how theory has contributed new perspectives; point out some directions I feel are as yet relatively unexplored; and, finally, make some comments about coupling with other levels of organization. This is meant as a theoretical complement to the excellent review and set of recommendations for future research by Ehleringer et al. (1986). What I discuss is limited by my own biases, including a blatantly terrestrial one. A comprehensive review of the area accessible to a general audience is contained in the January 1987 issue of *BioScience*. The most exhaustive compilation of research in the area to date is the series of books edited by Lange et al. (1981, 1982, 1983). Relatively few mathematically oriented books have appeared, but those containing some relevant material include Thornley (1976), DeWit (1978), Rose and Charles-Edwards (1981), Charles-Edwards (1981), Jean (1984), and Gross and Miura (1986). On the biophysical end, the books by Gates (1980) and Nobel (1983) are standards. For a fine collection of papers that take an economic, cost-benefit approach to energy capture and utilization by plants, see Givnish (1986a).

Generally, I have found that two quite different viewpoints prevail in plant physiological studies. On the one hand, ecologists approach problems from an evolutionary perspective and consider physiology as a means to carry out adaptations necessitated by selective forces. Those with an agronomic approach, on the other hand, are more typically reductionist in that they are primarily concerned with the "how" of direct hormonal control of physiology and not with the ultimate questions of how that hormonal control arose. In fact, ecologists also typically view hormonal control as fixed, and then try to explore its ecological role. To a certain extent, these differing viewpoints have produced a schism in the field. Much of what is published under the rubric of plant physiology deals with specific details of biochemical control of physiology, and recent emphasis on the cellular and molecular levels has relegated whole-plant physiology to a backwater, making it a retrospective rather than a predictive field (Kramer 1986). A great deal might be gained if ecologists recognized that their assumptions about the fixity of hormonal control are often overly restrictive and if physiologists took a more holistic view of plant functioning. In addition, ecophysiology has relatively recently become much more instrumentation oriented. The electronic revolution has made both field and lab measurements of certain physiological processes affordable even to those with minimal financial resources.

These trends have allowed us to learn a great deal about particular details of physiology, and they have produced a basic data set on physiological responses to a reasonably large fraction of the world's habitats. At the same time, however, the integration of our knowledge at the detailed biochemical level to investigate whole-plant phenomena is very inadequate. Modern physiology seems to have drifted far from practical applications to crop and forest population. This is not to say that the questions investigated are not of interest in their own right, nor that they will not eventually be important in addressing practical problems, but rather that currently most ecophysiologists, agronomists, and foresters make minimal use of recent plant physiology research. These difficulties are very much tied to the current limitations of reductionist approaches at longer time and larger spatial scales.

THE BASIC THEORETICAL APPROACHES

The central questions of ecophysiology concern (1) plant form; (2) response of metabolic processes—including photosynthesis, respiration, transpiration, and translocation—to environments within an individual's life span; (3) inferred population-level responses to environment over many generations; (4) partitioning of resources among plant parts and for defense from herbivores; and (5) interactions at other levels, including host-parasite relationships, ecosystem productivity and nutrient cycling analysis, and agricultural system analysis. These questions are, of course, not independent of each other. It is, however, often possible to effect a separation by recognizing that some processes take place on significantly different time scales than others. Thus, questions of metabolic responses on a daily time scale, or acclimation responses over weeks, consider the genetic makeup of the population as fixed, separating the questions of ecotypic differentiation from those of short-time responses of physiology. The following sections will briefly review the theoretical approaches to these questions.

Plant Form

The branched architecture of plants and the spiral patterns of phyllotaxis have long been topics of interest among botanists. A variety of mathematical theories have been proposed to both describe and provide a mechanistic basis for understanding phyllotaxis (Jean 1984). The theories are based on quite different physiological mechanisms, including those that maintain that primordia (1) grow until they are "pressured" by adjacent primordia (Adler 1977); (2) arise as a gap-filling process in regions of some minimal size (Adler 1975); and (3) produce a morphogen that acts as an inhibitor of new primordia, with reaction-diffusion equations describing the field of inhibition and predicting where new primordia will arise (Thornley 1976). Even the fairly elaborate mathematical models for these alternative physiological mechanisms do not produce testable hypotheses that would allow experiments to differentiate between them.

Much work has been done on the simulation of branching patterns to determine the general patterns one might expect under alternative design constraints (Bell, Roberts, and Smith 1979), including maximizing effective leaf area (Honda and Tomlinson 1982). Niklas (1986) has compared the conflicting requirements for light interception, biomechanical constraints of an upright form, hydraulic constraints for water supply, and reproductive display and maintenance. In addition to yielding fascinating comparisons to data on the evolution of architecture, these analyses show that there is no single "optimal" form: because of the nonlinear relationships between the constraints, the form predicted for a particular habitat varies with environmental conditions. Even with linear main effects, the multiplicity of interactions and constraints can preclude a single optimal form and hence promote diversity (Horn 1981). Indeed, the diversity of plant form evident in any single habitat argues that either form is only weakly coupled to mild environmental differences, that the models do not accurately take into account the alternative selective pressures on plant form, or that the phenotypic optimization approach of these models is inappropriate due to their neglect of the historical nature of the evolutionary process. Nevertheless, these models, especially when coupled with biophysically realistic analyses of environmental effects on basic physiological processes, provide a template to check overall patterns of plant form among differing environments.

Metabolic Processes

Many sophisticated mathematical models have been derived to analyze the response of basic plant metabolic processes to differing environmental conditions. These generally include simplifications of the known complexities of the biochemistry involved, and at some level even the most reductionist models are empirical. The vast majority of work has been done on static models that assume the plant is sitting in constant environmental conditions. In considering diurnal variations, the assumption is that the physiology instantaneously tracks the environment by moving among the associated steady states. This approach is reasonable only if the dynamics of physiological response operate much more rapidly than the environmental variations driving the response. In most natural conditions, I believe this assumption is not justified and the accuracy of calculations based on steady-state assumptions needs to be checked by reference to dynamic models that have been carefully validated. It would be useful to delineate the types of habitat in which the physiology can keep pace with environmental changes, making it possible to specify when dynamic models are necessary. Such work on light and photosynthesis is being carried out by Chazdon (1988) and Pearcy, Chazdon, and Kirschbaum (1987).

Dynamic responses may be investigated on physiological, acclimation, and evolutionary time scales (Gross 1986). The physiological time scale concerns variations in environmental components within a day, and I believe the theory can be developed in a straightforward manner once an adequate data base is established from lab and field measurements. In addition to allowing us to more accurately estimate ecologically important properties such as carbon gain in varying environments, such measurements should provide new insights into the basic physiology of the processes involved. The acclimation time scale concerns changes in physiology and anatomy that occur throughout the life span of the individual, including developmental changes, production of new branches and leaves, and the senescence of plant parts. This couples the dynamics of physiological changes (e.g., stomatal density and mesophyll thickness can change within a leaf because of light changes during development) to the demography of plant parts (e.g., the initiation of new leaves with a differing physiology) (Bazzaz 1984). On this time scale it may be useful to consider quasi-independent plant parts as individuals competing for resources, which premise forms the basis for much of the source-sink models developed to date. To be truly successful, however, we need more information on the mechanisms involved in the shedding of plant parts. Theoretical approaches on this time scale have been limited (but see Gross 1984 for an application to photosynthetic capacity, and the section below on allocation patterns). The evolutionary time scale concerns ecotypic differentiation of physiological traits and alternative physiological solutions to environmental constraints observed between taxa. Life-history theory is applied on this scale and one cannot really divorce the population from the individual here (Bazzaz et al. 1987). To date, however, even the most complete life-cycle models used to analyze alternative life histories include very little physiological detail (Caswell 1986). Linking the population-level parameters of fecundity and mortality to measurable physiology remains a very open area.

Of all the metabolic processes in plants, photosynthesis has been the one most intensively modeled. The starting point is often some simplification of the biochemical pathways for carbon assimilation, coupled with assumptions about stomatal behavior, to consider the effects of light, temperature, carbon dioxide, and humidity on net photosynthetic rates. A compendium of models is given by Hesketh and Jones (1980). These models are inherently steady-state and are designed mainly to grapple with the problem of how to combine the effects of the many environmental variables that control uptake. They have been extensively used in many physiologically based crop growth models and as submodels in ecosystem simulations (Reynolds and Acock 1985).

Models that attend to more biochemical and photochemical details (Farquhar, von Caemmerer, and Berry 1980; Farquhar and von Caemmerer 1982) have been instrumental in focusing attention on the importance of the intercellular carbon dioxide concentration in coupling stomatal function with the biochemical photosynthetic pathways. The basic approach is to utilize detailed biochemical models to specify the functional relationship,

$$A = f(p_i), \tag{1.1}$$

between net carbon assimilation rate A per unit leaf area and the intercellular partial pressure of carbon dioxide, p_i. The function $f(\cdot)$ contains a variety of parameters related to enzyme activation and sizes of various pools of metabolites, as well as environmental inputs such as light and partial pressure of oxygen. The result is a concave function of p_i, similar to a Michaelis-Menton curve, though with sharp transitions when alternative biochemical processes are limiting. This is then coupled to stomatal conductance, since by definition conductance g is such that

$$A = \frac{g(p_a - p_i)}{P} \tag{1.2}$$

where p_a is the partial pressure of carbon dioxide external to the leaf, and P is atmospheric pressure. Since, as a function of p_i, (1.1) is concave and (1.2) is linear with negative slope, upon setting (1.1) equal to (1.2) there is a unique solution for p_i. This solution, obtained numerically, gives the equilibrium p_i and A values for given leaf and environmental conditions. By pointing out the importance of the relationship (1.1), this modeling effort has led to changes in the way experimentalists carry out their observations. Together with advances in instrumentation over the last decade, particularly the availability of good mass flow meters, such modeling has led many researchers to measure the relationship (1.1) rather than to measure assimilation as a function of light or of other environmental factors, which were the prevalent measurements taken prior to the Farquhar-von Caemmerer approach. This modeling was among the first to integrate appropriately a variety of physiological functions, and ecophysiologists use it as a basis to couple detailed physiology with conditions in natural environments.

Another modeling approach that has had great influence on ecophysiology concerns the control of stomatal conductance (Cowan 1977, 1982). The goal is to determine how g in (1.2) behaves as a function of environmental variables as well as how it correlates with assimilation, since g is viewed as an externally determined variable in the Farquhar–von Caemmerer model. Here calculus of variations is applied to determine optimal patterns of stomatal opening. The objective is to minimize water loss, subject to the constraint that integrated carbon assimilation is maintained at some fixed level, presumably that which will assure adequate photosynthate production to meet plant needs for growth and reproduction. The mathematical form of the model involves the assumption that transpiration rate per unit leaf area, E, is an implicit function of A, and that each of these depend on time and location on the leaf. Then, if s is the space variable (one-dimensional here), the requirement is that

$$\int_0^T \int_0^S [E(A,s,t) - \lambda A]\, ds\, dt \tag{1.3}$$

be minimized, where λ is a Lagrange multiplier. The solution implies that

$$\left(\frac{\partial E}{\partial A} \right)_{s,t} = \lambda. \tag{1.4}$$

Here λ may be viewed as the benefit of carbon gain relative to the cost of water loss. Since both A and E may be measured, by varying environmental conditions (1.4) may be tested, though there are some difficulties related to constraints on stomatal control that are not taken into account in this simple framework (Cowan 1986). Alternative models for the trade-offs between carbon gain and water loss are closely related to this model (Givnish 1986b).

Considerably less theoretical work has been done on other physiological processes. Our most detailed knowledge is at the level of individual leaf function (Pearcy et al. 1987). Respiration remains an area in which lack of data and the confounding of dark and light respiration make theoretical developments quite difficult (Penning de Vries 1983). Maintenance respiration is normally assumed to be just proportional to biomass, with some temperature dependence, and the respiratory load of below-ground material as well as any associated mycorrhizae is relatively unexplored. Patterns of fluid flow within a plant have been investigated with a variety of models (Rand 1983), though with rather little concern for the detailed operations of the associated conductive tissue. Elaborate biomechanical models of stomatal behavior (Delwiche and Cooke 1977; Sharpe and Wu 1978), often specifically constructed to mimic stomatal oscillations, have thus far led to rather little in the way of ecological insight, though they are mathematically interesting. Indeed, I believe a more feasible approach is to consider a "statistical mechanics" of stomata taking account of the coupling between stomata that can produce spatially structured patterns across a leaf surface (Rand and Ellenson 1986).

Biophysical models allow the coupling of the many physical processes which affect plant function, with particular emphasis being given to energy-balance models. These models can predict how alterations in wind, ambient temperature and radiation, and humidity affect heat loads on organisms of any given shape. They have been particularly useful in predicting trends in leaf size and shape across habitats (Parkhurst and Loucks 1972; Givnish 1979). They serve as the basis for understanding plant-atmosphere interactions (Grace 1983) and allow the coupling of detailed within-canopy radiation models (Norman 1980) with leaf distribution models to produce whole canopy assimilation estimates (Baldocchi and Hutchison 1986). Despite their utility, these approaches can provide only a fairly crude understanding of the biophysical limitations to growth in a particular habitat, which is inadequate to explain the great variability in leaf shape and size in many habitats. Although there are some fine and detailed models of simple canopies, they ignore much of the variation in both environment and leaf physiology within canopies, often by simply breaking the canopy into sunlit and nonsunlit fractions.

Allocation Patterns

Knowledge about plant allocation patterns is largely empirical, and relatively little is known about their control (Pearcy et al. 1987). Because a limited amount of available resources (photosynthate, water, and nutrients) must be apportioned between alternative demands, there are trade-offs, and two alternative approaches have been taken to analyze them. The first is a cost-benefit analysis that takes biomass or a limited nutrient as a currency to measure the allocation pattern. A key assumption is that increasing photosynthetic capacity will lead to increased growth, though there is surprisingly little data to support this. Growth is strongly correlated with total light interception capacity and is typically limited by environment, not physiology (Kramer 1986). The second approach is to construct mechanistic compartment models for whole plant growth that break a plant into roots, shoot, leaves, fruit, etc. Flows of nutrients between compartments are driven by sink and source strengths, and partitioning of new material is governed by a goal-seeking assumption that is set a priori—for example, to maintain a fixed C : N ratio or fixed root : shoot ratio (Thornley and Johnson 1986). This structure is typically applied in physiologically based crop-growth models where the model predictions can be readily tested against data.

An alternative to compartment-type models is to assume the existence of underlying organizing principles of evolutionary origin that specify the growth form in any particular environment. The cost-benefit analysis here takes the form of an optimal control problem, first analyzed by Cohen (1971). The chosen optimization criterion is usually some measure of reproduction, and elaborations consider random season lengths, varying environments, herbivory, and several vegetative compartments (Roughgarden 1986). This approach, though capable of producing allocation patterns similar to those observed (i.e., bang-bang allocation of resources in annuals), is almost totally lacking in the physiological detail necessary to couple its parameters to measurable aspects of any particular species or group of species. Nevertheless, the models do produce reasonably accurate portrayals of the general balance that seems to exist in plant allocation, and they mimic the types of adjustments that are observed to occur when there are imbalances in the availability of different resources (Chapin et al. 1987). Detailed validation for particular taxa cannot be attempted until the models become more physiologically realistic, but the models *can* provide hypotheses about trends across taxa and environmental conditions. The approach suffers all the difficulties associated with optimization schemes, but see Givnish (1986a) for arguments as to why the adaptationist program has proven so useful here.

There are a host of open theoretical questions regarding allocation of resources for defense from herbivores. Models that relate plant quality to herbivore population dynamics are few (Edelstein-Keshet 1986) but represent a first step toward producing a physiologically based approach. Such an approach involves a partial differential equation for the change in plant quality through time, and represents only one of many potential ways to tie physiology to population-level models by using a relevant physiological variable to structure the population (Metz and Diekmann 1986). This approach still needs to be used with a control-type problem to produce, for example, a cost-benefit analysis related to plant apparency. Gulmon and Mooney (1986) have already proposed a starting framework for a theory of allocation to defense, particularly as it relates to resource availability. Remaining theoretical questions involve spatial and temporal patterns of allocation to defensive compounds and the multiple constraints that act on such allocations (Bazzaz et al. 1987).

Systems and Future Directions

Physiological models are combined to serve as the basis for complex systems models of both natural and managed ecosystems (Reynolds and Acock 1985).

Systems models are typically structured around particular crops or natural habitats, though there have been recent attempts to construct a generic model (Reynolds et al. 1986). These mechanistically based models have served to reveal areas in which our ignorance limits the model's usefulness, and they conveniently join the vast array of physiological processes operating in these systems. But they do have their limitations. They cannot compensate for our lack of expertise in certain areas, such as how to handle the interactions of multiple environmental stresses (Chapin et al. 1987). Systems models typically contain hundreds of parameters whose values we can only approximate from our knowledge of physiology, and we know little about them regarding variation between individuals or species. Even the most elaborate crop growth models are not yet as accurate in yield prediction as relatively simple regression models. On the other hand, the complex mechanistic models serve as the only means to track the dynamics of the components of ecosystems, and thereby determine appropriate control measures. Indeed, this is one of the current uses of detailed crop models, particularly with respect to irrigation, fertilization, and pesticide scheduling. Regression approaches lack this capability because they are strictly limited by the data set used in their construction.

At certain levels, it may simply be inappropriate to include the details of physiology. Forest-stand simulation models appear to reflect forest composition changes realistically over periods of centuries by tracking individual trees through their lifespan (Shugart 1984). Yet their physiological component is extremely naive, causing some physiologists to lambast the approach. This disagreement occurs in part because the time scales that are of interest to physiologists are much shorter than the centuries that this simulation approach is designed to work on. Of course, the lack of knowledge about physiology limits the types of questions for which the simulation approach is appropriate. It would be absurd, for example, to attempt to apply it in a rigorous way to predict effects of atmospheric carbon dioxide increases on forest successional patterns, since that would necessitate making a priori assumptions about how these atmospheric changes would differentially affect the component species of the system. Despite the lack of physiological detail in these individual-based models, the approach provides one of the few available means to investigate how altering physiological characteristics of component species will affect community-level processes (Huston and Smith 1987).

A further remark regarding the carbon dioxide question is relevant. One argument in support of continuing physiological studies about the direct effects of enhanced carbon dioxide emphasizes that the data are necessary to predict worldwide effects of the predicted increases. It may well be that this problem is impossible to solve. That is, no matter how much effort we exert to understand physiological responses to enhanced carbon dioxide, we cannot possibly answer the questions posed by policymakers because of the complexity of applying even our limited knowledge at the population and community levels. In this case, although one can make an argument that the basic physiological questions are inherently interesting in their own right, completely different top-down methods may be much more appropriate to carry out a sort of risk analysis on the predictions demanded by politicians. A reductionist approach based on physiological details may be so incomplete that it is irrelevant, and it may even be counterproductive by taking attention and limited funding away from approaches more appropriate for the policy questions being posed.

Many areas of plant physiological ecology are still ripe for further theoretical development. Some of these were mentioned above. With relatively few exceptions (Hay 1986; Koehl 1986), many of the questions posed above have been unanswered for aquatic plants, for example. Another open area involves the application of the theory of evolutionarily stable strategies to partitioning among plant parts and to competition between individuals. A very crude example of this area is given in Riechert and Hammerstein (1983) in regard to rooting behaviors. There is also a need to develop a general theory of plant epidemiology. Though there has been a tremendous burst of activity in epidemiological modeling over the last decade, it has not been reflected in plant studies. Indeed, considering the severe economic and ecological effects of plant diseases, it is unfortunate that much of plant epidemiology is tied to specific agronomic situations. Although theories developed for animals do not apply directly to plants because of the importance of spatial effects and the display among populations of a continuum of resistance levels to any particular pathogen, models developed for macroparasites (Hassell and May, chapter 22) may be quite useful in plant situations. Coupling the physiological response to pathogen infection with the demographics of pathogens presents a fascinating and ultimately highly applicable area of theory. Gilligan (1985) has reviewed the state-of-the-art with regard to crop disease models, and a recent series of papers by van den Bosch, Zadoks, and Metz (1988a,b) covers spatial aspects of crop disease spread.

The general problems of scaling up, from our relatively good understanding of processes on a leaf-level, short-time scale to the whole canopy and plant, center around what can be reasonably ignored on the scales of the questions being posed. Indeed, the hope is that we need not consider much of the physiological detail when the focus is on population and community-level interactions. Determining how much physiology can be safely ignored in a particular problem is still debatable. One hope is that detailed physiological models will lead the way to more appropriate holistic descriptions of natural systems than are currently available. This uses reductionism to scale up rather than down by giving us clues as to how representative particular models at the population and community level are when used across different environmental circumstances. The idea is that physiologically based models not only specify the appropriate form for more empirical macrodescriptors of system behavior at larger scales, but also provide means to test the robustness of these descriptors across different natural systems. The advances in our knowledge of molecular control of physiology and its genetic manipulation must be explored from an ecophysiological viewpoint before we can hope to ascertain community-level effects of the release of manipulated organisms. A physiological perspective is essential to let us know which manipulations will be successful agronomically as well as to allow us to evaluate what the long-term systems effects might be. In the hierarchy of natural systems, ecophysiology links cellular and biochemical phenomena to population and community-level processes, and this ability to consider the implications of vastly different scales adds to the exciting prospects in the field. Besides this coupling between scales, ecophysiology has its own interesting questions to address, independent of its utility to transfer information upscale to the population and community levels.

ACKNOWLEDGMENTS

I thank Henry Horn for an incisive reading of an earlier version of this paper, which led to substantive changes. My understanding of many of the topics discussed here has been greatly improved through extensive conversations with Bob Pearcy, and I thank him for his friendship, patience, and tolerance of the naive questions of a theoretician.

REFERENCES

Adler, I. 1975. A model of space filling in phyllotaxis. *J. Theor. Biol.* 53:435–44.

Adler, I. 1977. The consequences of contact pressure in phyllotaxis. *J. Theor. Biol.* 65:29–77.

Baldocchi, D. D., and B. A. Hutchison. 1986. On estimating canopy photosynthesis and stomatal conductance in a deciduous forest with clumped foliage. *Tree Physiol.* 2:155–68.

Bazzaz, F. A. 1984. Demographic consequences of plant physiological traits: Some case studies. In R. Dirzo and J. Saukhan, eds., *Perspectives on Plant Population Biology*, pp. 324–46. Sinauer, Sunderland, Mass.

Bazzaz, F. A., N. R. Chiariello, P. D. Coley, and L. F. Pitelka. 1987. Allocating resources to reproduction and defense. *Bioscience* 37:58–67.

Bell, A. D., D. Roberts, and A. Smith. 1979. Branching patterns: The simulation of plant architecture. *J. Theor. Biol.* 81:351–75.

Caswell, H. 1986. Matrix models and the analysis of complex plant life cycles. In Gross and Miura (1986), pp. 171–233.

Chapin, F. S. III, A. J. Bloom, C. B. Field, and R. H. Waring. 1987. Plant responses to multiple environmental factors. *Bioscience* 37:49–57.

Charles-Edwards, D. A. 1981. *The Mathematics of Photosynthesis and Productivity.* Academic Press, London.

Chazdon, R. L. 1988. Sunflecks and their importance to forest understory plants. *Advances in Ecological Research* 18. In press.

Cohen, D. 1971. Maximizing final yield when growth is limited by time or by limiting resources. *J. Theor. Biol.* 33:229–307.

Cowan, I. R. 1977. Stomatal behaviour and environment. *Adv. Bot. Res.* 4:1176–1227.

Cowan, I. R. 1982. Regulation of water use in relation to carbon gain in higher plants. In Lange et al. (1981–83), vol. 12B, pp. 589–613.

Cowan, I. R. 1986. Economics of carbon fixation in higher plants. In Givnish (1986a), pp. 133–70.

Delwiche, M. J., and J. R. Cooke. 1977. An analytical model of the hydraulic aspects of stomatal dynamics. *J. Theor. Biol.* 69:113–41.

DeWit, C. T. 1978. *Simulation of Assimilation, Respiration and Transpiration of Crops.* Wiley, New York.

Edelstein-Keshet, L. 1986. Mathematical theory for plant-herbivore systems. *J. Math. Biol.* 24:25–58.

Ehleringer, J. R., R. W. Pearcy, and H. A. Mooney. 1986. Recommendations of the workshop on the future development of physiological ecology. *Bull. Ecol. Soc. Amer.* 67:48–58.

Farquhar, G. D., and S. von Caemmerer. 1982. Modelling of photosynthetic response to environmental conditions. In Lange et al. (1981–83), vol. 12B, pp. 549–87.

Farquhar, G. D., S. von Caemmerer, and J. A. Berry. 1980. A biochemical model of photosynthetic CO2 assimilation in leaves of C3-species. *Planta* 149:78–90.

Gates, D. M. 1980. *Biophysical Ecology.* Springer-Verlag, New York.

Gilligan, C. A. 1985. *Mathematical Modelling of Crop Diseases.* Academic Press, London.

Givnish, T. 1979. On the adaptive significance of leaf form. In O. T. Solbrig, S. Jain, G. B. Johnson, and P. H. Raven, eds., *Topics in Plant Population Biology,* pp. 375–407. Columbia University Press, New York.

Givnish, T. J., ed. 1986a. *On the Economy of Plant Form and Function.* Cambridge University Press, Cambridge, England.

Givnish, T. J. 1986b. Optimal stomatal conductance, allocation of energy between leaves and roots, and the marginal cost of transpiration. In Givnish (1986a), pp. 171–213.

Grace, J. 1983. *Plant-Atmosphere Relationships.* Chapman and Hall, London.

Gross, L. J. 1984. On the phenotypic plasticity of leaf photosynthetic capacity. *Lecture Notes in Biomath.* 52:2–14.

Gross, L. J. 1986. Photosynthetic dynamics and plant adaptation to environmental variability. In Gross and Miura (1986), pp. 135–70.

Gross, L. J., and R. M. Miura, eds. 1986. *Some Mathematical Questions in Biology—Plant Biology.* American Mathematical Society, Providence.

Gulmon, S. L., and H. A. Mooney. 1986. Costs of defense and their effects on plant productivity. In Givnish (1986a), pp. 681–698.

Hay, M. E. 1986. Functional geometry of seaweeds: Ecological consequences of thallus layering and shape in contrasting light environments. In Givnish (1986a), pp. 635–66.

Hesketh, J. D., and J. W. Jones, eds. 1980. *Predicting Photosynthesis for Ecosystems Models.* CRC Press, Boca Raton, Fla.

Honda, H., and P. B. Tomlinson. 1982. Two geometrical models of branching of botanical trees. *Ann. Bot.* 49:1–11.

Horn, H. S. 1981. Some causes of variety in patterns of secondary succession. In D. C. West, H. H. Shugart, and D. B. Botkin, eds., *Forest Succession: Concepts and Application,* pp. 24–35. Springer-Verlag, New York.

Huston, M., and T. Smith. 1987. Plant succession: Life history and competition. *Amer. Natur.* 130:168–98.

Jean, R. V. 1984. *Mathematical Approach to Pattern and Form in Plant Growth.* Wiley, New York.

Koehl, M.A.R. 1986. Seaweeds in moving water: Form and mechanical function. In Givnish (1986a), pp. 603–34.

Kramer, P. J. 1948. Plant physiology in forestry. *J. Forestry* 46:918–21.

Kramer, P. J. 1986. The role of physiology in forestry. *Tree Physiol.* 2:1–16.

Lange, O. L., P. S. Nobel, C. B. Osmond, and H. Ziegler, eds. 1981–83. *Physiological Plant Ecology I–IV. Encyl. Plant Physiol.,* vol. 12A–D. Springer-Verlag, Berlin.

Metz, J.A.J., and O. Diekmann. 1986. *The Dynamics of Physiologically Structured Populations.* Springer-Verlag, Berlin.

Niklas, K. J. 1986. Computer simulations of branching-patterns and their implications on the evolution of plants. In Gross and Miura (1986), pp. 1–50.

Nobel, P. S. 1983. *Biophysical Plant Physiology and Ecology.* Freeman, San Francisco.

Norman, J. M. 1980. Interfacing leaf and canopy light interception models. In Hesketh and Jones (1980), pp. 49–67.

Parkhurst, D. F., and O. L. Loucks. 1972. Optimal leaf size in relation to environment. *J. Ecol.* 60:505–37.

Pearcy, R. W., R. L. Chazdon, and M.U.F. Kirschbaum. 1987. Photosynthetic utilization of lightflecks by tropical forest plants. In J. Biggens, ed., *Progress in Photosynthesis Research,* vol. 4, pp. 257–60. Martinus Nijhoff Publishers, Dordrecht, Netherlands.

Pearcy, R. W., O. Bjorkman, M. M. Caldwell, J. E. Keeley, R. K. Monson, and B. R. Strain. 1987. Carbon gain by plants in natural environments. *Bioscience* 37:21–29.

Penning de Vries, F.W.T. 1983. Modeling of growth and production. In Lange et al. (1981–83), vol. 12D, pp. 117–50.

Rand, R. H. 1983. Fluid mechanics of green plants. *Ann. Rev. Fluid Mech.* 15:29–45.

Rand, R. H., and J. L. Ellenson. 1986. Dynamics of stomate fields in leaves. In Gross and Miura (1986), pp. 51–86.

Reynolds, J. F., and B. Acock. 1985. Predicting the response of plants to increasing carbon dioxide: A critique of plant growth models. *Ecol. Model.* 29:107–29.

Reynolds, J. F., D. Bachelet, P. Leadley, and D. Moorhead. 1986. Assessing the effects of elevated carbon dioxide on plants: Towards the development of a generic plant growth model. DOE report 28 in series Response of Vegetation to Carbon Dioxide.

Riechert, S. E., and P. Hammerstein. 1983. Game theory in the ecological context. *Ann. Rev. Ecol. Syst.* 14:377–409.

Rose, D. A., and D. A. Charles-Edwards, eds. 1981. *Mathematics and Plant Physiology.* Academic Press, London.

Roughgarden, J. 1986. The theoretical ecology of plants. In Gross and Miura (1986), pp. 235–67.

Sharpe, P.J.H., and H-I. Wu. 1978. Stomatal mechanics: Volume changes during opening. *Plant, Cell and Environ.* 1:259–68.

Shugart, H. H. 1984. *A Theory of Forest Dynamics.* Springer-Verlag, New York.

Thornley, J.H.M. 1976. *Mathematical Models in Plant Physiology.* Academic Press, New York.

Thornley, J.H.M., and I. R. Johnson. 1986. Modelling plant processes and crop growth. In Gross and Miura (1986), pp. 87–133.

van den Bosch, F., J. C. Zadocks, and J.A.J. Metz. 1988a. Focus expansion in plant disease. I: The constant rate of focus expansion. *Phytopathology* 78:54–58.

van den Bosch, F., J. C. Zadoks, and J.A.J. Metz. 1988b. Focus expansion in plant disease. II: Realistic parameter-sparse models. *Phytopathology* 78:59–64.

Chapter 2

Individual Behavior and the Procurement of Essential Resources

H. RONALD PULLIAM

The relationship between behavior and the availability of resources is a central part of behavioral ecology. Accordingly, many questions asked by behavioral ecologists deal with either the procurement or allocation of resources. In the case of procurement, the resources in question may be food, mates, space, or refuges and the individual in question must decide how to acquire them. On the other hand, questions concerning sex ratio, life-history strategy, and helping behavior deal with the allocation of resources such as nutrients, energy, and time that are already at the disposal of the individual in question. In this paper, I discuss only the theory of resource procurement, though the basic philosophy and theoretical approach is similar for questions dealing with resource allocation and, in many cases, procurement and allocation must be considered simultaneously.

Many of the original practitioners of behavioral ecology were ecologists interested in behavioral mechanisms of population regulation and species interaction. More recently, the discipline has attracted the interest of other behaviorists, particularly ethologists and behavioral psychologists, interested more in behavior for its own sake. Practitioners of behavioral ecology can also be divided according to their theoretical approach. Most behavioral ecologists routinely use optimization models to guide their research and interpret their results. Other behavioral ecologists employ more descriptive or mechanistic models to summarize behavior and predict the consequences of behavior for higher-level phenomena such as spatial distribution, population dynamics, and species interactions (e.g., Hassell and May 1985; Pulliam 1987; Turchin 1986; and Kareiva 1987).

Those behavioral ecologists who make extensive use of optimization theory have commonly used static optimization procedures, including game theory, and in recent years have begun to employ dynamic programming, optimal control theory, and the theory of dynamic games. Whether or not, in the long run, this optimization approach proves useful for a better understanding of the fine details of behavior remains to be seen. Nonetheless, current evidence overwhelmingly indicates that the approach provides a means of predicting general trends of behavior at the coarser level of resolution of interest to most ecologists. Accordingly, many ecologists are beginning to explore how to incorporate the predictions of behavioral ecology into more general ecological theories of population dynamics and species interaction (e.g., Rosenzweig 1981, 1985; Holt 1985; Pulliam 1988).

In this paper, I consider, in general, the relationship between resource distribution and patterns of resource procurement and consumption. In particular, I ask, "Can the spatial distribution of resources be described in a manner sufficient to allow accurate prediction of the distribution and success of consumers." In the case of predator (as consumer) and prey (as resource), the abundance and distribution of prey is strongly influenced by where consumers feed. For example, during the nonreproductive season, changes in the abundance and distribution of seeds is largely determined by where granivores feed and how much they eat. In such situations, the feeding behavior of the consumer population is largely determined by the abundance and distribution of prey and, in turn, changes in the prey population are mostly determined by the behavior of the consumer population. Thus, by iteration, answering the question posed above allows one to predict changes in the spatial and temporal distribution and abundance of both prey and consumer populations.

The approach of this paper will be to review briefly the theoretical literature pertaining to questions of resource procurement, to give examples of empirical tests of this theory, and to discuss current directions of both theoretical and empirical research in this area. Though the primary concern of the paper will be how the abundance and distribution of resources influences individual behavior, I shall also consider how behavior in turn influences the distribution of resources. As mentioned above, in the case of foraging behavior, this sets the stage for a consideration of the role of individual behavior in population dynamics and species interactions.

FUNCTIONAL RESPONSE AND MUTUAL INTERFERENCE

Solomon (1949) and later Holling (1959, 1961, 1966) distinguished between the functional response, defined as the relationship between the number of prey consumed per predator and prey density, and the numerical response, defined as the relationship between the number of predators and prey density. The functional response is, in large part, due to the changes in predator searching efficiency that accompany changes in prey density. Such changes in searching efficiency are well documented for many kinds of organisms (Hassell 1978) and, accordingly, the functional response forms an integral part of modern predator-prey theory.

The now-standard way of modeling the rate of intake as a function of resource (prey) density is to divide the total time required, per prey consumed, into search- and pursuit-time components. Pursuit time (τ), defined as the total time spent pursuing and handling prey once they are encountered, is usually considered constant as long as prey are not divisible. Search time is usually modeled by assuming that prey are randomly distributed and are located by a random (Poisson) search process. Thus, if λ denotes the expected number of prey located per unit search time, then the expected search time per prey is $1/\lambda$. Common sense suggests that the long-term feeding rate should be the mass of prey consumed per encounter divided by the sum of search time and pursuit time ($\tau + 1/\lambda$). Use of the renewal theorem of stochastic processes (Feller 1968; Charnov 1976; Clark and Mangel 1986) shows indeed that the asymptotic mean rate of intake is given by

$$F(n,x) = B/[\tau + 1/\lambda(n,x)], \qquad (2.1)$$

where B is the mass of prey consumed per prey located, n is prey (or resource) density, and x is predator (or consumer) density.

When prey are very abundant, search time approaches zero and eq. (2.1) describes the Holling type I functional response. When encounter rate is directly proportional to prey abundance, so $1/\lambda = 1/ax$, eq. (2.1) describes the Holling type II functional response, where a is the effective area searched per unit search time. For the Holling type III functional response, the parameter a is an increasing function of prey density, and the intake rate is sigmoidal with increasing prey density. It is worth noting that, in practice, type II and type III functional responses are extremely difficult to distinguish given the noisy data normally available.

Searching efficiency may change with predator density as well as with prey density. In most models, the influence of predator density is modeled as mutual interference (e.g., Beddington 1975; Hassell 1978; Rogers and Hassell 1974). In eq. (2.1) this amounts to making the number of prey located per unit of search time (λ) a decreasing function of predator density (n). However, under some circumstances, intake rates may increase in the presence of other predators (see Clark and Mangel 1986), as demonstrated for some social species. One mechanism leading to such mutual enhancement is the reduced time spent scanning for predators, based on mutual vigilance (Pulliam and Caraco 1984).

Clearly, if animals choose a feeding location based, in part, on the feeding rate that can be achieved, they may leave crowded areas because of interference. However, the decision to leave a crowded area may result in poorer success if there is no alternative area where a higher feeding rate can be achieved. In his seminal 1959 paper, Holling clearly recognized the importance of considering the density and quality of alternative foods. He pointed out that an "increase in alternate foods decreases predation by dilution of the functional response, but increases predation by promoting a favorable numerical response" (p. 309). Holling's study of mammal predation on European pine sawflies was conducted on a rather homogeneous pine forest; thus his emphasis was more on alternative foods within the habitat and not so much on alternative habitats per se. In what follows, I shall discuss how, starting with Holling's concepts of functional response and mutual interference, optimal foraging theory can be used to predict the numerical response and patterns of habitat usage.

APPROACH OF OPTIMAL FORAGING THEORY

The preceding section illustrates the use of *non*optimization models to describe the dynamics of predator-prey interactions. Much of modern optimal foraging theory (see Stephens and Krebs 1987 for a recent review) focuses on the choice of foods within a multiprey environment and thus is especially relevant to predicting the functional response, discussed above. Food choice in a multiprey environment can be modeled without the use of optimization procedures by measuring food preferences and incorporating them into the approach outlined above; however, optimization models allow the possibility of predicting preferences and, therefore, diets a priori.

The most developed and widely tested foraging models are concerned either with the choice among randomly dispersed prey in a homogeneous environment or the choice among depletable patches of prey. In the latter case, the patches are usually assumed to be small and locally abundant so that a forager might visit many patches during a single foraging bout. Stephens and Krebs (1987) distinguish between a prey and a patch by stating "a prey yields a fixed amount of energy and requires a fixed amount of time to handle it . . . however, the forager controls the time spent in, and hence, the energy gained from a patch" (p. 14). Most models of patch choice (e.g., the marginal value theorem) calculate the asymptotic rate of intake based on visiting many patches. I distinguish between patch selection and habitat selection by stating that many patches are visited during a foraging bout but that only one habitat is visited. Thus, choice of a habitat determines what patches are available during a foraging bout just as choice of a patch determines what prey are available in that patch. A forager may make many repeated visits (i.e., many foraging bouts) to the same habitat and may therefore use the same habitat for an entire season or lifetime. In practice, however, the distinction between patch choice and habitat choice becomes blurred and, for many purposes, the models for habitat choice and patch choice are the same.

Most optimization models of food and habitat selection have assumed that a forager chooses whichever option maximizes its chance of survival and/or reproduction. Often this is assumed to be the option that maximizes the rate of intake and/or minimizes total foraging time. A much more general approach incorporates both current survival probability and future reproductive success (Houston et al. 1988 1986; Mangel and Clark 1986). This approach requires the specification of the expected reproductive success following any feasible sequence of behavior. A relevant interval of time, such as a day or season, is chosen, and a reward function, $R(\mathbf{x})$, specifying expected future reproduction, is defined on the animal's state, $\mathbf{x}$, at the final time, T. To determine the optimal behavior sequence from time t to the final time T, a function $H(\mathbf{x}, a, t)$ specifies the expected future reproductive success at time $t + 1$, given the animal is in state $\mathbf{x}$ and performs action a at time t. Since maximizing $H(\mathbf{x}, a, t)$ when time $t = T - 1$ gives $R(\mathbf{x})$ for the final time T, the optimal action for any previous time can be found by starting at $T - 1$ and working backwards.

Since the consequences of habitat selection depend on the choices made by other individuals, habitat selection is often modeled as a game (Caraco 1987; Caraco and Pulliam 1984). The Nash solution (or Evolutionarily Stable Strategy) for the habitat selection game occurs when individuals are distributed among habitats in such a manner that no individual can do better by changing habitats. Furthermore, since the habitat choice made at one time may constrain the options available in the future, the game is a dynamic one, and optimal control theory or dynamic programming may be required to find optimal solutions. In practice, however, habitat selection has usually been modeled as a static game, and the dynamic aspects of how the choices made by one individual influence the future options of other individuals have rarely been considered.

Though optimal habitat choice has been studied by behavioral ecologists, it is not as well developed either theoretically or experimentally as are other areas of foraging theory. Some models of habitat selection (e.g., Fretwell's Ideal Free Distribution) assume that the animal has perfect knowledge of the alternative habitats available and is able to choose whichever is best. More recent models have assumed animals must gain information about available alternatives and may be prevented from occupying the best habitat because of dominance or other social constraints (Clark and Mangel 1984). Also, many current models do not explicitly consider the time and energy costs incurred in moving from one habitat to another (but see Rosenzweig 1981). A challenge to the future development of the theory of habitat selection, and of foraging theory in general, is to incorporate more realistic constraints into models without making them mathematically inaccessible to empiricists interested in testing them.

HABITAT CHOICE AND NUMERICAL RESPONSE

The numerical response is essential to an understanding of predator-prey dynamics because the total consumption of prey depends on the product of the functional response (prey consumed per predator) and numerical response (number of predators). Changes in the number of predators may be due either to birth and death of individuals or to the movement of individuals. Changes in local density due to dispersal have proven difficult both to model and to measure, though they may be critical to understanding the dynamics of animal populations in spatially heterogeneous environments. Theoretical models of habitat selection, when applied to an entire population, may allow the accurate prediction of habitat distribution, or, from the perspective of local predator density, of numerical response to local prey abundance.

In a spatially heterogeneous environment, any habitat can be characterized by the rate of consumption that can be achieved there. Another way of saying this is that each type of homogeneous habitat has its own functional response. If mobile predators are free to choose, they might tend to congregate in the habitats where their feeding rate is highest. However, if mutual interference reduces individual feeding rates in the high-quality areas, some individuals might benefit by emigrating to less crowded areas.

The "ideal free distribution" (Fretwell 1972) is a distribution of individuals among habitats that results in no individual being able to achieve a higher feeding rate by moving to another habitat. As pointed out by Milinski (1979) and Pulliam and Caraco (1984), the ideal free distribution corresponds to an evolutionarily stable strategy, or ESS (Maynard Smith 1976). An ESS is a strategy (pure or mixed) which, if adopted by a population, is not susceptible to invasion by any alternative strategy.

In several field and laboratory studies, the ideal free distribution has been shown to predict satisfactorily the equilibrium distribution of individuals among available habitats. Among the first and best-known tests of this model are the experiments of Milinski (1979), who offered stickleback fish a choice between feeding areas where food was delivered at different rates. As predicted by the ideal free distribution, the fish consistently distributed themselves in the ratio of the patch profitabilities. Somewhat similar studies with insects choosing mates or breeding sites (Parker 1979; Whitham 1980), amphibians choosing mates (Davies and Halliday 1979), and birds choosing feeding sites (Harper 1982) all support the notion that animals approximately distribute themselves among available sites in such a way that no individual could improve the quality of resource (e.g., food or mate) it obtains by changing sites.

The ideal free distribution assumes that individual consumers are "free" to choose either area and that all individuals have identical average intake rates. These two assumptions are true only if there is no dominance or resource defense. The same general approach can be used when there are dominance-related differences in intake rate, as shown in figure 2.1. Here individuals differ in their ability to obtain access to food (or other resources), and some individuals achieve higher average rates of intake than others. All individuals are expected to be found only in one area if the intake rate of the lowest ranked individual when

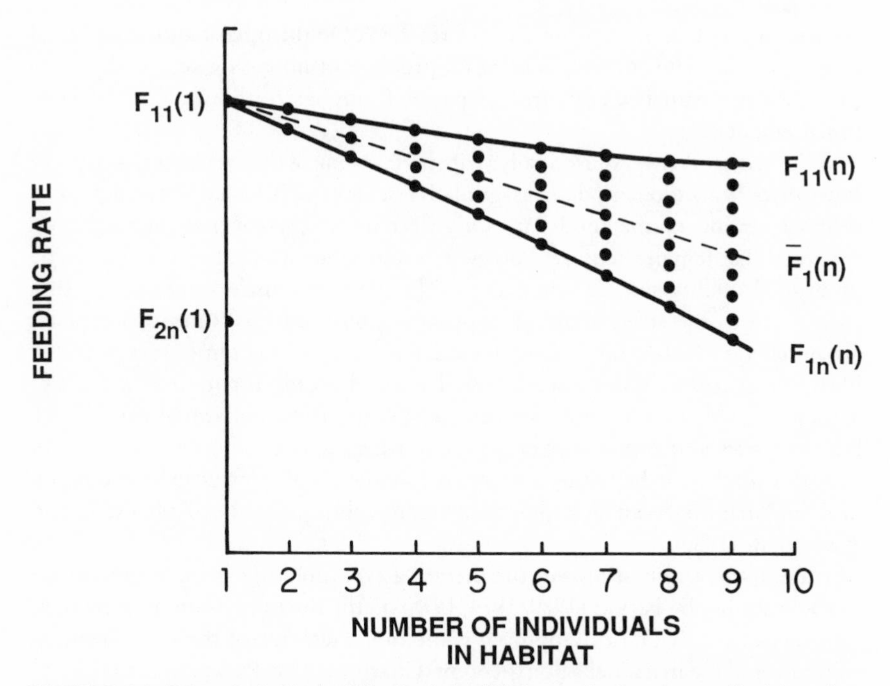

FIGURE 2.1. When individuals differ in their abilities to obtain access to food (or other resources), some individuals achieve higher average rates of intake than others. F_{11} (1) is the feeding rate available to the most dominant individual when it is alone in habitat 1, while F_{11} (n) is this individual's feeding rate in habitat 1 when there are n individuals present in the habitat. The most subordinate individual (i.e., the nth individual) will stay in habitat 1 only if the feeding rate it obtains there is greater than that available to it in habitat 2; that is, it will stay as long as F_{1n} $(n)>F_{2n}$ (1).

all individuals are together exceeds the expected rate of intake that individual could achieve by moving to an alternative area. In any case, the expected distribution of individuals among areas is such that no individual can increase its rate of intake by moving. As shown by Pulliam and Caraco (1984), no such Nash solution exists in some situations, and the individuals may continue to move back and forth between the available areas in a kind of round robin, or an equilibrium may be achieved by cooperation (i.e., the Pareto solution to an iterated game; see Axelrod and Hamilton 1981 and Pulliam, Pyke, and Caraco 1982).

EVOLUTIONARILY STABLE MATING STRATEGIES

Mating decisions can be viewed as involving both the procurement and allocation of resources. If mates are relatively scarce or otherwise hard to come by, mating becomes a problem of securing a scarce resource. On the other hand, if potential mates are abundant, mating becomes a problem of allocating scarce resources (time, energy, gametes, etc.). In this paper, I emphasize the problem of mate procurement.

The Orians-Verner-Wilson polygyny model is one of the very best examples of how an optimization model has guided research in behavioral ecology. The original version of the model by Orians (1969) envisions that males defend resources that females require. Polygyny results when a female chooses to settle on a good quality site already occupied by another female rather than use a poorer site as the sole female. The optimization model assumes that females always choose the best site currently available much in the same ways as in the ideal free distribution discussed above. The equilibrium distribution of females among available sites is evolutionarily stable only if no individual can change breeding sites and expect higher reproductive success. Numerous examples of how this model has been the conceptual foundation of fruitful field studies on birds and mammals can be found in the recent volume edited by Rubenstein and Wrangham (1986).

Among the very best tests of the theory of evolutionarily stable strategies are the field studies by Parker (1970, 1974, 1978) on the mating system of dung flies. In a very clever theoretical approach combining elements of the ideal free distribution and the marginal value theorem (Charnov 1976), Parker predicts where male dung flies should search and how long they should search in one patch before leaving to look elsewhere. He demonstrates that the payoffs of various mating strategies available to the males are frequency dependent, that is, the payoff depends on how many other males adopt the same strategy. Accordingly, he predicts a mixed ESS with males spending part of their time in very crowded but otherwise high quality patches and part of their time in poorer but less crowded patches. His very careful field tests confirm the model predictions and show the utility of the ESS approach.

CONSIDERATION OF SPATIAL AND TEMPORAL SCALE

If animals choose a habitat based on the feeding rate or reproductive success that can be achieved there, models of population growth and regulation may be built using submodels of habitat choice and distribution. At least in theory, the habitat distribution can be predicted based on the distribution of resources. In turn, the habitat distribution of consumers gives information needed to predict changes in the distribution of resources and, again in turn, changes in the distribution of resources allow prediction of the change in consumer distribution, and so on. Furthermore, if the basis of habitat distribution is the expected survival and reproductive value of consumers using that habitat, the models of habitat distribution inherently contain information about the demographic consequences of habitat selection.

To date, most of the use of game theory for modeling the habitat distributions of animals has focused on predicting the equilibrium distributions of individuals among habitats using the concept of the Nash equilibrium. In applying such habitat distribution models to models of population dynamics, several authors (e.g., Rosenzweig 1981; Holt 1985; and Pulliam 1988) have assumed that behavioral events such as habitat selection occur much more rapidly than do the demographic events involved in population regulation, so that the habitat distribution of individuals is usually considered to be in equilibrium with the current distribution of resources. In reality, this may not always be the case, particularly if the time scale over which individuals assess and move between habitats is on the same temporal scale as that of demographic events. For example, if birds breed only once a year and choose a habitat only once a year during a spring dispersal period, then the distribution of individuals between habitats cannot be assumed to track within-year changes in food abundance.

Models of habitat distribution may also need to take spatial scaling into account. For example, a checkerboard of one-hectare plots of oak and pine forest may allow individual mobile animals with home ranges of a few hectares quickly to detect changes in the availability of food or breeding sites in adjacent habitats, whereas the same animals in a landscape of 100-hectare checker squares would only very slowly take advantage of such opportunities. Models of habitat selection with rate constants determining the rate of approach to an equilibrium spatial ditribution may, therefore, capture much more of the dynamics of habitat selection in a changing, heterogeneous environment than do idealized models that assume instantaneous equilibrium between consumers and their resources. Making the rate constants dependent upon the actual amount of contact between adjacent habitats in natural landscapes may allow an accuracy of prediction useful to the management of real populations.

PREDATION HAZARD AND HABITAT DISTRIBUTION

So far, I have discussed habitat distribution entirely in terms of the access to resources, mostly food and mates, in those habitats. A number of recent experiments have convincingly demonstrated that predators play an important indirect role in the habitat distribution of prey. For example, Werner et al. (1983) demonstrated that bluegill sunfish choose the habitat where their feeding rate is highest in the absence of predators, but prefer a less profitable habitat when predatory bass are in the otherwise better habitat. In my own studies of white-throated sparrows, I have found that individual birds will accept approximately a 20% reduction in their intake rate in order to stay a half meter closer to a brush pile where they are safe from predators (H. R. Pulliam and M. Dodd, in prep.). Other studies with insects (Sih 1982), amphibians (Petranka 1983), rodents (Kotler 1984; Price 1986), and birds (Ekman 1986) suggest that predation is an important factor influencing the habitat preferences of many mobile organisms.

The most thorough theoretical treatment, to date, of the conflicting needs of achieving high food intake and avoiding predators is due to Gilliam (1982; also see Werner and Gilliam 1984; Mangel and Clark 1986; and Stephens and Krebs 1987). The Gilliam model uses optimal control theory to explore how predation hazard might influence habitat choice by animals attempting to maximize their lifetime net reproductive rate.

A very simplified model of habitat choice under a trade-off between predation hazard and feeding rate is as follows. Assume an animal must acquire R grams of food in order to avoid starvation. The animal has the choice of feeding in habitat 1 where its intake rate is f_1 or habitat 2 where its intake rate is f_2. The time (T) required for the animal to achieve its requirement of R grams is thus R/f_1 for habitat 1 and R/f_2 for habitat 2. Now assume that the instantaneous predation hazard in habitat 1 is u_1 and in habitat 2 is u_2, so that, for small uT, the probability of predation in time T is u_1T and u_2T, respectively. Clearly then, assuming that the animal has enough time to find the required amount of food in either habitat and that it is subject to predation only while foraging, the likelihood of predation is lower if the animal chooses the habitat with the lower ratio of u/f.

The most direct test of the prediction that animals should choose a habitat based on the rule "minimize u/f" is an experiment by Gilliam and Fraser (1987). They offered juvenile creek chubs a choice between feeding sites in a natural stream that differed in both the number of predators (adult chubs) and the density of prey (tubifex worms). They measured both the functional response, relating feeding rate (f) to tubifex worm density, and the predation hazard, relating probability of death per unit time (u) to predator density. They then chose combinations of prey and predator densities that differed in the ratio u/f. The juvenile chubs showed a strong preference for whichever habitat had the lowest u/f ratio. More experiments of this kind will be necessary to determine how precisely animals assess feeding rate and predation hazard, but the Gilliam and Fraser experiments already suggest that a very simple model may predict the general patterns of habitat choice with predation hazard.

UTILITY OF THE OPTIMIZATION APPROACH

Despite the fact that optimization models of animal behavior have proven useful for both the design and interpretation of experiments and field observations, the approach has been criticized on the grounds that the proposition that evolution maximizes fitness is inherently circular and therefore unfalsifiable (Gould and

Lewontin 1979; Ollason 1987). However, as has been pointed out by many others (Maynard Smith 1978; Stephens and Krebs 1987; Krebs and Davies 1984), the research program does not attempt to prove that evolution maximizes fitness. Rather than test the proposition that evolution maximized fitness in the past, behavioral ecologists attempt to use the proposition that current behavior maximizes expected fitness to generate predictions about how animals will choose among the options available to them. It is clear that, when presented with the sort of options faced in their natural environment, animals do not choose at random but are rather more likely to perform the actions leading to higher expected future reproductive success.

It is true that, when animals fail to behave as predicted, behavioral ecologists often have difficulty determining whether the experiment was not designed properly or if the model failed to predict the optimal behavior. This is analogous to the problem in any science of determining whether anomalous results are due to a faulty experiment or a faulty model. The assumption that behavior maximizes fitness is part of any specific optimization model of behavior, and if the experimenter concludes the experiment was conducted properly, then he must also conclude the fault lies with the model. However, to conclude that one particular model is in fault is not justification to abandon all use of optimization models. Only if repeated attempts to rectify the model fail to predict the behavior at the desired level of resolution should the optimization approach be abandoned. To date, most experimentalists have been satisfied that optimization models of behavior explain enough of reality to justify their continued use.

Much of the success of foraging theory to date hinges on two facts: (1) the theoretical models are built on realistic assumptions about how animals locate and consume food; and (2) the parameters of the theoretical models are operationally defined and can be measured in field and laboratory experiments. The continued success of the field requires that theorists continue to work closely with experimentalists so that the models do not lose touch with reality, as has often happened in other areas of ecology.

ACKNOWLEDGMENTS

Joel Cohen, Peter Kareiva, Bob May, Marc Mangel, and Glenn Reynolds all read an earlier version of the manuscript and made many helpful suggestions.

REFERENCES

Axelrod, R., and W. D. Hamilton. 1981. The evolution of cooperation. *Science* 211:1390–96.

Bedington, J. R. 1975. Mutual interference between parasites or predators and its effect on searching efficiency. *J. Anim. Ecol.* 44:331–40.

Caraco, T. 1987. Foraging games in a random environment. In A. Kamil, J. R. Krebs, and H. R. Pulliam, eds., *Foraging Behavior.* Plenum Press, New York.

Caraco, T., and H. R. Pulliam. 1984. Sociality and survivorship of animals exposed to predation. In P. N. Price, C. N. Slobodchikoff, and W. S. Gand, eds., *A New Ecology: Novel Approaches to Interactive Systems.* Wiley, New York.

Charnov, E. L. 1976. Optimal foraging, the marginal value theorem. *Theor. Pop. Biol.* 9:129–36.

Clark, C. W., and M. Mangel. 1984. Foraging and flocking strategies: Information in an uncertain environment. *Amer. Natur.* 123:626–41.

Clark, C. W., and M. Mangel. 1986. The evolutionary advantages of group foraging. *Theor. Pop. Biol.* 30:45–75.

Davies, N. B., and T. R. Halliday. 1979. Competitive mate searching in common toads, *Bufo bufo. Anim. Behav.* 27:1253–67.

Ekman, J. 1986. Tree use and predator vulnerability of wintering passerines. *Ornis Scand.* 17:261–67.

Feller, W. 1968. *An Introduction to Probability Theory and its Applications,* 3d ed., vol. 1. Wiley, New York.

Fretwell, S. D. 1972. *Populations in a Seasonal Environment.* Princeton University Press, Princeton, N.J.

Gilliam, J. F. 1982. Habitat use and competitive bottlenecks in size-structured fish populations. Ph.D. dissertation, Michigan State University, East Lansing, Mich.

Gilliam, J. F., and D. F. Fraser. 1987. Habitat selection under predation hazard: Test of a model with foraging minnows. *Ecology* 68:1856–62.

Gould, S. J., and R. C. Lewontin. 1979. The spandrels of San Marco and the Panglossian paradigm: A critique of the adaptationist programme. *Proc. Roy. Soc. Lond.* (B). 205:581–98.

Harper, D.G.C. 1982. Competitive foraging in mallards: "Ideal free" ducks. *Anim. Behav.* 30:575–84.

Hassell, M. P. 1978. The dynamics of arthropod predator-prey systems. Princeton University Press, Princeton, N.J.

Hassell, M. P., and R. M. May. 1985. From individual behaviour to population dynamics. In R. M. Sibly and R. H. Smith, eds., *Behavioural Ecology: Ecological Consequences of Adaptive Behaviour,* Blackwell Scientific, Oxford.

Holling, C. S. 1959. Some characteristics of simple types of predation and parasitism. *Can. Entomol.* 91:385–98.

Holling, C. S. 1961. Principles of insect predation. *Ann. Rev. Entomol.* 6:163–82.

Holling, C. S. 1966. The strategy of building models of complex ecological systems. In K.E.F. Watt, ed., *Systems Analysis in Ecology,* pp. 195–214. Academic Press, New York.

Holt, R. D. 1985. Population dynamics in two-patch environments: Some anomalous consequences of optimal habitat selection. *Theor. Pop. Biol.* 28:181–208.

Houston, A., C. Clark, J. McNamara, and M. Mangel. 1988. Dynamic models in behavioural and evolutionary ecology. *Nature* 332:29–34.

Kareiva, P. 1987. Habitat fragmentation and the stability of predator-prey interactions. *Nature* 326:388–90.

Kotler, B. P. 1984. Risk of predation and the structure of desert rodent communities. *Ecology* 65:689–701.

Krebs, J. R., and N. B. Davies, eds. 1984. *Behavioral Ecology: An evolutionary approach,* 2d ed. Blackwell Scientific, Oxford.

Mangel, M., and C. W. Clark. 1986. Towards a unified foraging theory. *Ecology* 67:1127–38.

Maynard Smith, J. 1976. Evolution and the theory of games. *Amer. Scient.* 64:41–45.

Maynard Smith, J. 1978. Optimization theory in evolution. *Ann. Rev. Ecol. Syst.* 9:31–56.

Milinski, M. 1979. An evolutionarily stable feeding strategy in sticklebacks. *Z. Tierpsychol.* 51:36–40.

Ollason, J. G. 1987. Foraging theory and design. In A. Kamil, J. R. Krebs, and H. R. Pulliam, eds., *Foraging Behavior.* Plenum Press, New York.

Orians, G. H. 1969. On the evolution of mating systems in birds and mammals. *Amer. Natur.* 103:589–603.

Parker, G. A. 1970. The reproductive behavior and the nature of sexual selection in *Scatophaga stercoraria* L. II. The fertilization rate and the spatial and temporal relationships of each sex around the site of mating and oviposition. *J. Anim. Ecol.* 39:205–28.

Parker, G. A. 1974. The reproductive behavior and the nature of sexual selection in *Scatophaga stercoraria* L. IX. Spatial distribution of fertilization rates and evolution of male search strategy within the reproductive area. *Evolution* 28:93–108.

Parker, G. A. 1978. Searching for mates. In J. R. Krebs and N. B. Davies, eds., *Behavioural Ecology: An Evolutionary Approach,* Blackwell Scientific, Oxford.

Parker, G. A. 1979. Sexual selection and sexual conflict. In M. S. Blum and N. A. Blum, eds., *Sexual Selection and Reproductive Competition in Insects,* pp. 123–66. Academic Press, New York.

Petranka, J. W. 1983. Fish predation: A factor affecting the spatial distribution of a stream-breeding salamander. *Copeia* 1983:624–28.

Price, M. V. 1986. Structure of desert rodent communities: A critical review of questions and approaches. *Amer. Zool.* 26:39–49.

Pulliam, H. R. 1987. On the evolution of density-regulating behavior. In P.P.G. Bateson and P. H. Klopfer, eds., *Perspectives in Ethology,* vol. 7: *Alternatives.* Plenum Press, New York.

Pulliam, H. R. 1988. Sources, sinks, and population regulation. *Amer. Natur.* 132. In press.

Pulliam, H. R., and T. Caraco. 1984. Living in groups: Is there an optimal group size? In J. R. Krebs and N. B. Davies, eds., *Behavioural Ecology: An Evolutionary Approach,* 2d ed. Blackwell Scientific, Oxford.

Pulliam, H. R., and M. Dodd. In prep. The influence of social dominance and distance to cover on feeding site selection by white-throated sparrows.

Pulliam, H. R., G. H. Pyke, and T. Caraco. 1982. The scanning behavior of juncos: A game-theoretical approach. *J. Theor. Biol.* 95:89–103.

Rogers, D. J., and M. P. Hassell. 1974. General models for insect parasite and predator searching behaviour: Interference. *J. Anim. Ecol.* 43:239–53.

Rosenzweig, M. L. 1981. A theory of habitat selection. *Ecology* 62:327–35.

Rosenzweig, M. L. 1985. Some theoretical aspects of habitat selection. In M. L. Cody, ed., *Habitat Selection in Birds,* pp. 517–39. Academic Press, New York.

Rubenstein, D. I., and R. W. Wrangham, eds. 1986. *Ecological Aspects of Social Evolution: Birds and Mammals.* Princeton University Press, Princeton, N.J.

Sih, A. 1982. Optimal patch use: Variation in selective pressure for efficient foraging. *Amer. Natur.* 120:666–85.

Solomon, M. E. 1949. The natural control of animal populations. *J. Anim. Ecol.* 18:1–35.

Stephens, D. W., and J. R. Krebs. 1987. *Foraging Theory.* Princeton University Press, Princeton, N.J.

Turchin, P. B. 1986. Modelling the effect of host patch size on Mexican bean beetle emigration. *Ecology* 67:124–32.

Werner, E. E., and J. F. Gilliam. 1984. The ontogenetic niche and species interactions in size-structured populations. *Ann. Rev. Ecol. Syst.* 15:393–425.

Werner, E. E., J. F. Gilliam, D. J. Hall, and G. G. Mittelbach. 1983. An experimental test of the effects of predation risk on habitat use in fish. *Ecology* 64:1540–48.

Whitham, T. G. 1980. The theory of habitat selection examined and extended using Pemphigus aphids. *Amer. Natur.* 115:449–66.

Chapter 3

Discussion: From Individuals to Populations

M.A.R. KOEHL

Processes operating at the level of individual organisms can determine the properties of populations, communities, and ecosystems. Theoretical studies can play an important role in advancing our understanding of the connection between organismal-level performance and patterns at the ecological level. In this paper I will first report our discussion of approaches to the area, including the roles of mechanistic versus phenomenological models, the interplay of theory and empiricism, and the usefulness of simple models and microcosm studies in understanding a complex world. I will then present a brief summary of examples of organismal-level analyses that have contributed to our understanding of ecological phenomena, and will conclude with a report of our discussion of directions for future research. This paper does not represent my personal view of this field, but rather attempts to report the range of opinions expressed during our meeting.

APPROACHES

Mechanistic versus Phenomenological Models: Why Should We Worry about an Individual When We Are Describing a Population?

Some models are phenemonological descriptions of a system, whereas other models seek to understand the essential processes governing components of the system and to build up from such a basic understanding of underlying mechanisms to an overall description.

A number of arguments can be made for focusing on phenomenological models (e.g., Rigler 1982). Using phenomenological regression predictions, one has some hope of getting answers in a timely fashion. For example, it is more reasonable to use such a model of birth and death rates to forecast the human population next year than it is to try to understand all the reasons that people decide to have children or to drive while drunk. Furthermore, often even elaborate mechanistic models do not match the data as well as some very simple phenomenological regression models (e.g., Gross, chapter 1). Why, then, should we worry about individual function or burden ourselves with the more cumbersome mechanistic models?

The limitations of phenomenological models render them inappropriate for certain types of analyses. Whenever we make a prediction using a phenomenological model, we implicitly assume (1) that conditions do not change, and (2) that the phenomena that go into the model adequately sample the causal pattern of interest. Therefore, phenomenological models are best used for making short-term predictions. For example, although it is sensible to use phemomenological descriptions of fertility rates to predict human population growth on the time scale of decades, it is necessary to understand important aspects of the physiology and behavior of individuals (such as how reproductive physiology is affected by factors like intervals of breast-feeding, or the interplay of workload and nutrition) to understand what may have held hunter-gatherer populations in balance for thousands of years. Similarly, regression predictions of the spread of a disease such as AIDS are useful for the next few years, but longer-term predictions should be made based on an understanding of the dynamics of the behaviors of individuals that transmit AIDS (May and Anderson 1987). Furthermore, conditions do not even have to change for phenomenological models to break down if a system is regulated chaotically—a system can be characterized by two regular patterns between which it can suddenly shift. Another major limitation to the use of phenomenological models is that they can be applied only to systems for which data are already available. Today man is contemplating perturbations to unique and irreplaceable ecosystems for which we have no data on which to base phenomenological models; in such cases mechanistic models have to be employed.

Although mechanistic models may not fit the data as well as phenomenological models and may be complicated and slow to provide answers, the development of mechanistic theories can lead to increased understanding of how a system works. A number of examples of how mechanistic models have provided ecological insights are reviewed by Schoener (1986). Note that one important lesson we have learned even from simple linear mechanistic models (such as the physical laws governing the motion of a pendulum) is that something that is rigidly determined can essentially be unpredictable if it is sensitive to initial conditions. Such mechanistic models point out the importance of history and stochastic processes in determining the course of events, even in systems for which we know the rules governing behavior.

A further discussion of the pros and cons of phenomenological versus mechanistic models can be found in Peters (1986) and Lehman (1986). Ideally these different approaches can reinforce each other. Phenomenology helps organize observations so that mechanistic laws can be formulated, and these in turn can explain the phenomenological rules. There are important problems that can best be addressed by one or by the other approach, depending on the particular question being studied.

What Can Organismal Biologists and Ecologists Do For Each Other?

Many mechanistic ecological models assume that particular processes at the organismal level are important in governing the behavior of a system at a larger level of organization, such as a population or community. Examples of such ecological modeling are discussed by Schoener (1986) in a recent symposium about mechanistic approaches in ecology (Price 1986). One important role of studies at the organismal level is that they can tell ecologists factors that can be ignored versus those that must be included when simplified models are developed. In addition, organismal studies can provide information for ecological theorists about the rates at which various processes occur, revealing those that are nonlinear. One important consequence of learning how mechanisms on the organismal level affect the properties of populations and communities is that such knowledge can provide a link between ecology and evolution. Conversely, we cannot understand the evolution of organismal-level traits (physiological, morphological, behavioral) without understanding the ecological context in which natural selection operates.

Interplay between Theory and Empiricism

The interaction of theory and empiricism should be a leapfrogging activity where theories lead to experiments that point out their limitations or disprove them, thereby stimulating new theories and experiments; our understanding of the way a system works is thus improved step by step. As physicists are well aware, the successful theory is one that leads to an understanding of why it is wrong and what should be done next. Although it should be a constructive activity and cause for celebration when a theory is found to be less than perfect, ecologists tend to denigrate theory when models are found to be flawed. We need to recognize that the successful theories are the ones that are kicked out, thereby leading to interesting experiments and new theories. A recent discussion of the psychological and philosophical aspects of the interplay of theory and empiricism in ecology can be found in Loehle (1987). Improving communication between theoretical and empirical ecologists, who often have quite different training, should lead to more fruitful interactions.

What can theorists do for empiricists? One important role of theory is to simplify complex ecological problems—to take the unmanageable and make it manageable. One of the many examples of such simplification is the concept used in some food-web models of a "trophic species," which may encompass a hundred taxonomic species with similar diets, and hence similar trophic roles in an ecosystem. By thus abstracting the essential elements of what is known about a system, theory can point out where to look for the next problem and can instruct empiricists to focus on research systems that are not overwhelmingly complex. Another way in which quantitative theorists can simplify the work of empiricists is by pointing out which of the myriad of parameters that could be measured are the ones most likely to have large effects on the process being studied. Optimization models can reveal whether the "price" of straying from some optimum behavior or morphology is large or trivial, and hence can alert experimentalists to traits on which to focus their attention. One other important job for theorists is to be very explicit about the whole battery of assumptions on which their models are based. Such clarification enables empiricists to judge whether or not a particular model is appropriate for the system they are investigating and permits them to avoid the trap of simply using the model in vogue at the time of their study.

What can empiricists do for theorists? One role for experimentalists is to point out to theorists the important unsolved empirical problems and the unexplained patterns observed in nature. Another job for empiricists is to use their knowledge of natural history to show theorists what they can and cannot ignore as they try to simplify and abstract a system. It behooves empiricists to demonstrate to theorists the relevance to ecological processes of organismal-level details, such as the relationship of nutritional physiology or thermoregulation to foraging behavior, and hence to community structure. A very important service that empiricists can perform is to test whether models have any relationship to what goes on in nature. One aspect of this empirical work is descriptive: the measurement of parameters to plug into models, and the comparison of predictions of models with observations in nature. We need to remember, however, that agreement of nature with the predictions of a theory does not necessarily mean that the theory is correct (e.g., Dayton 1973). Another aspect of empirical evaluation of theory is experimental research: specific hypotheses are tested by controlled, manipulative experiments conducted either in laboratory microcosms or in the field. If theorists want empiricists to test their models, they should build models with *measurable* parameters and *testable* hypotheses. Empiricists need to let theorists know what is "doable," and theorists should familiarize themselves with how empirical work is done and with the difficulties of experimental ecology (e.g., Platt 1964; Quinn and Dunham 1983; Hurlbert 1984).

The key to the successful interplay of theory and empiricism is communication. It is encouraging that some ecologists do both theoretical and empirical research, and that close collaborations have developed between a number of modelers and experimentalists.

Simple Models of a Complex World: Theories and Microcosm Experiments

Rather than become overwhelmed by the bewildering complexity of nature, ecologists can develop simplified theories or conduct experiments in manageable microcosms. The hope is that such simple systems may reveal basic principles valid both for and within more complex systems. Experiments using microcosms offer a number of advantages: they can be designed to reproduce the assumptions of a model being tested, and they can be conducted in replicate with appropriate controls. The small sizes and short generation times of organisms such as insects or bacteria make them appealing subjects for microcosm studies. Furthermore, modern microbiological techniques permit us to know a great deal about the genetic variability between individuals in populations of micro-organisms we can use in our experiments. Nonetheless, we must remember that the kinds of questions that we can ask of such organisms are different from those we must ask of organisms with larger bodies, longer lives, and more flexible behavioral repertoires, such as vertebrates.

Although we certainly should first measure processes in and develop theory for simple systems before moving on to more complicated ones, we must keep a number of questions and precautions in mind. If bottle experiments—for example, of predator-prey relations—are made simple enough to mimic the assumptions of a model they are testing, isn't such laboratory ecology little more than analog computing? Such experiments are certainly useful to test models, but are they so unrealistic that they tell us little about nature? Microcosms are by definition smaller than natural systems; how do we decide what scale (spatial and temporal) is appropriate to test theories about processes occurring in natural systems (see Giesy 1978; Powell, chapter 11, this volume; Steele, chapter 12, this volume)? To what extent can the unpredictability encountered in nature be sensibly replicated in controlled experiments? Good lab work must be based on a sensitive appreciation of natural history; otherwise it is all too easy to miss crucial elements of the real system when we bring it into the lab for study. We must also remember that what is considered "simple" is strongly conditioned by the background and objectives of the scientist. For example, a thermal spring community of twenty-five species that seems appallingly messy to a microbial physiologist can appear delightfully simple to a field ecologist.

It is important that we pursue sensible answers to these questions: there is a very specific and growing demand for understanding what we need to do as pilot studies in microcosms prior to the release of genetically engineered organisms.

WHERE ARE WE NOW?

Information about organismal-level functions (e.g., behavior, physiology, biomechanics) has proved useful in understanding ecological processes, and vice versa. Gross and Pulliam (chapters 1 and 2) review a number of studies that illustrate the fruitful interplay between theory and empiricism at this interface between the organismal and ecological levels of organization. In this report I will briefly mention (with a few leads into the literature) some additional areas of research brought out in our discussions, but not covered in those papers, that exemplify the contributions of theoretical work.

Plant Pysiological and Biophysical Ecology

Gross (chapter 1) reviews the physiological ecology of terrestrial plants, but not aquatic ones. Examples of the usefulness of organismal-level physiological information to the development of community-level models of phytoplankton can be found in Tilman (1982) and Powell and Richerson (1985). Discussions of the importance of the physics of nutrient flux at the surfaces of individual cells to questions about the productivity of lakes and oceans can be found, for example, in Jackson (1980) and Lehman (1984). Empirical and theoretical work on the biophysical ecology of marine macrophytes is reviewed by Koehl (1985).

Behavioral Ecology

Aspects of the interface between animal behavior and ecology not covered by Pulliam (chapter 2) are reviewed in Krebs and Davies (1984) and Rubenstein and Wrangham (1986), and a recent synthesis of optimal foraging models is given in Stephens and Krebs (1987).

Animal Biophysical and Physiological Ecology

Another field rich in examples of the interplay of theory and empiricism is the study of the interface between the physical and physiological performance of animals and their ecological function. Rather than attempt to review this growing field, I will merely mention a few examples here.

Analyses of the biophysics of heat and water regulation reveal when and where particular animals can be active, and hence point out constraints on habitat use, on ecological interactions such as competition and predation (e.g., Porter et al. 1975; Heinrich 1979), and on reproductive strategies (e.g., Kingsolver 1983). Conversely, models that incorporate the ecological roles of animals can provide insights about organismal-level function. such as which animals should be expected to thermoregulate and which should not (e.g., Huey and Slatkin 1976).

Our understanding of foraging ecology and habitat use by animals has also been expanded considerably by analyses of the metabolic costs of various activities (e.g., Heinrich 1979), and by studies of nutritional physiology and chemical defenses (e.g., Rosenthal and Janzen 1979; Crawley 1983; Hubbel and Howard 1984).

Biomechanics (see, e.g., Wainwright et al. 1976; Vogel 1981; Alexander 1983; Denny 1984) also provides ecologists with information about physical constraints on the ecological performance of organisms. One obvious example of the ecological importance of mechanical processes is the role of physical disturbance in structuring many communities (e.g., Sousa 1984). Examples for rocky shore communities of mechanistic studies of disturbance that involve both theory and empirical work include biomechanical studies at the organismal level (e.g., Koehl 1977; Denny, Daniel, and Koehl 1985), which reveal mechanisms responsible for the differences in susceptibility of various organisms to removal by waves, and studies at the community level, which focus on the ecological consequences of this removal (e.g., Paine and Levin 1981). Biomechanics has also shed light on other ecological questions. For example, flight aerodynamics provide a mechanistic explanation for the patterns of hummingbird foraging in habitats at different altitudes (Feinsinger et al. 1979). Similarly, a mathematical model of the biomechanics of nectar feeding led to predictions of strategies of foraging by hummingbirds and of nectar production by plants (Kingsolver and Daniel 1983). An analysis of the mechanics of silk webs proved necessary to understand habitat use and foraging by spiders in a tropical forest (Craig 1987). Similarly, patterns in habitat use and foraging by fish of different body forms can be related to the biomechanics of swimming (e.g., Webb 1984). Conversely, ecological analyses can make sense of otherwise puzzling biomechanical features. For example, the abundance on wave-swept reef crests of a species of coral with a "bad" (e.g., breakage-enhancing) mechanical design was explained by an ecological study that showed breakage to be an important mode of asexual reproduction and dispersal for this species (Tunnicliffe 1981).

A combination of mathematical modeling and empirical measurements have also been used to study the physical environments of organisms (e.g., Monteith 1973; Gates 1980; Okubo 1980; Nowell and Jumars 1984; and Denny 1988). At the ecosystem level, such information is necessary to analyze the flux of various substances into and out of the components of a system (e.g., slowed currents limit nutrient supply in a kelp forest [Jackson and Winant 1983]; boundary layer hydrodynamics determine mass transport to and from benthic communities [Jumars and Nowell 1984]; turbulence affects the flux of nutrients driving phytoplankton productivity in the ocean [Lewis et al. 1986]). At the biogeographic and community levels, analyses of fluid motions in the environment are sometimes necessary to understand the large- and small-scale spatial patterns of distribution of organisms that disperse by propagules such as wind-borne seeds or current-borne larvae (e.g., Scheltema 1975; Eckman 1983; Shanks 1985). Obviously the physics of water motion is also critical in producing patterns of distribution of phyto- and zooplankton (e.g., Denman and Powell 1984; Mackas, Denman, and Abbott 1985).

WHAT NEXT?

A number of directions for future research at the interface between organismal biology and ecology were discussed. Although our discussion ranged from everyone's favorite philosophies of science to each participant's pet research topic, I will attempt a concise summary.

Techniques for the Future

New techniques often pave the way for scientific progress, opening up problems that previously were not accessible to incisive research. Several empirical and theoretical tools were mentioned that may prove useful in the near future. On the empirical side, new molecular techniques that enable us to measure the genetic structure of field populations open exciting possibilities for investigating questions at the interface between population ecology and population genetics (as do the microbiological techniques mentioned above in the context of microcosm experiments). On the theoretical side, behavioral ecologists might borrow the models of irrational choice now being developed by economists (Tversky and Kahneman 1981). Artificial intelligence languages should be more useful than the more usual high-level computer languages for modeling the behavior of individuals aggregated into populations of interacting units. In the coming years there is likely to be extension of ESS and game theory modeling to include dynamical game theory (e.g., Brown and Vincent 1987). Similarly, relatively simple optimization models are giving way to more complex ones; work in dynamic optimization is one direction, and modeling that includes the effects of optimization under uncertainty is another (e.g., Mangel and Clark 1983, 1986, 1988).

Questions on Which to Focus in the Future

In spite of the temptation to find problems suited to the latest flashy technique, there was a consensus that we need to focus more deliberately on asking the right questions. We should let important unsolved biological problems, rather than appealing research tools, be our guide for future research. While some participants stressed the importance of focusing future research on specific issues important to mankind (such as the release of genetically engineered organisms, spread of diseases, or destruction of ecosystems), others argued that the answers to these practical problems, as well as to more basic academic questions, still hinge on gaining a better understanding of what determines the abundance and distribution of organisms. It appears that a multiplicity of approaches is in order.

POPULATION REGULATION

One critical area for future investigation remains the problem of what factors regulate populations. Rather than focus on resources, as we have tended to do in the past, we should be more explicit about the demographic consequences of behavioral, physiological, and biophysical mechanisms that affect survivorship and reproduction, and that determine the distribution of individuals among habitats.

WORLD DOMINANCE BY CERTAIN ORGANISMS

An interesting phenomenon, well known to biogeographers, is that there are certain groups of organisms that arise (sometimes rather quickly) in the geological record and literally sweep the world. This is an ecologically important phenomenon because such organisms tend to wipe out certain pre-existing groups but to co-exist with others, for reasons we do not yet understand. What are the general traits, if any, that contribute to world dominance by certain groups of organisms?

VARIABILITY

A number of the directions of future research that we discussed share the common theme of incorporating variability (within a population, within a habitat, or of an individual) into ecological models. In analyzing ecosystems, communities, or populations, when should we retain the diversity of species or individuals, and when can we simplify?

It is crucial to recognize that the average dynamical behavior of a nonlinear system can be very different from the dynamical behavior of the average (e.g., May 1986). Therefore individual differences can be quite important, and organismal-level information about the degree of variability within a population (or species or community) will become increasingly necessary.

Not only should we pay more attention to how population dynamics are affected by individual variability, but we should also investigate mechanisms (such as those resulting from breeding systems or environmental heterogeneity) that maintain variability in populations. Optimization models predict a "best" phenotype for a habitat, and hence cannot account for the variation seen within a population (or between different species utilizing the same habitat) (e.g., Gross, chapter 1). In the future, such models should take into account environmental variations in space and time to explore the relationship between genetic variability and the predictability of particular habitat types. A compendium of the sorts of mathematical models that could lead to maintenance of variation would be very useful at this time.

Habitat variability in space and time is also a central feature of several other directions of future research. For example, when modeling the population dynamics of plants or sessile animals, we need to learn whether predictions can be made without detailed knowledge of the neighborhood (biotic and abiotic) around individuals. In addition, we should explore the extent to which the local population abundance of motile or sessile organisms is determined by characteristics of the resources available on a site versus characteristics of the mosaic of habitats in which the site is embedded (e.g., Pulliam, chapter 2). Future research should also explore factors that produce and maintain the spatial structure of populations in variable environments. Although dispersal (by seeds, larvae, or motile individuals) in patchy environments is no doubt a crucial factor in determining population dynamics (e.g., Roughgarden and Iwasa 1986) and is an important feature of life-history strategies (e.g., Jackson and Strathmann 1981), we should recognize the tremendous technical challenge posed by empirically studying this phenomenon in nature.

Variability within one individual (both phenotypic plasticity and learned behavioral change) represents another important topic for future research. What are the laws, if any, governing the relationship between phenotypic plasticity and genotypic variability in natural populations (e.g., Travis and Mueller, chapter 7)? How might genotype-environment interactions affect plasticity (e.g., Via and Lande 1985)? Some kinds of animals and some types of behaviors are modified by experience (learning) and others are not. Does the time-course of environmental events relate to this difference in flexibility? Dynamic game theories that consider mixed strategies versus coalition strategies or pure strategies might be employed to tell us those circumstances under which an individual ought to be flexible versus rigid in its behavior. At present such models are deterministic (e.g., Brown and Vincent 1987); the role of stochastic events might be incorporated in the future. Another aspect of variability within individuals that should be explored is the population-level consequences of such plasticity in behavior, morphology, or physiology. For example, in considering ESS models, is the effect of each individual doing particular behaviors with a certain probability the same as the effect of a population composed of certain proportions of individuals that each specialize in one behavior?

OPTIMIZATION MODELS

A good deal of attention has been paid by theorists to organisms as entities that behave according to some sort of "optimality." One problem with this approach has been the difficulty of determining what is being optimized. For example, when considering optimal foraging, detailed questions of nutrition must be dealt with rather than simple caloric intake. Another problem with the optimality approach has been elucidating the constraints that limit the optimization process: for both plants and animals, we need a systematic exploration of the way in which physiology and physics constrain optimization.

An important direction for future research is to seek organizing principles of communities that are predictable from a knowledge of the degree to which behavior is optimized as a function of the density of a population.

PLANT ECOLOGY

Several specific areas of future research were mentioned for plant ecology. One is the coupling of multiple environmental factors as they affect an individual's performance. Once such coupling is better understood, we can move on to a more unified quantitative theory that can be related to models of plant geographic distribution or community composition.

Another area of future activity in plant ecology concerns the allocation of resources within individuals (e.g., Bloom et al. 1985). Analysis of the control of root-shoot allocation by plants in various environmental complexes should provide us with one simple pattern we can use to address questions at several levels of organization, from individual growth to competition and succession. Future modeling efforts should incorporate the roles of storage and of reproduction in mediating such allocation patterns. We should also explore the population- and community-level consequences of patterns of carbon and nitrogen allocation by plants into various products and defensive chemicals in different types of habitats.

Interface between Ecology and Evolution

Basic evolutionary processes shape the properties of individuals, which add up to the dynamical properties of populations. For too long, organismal biologists have tended to focus on the relationship between individual performance and evolution, while ecologists have tended to focus on the dynamics of populations without asking how the behavior of individuals affects population parameters. An overarching direction for future research should be the explicit relationships between individual performance and population dynamics. We must deal with the genetics as well as the behavior of populations if we are to understand how individuals determine the ways in which populations respond to change.

CONCLUSIONS

There appears to be a growing recognition that processes that occur at the level of individuals can form the basis for constructing a theoretical framework with which to interpret the properties of populations or communities.

Biological problems, rather than research techniques, should drive the directions of future investigations at this interface between ecology and organismal biology. There is considerable room for theoretical work of a great variety of kinds (ranging from simple phenomenological models to complicated mechanistic ones), as well as for empirical work of various sorts (ranging from insightful natural-history observations to manipulative field and laboratory experiments). If we are to make headway in this field, it is important to enhance communication between theorists and empiricists, and between ecologists, population geneticists, and organismal biologists.

REFERENCES

Alexander, R. McN. 1983. *Animal Mechanics,* 2d ed. Blackwell Scientific, Oxford.

Bloom, A. J., F. S. Chapin III, and H. A. Mooney. 1985. Resource limitation in plants—An economic analysis. *Ann. Rev. Ecol. Syst.* 16:363–92.

Brown, J. S., and T. L. Vincent. 1987. A theory for the evolutionary game. *Theor. Pop. Biol.* 31:140–66.

Craig, C. L. 1987. The ecological and evolutionary interdependence between web architecture and web silk spun by orb weaving spiders. *Biol. J. Linn. Soc.* 30:135–62.

Crawley, M. 1983. *Herbivory: The Dynamics of Animal-Plant Interactions.* University of California Press, Berkeley.

Dayton, P. K. 1973. Two cases of resource partitioning in an intertidal community: Making the right prediction for the wrong reason. *Amer. Natur.* 107:662–70.

Denman, K. L., and T. M. Powell. 1984. Effects of physical processes on planktonic ecosystems in the coastal ocean. *Oceanogr. Mar. Biol. Ann. Rev.* 22:125–68.

Denny, M. W., ed. 1984. Symposium: Biomechanics. *Amer. Zool.* 24:3–134.

Denny, M. W. 1988. *Biology and the Mechanics of the Wave-Swept Environment.* Princeton University Press, Princeton, N.J.

Denny, M. W., T. Daniel, and M.A.R. Koehl. 1985. Mechanical limits to the size of wave-swept organisms. *Ecol. Monogr.* 55:69–102.

Eckman, J. E. 1983. Hydrodynamic processes affecting benthic recruitment. *Limnol. Oceanogr.* 28:241–57.

Feinsinger, P., R. K. Colwell, J. Terborgh, and S. B. Chaplin. 1979. Elevation and the morphology, flight energetics, and foraging ecology of tropical hummingbirds. *Amer. Natur.* 113:481–97.

Gates, D. M. 1980. *Biophysical Ecology.* Springer-Verlag, New York.

Gates, D. M., and R. B. Schmerl, eds. 1975. *Perspectives in Biophysical Ecology.* Springer-Verlag, New York.

Giesy, J. P., ed. 1978. Microcosms in Ecological Research. *D.O.E. Symp.* 52, Augusta, Georgia, Conf. 781101, NTIS.

Heinrich, B. 1979. *Bumblebee Economics*. Harvard University Press, Cambridge, Mass.

Hubbell, S. P., and J. J. Howard. 1984. Chemical leaf repellency to an attine ant: Seasonal distributions among potential host plant species. *Ecology* 65:1067–76.

Huey, R. B., and M. Slatkin. 1976. Cost and benefits of lizard thermoregulation. *Quart. Rev. Biol.* 51:363–83.

Hurlbert, S. H. 1984. Pseudoreplication and the design of ecological field experiments. *Ecol. Monogr.* 54:187–211.

Jackson, G. A. 1980. Phytoplankton growth and zooplankton grazing in oligotrophic oceans. *Nature* 248:439–41.

Jackson, G. A., and R. R. Strathmann. 1981. Larval mortality from offshore mixing as a link between precompetent and competent periods of development. *Amer. Natur.* 118:16–26.

Jackson, G. A., and C. D. Winant. 1983. Effects of a kelp forest on coastal currents. *Cont. Shelf Res.* 2:75–80.

Jumars, P. A., and A.R.M. Nowell. 1984. Fluid and sediment dynamic effects on benthic marine community structure. *Amer. Zool.* 24:45–55.

Kingsolver, J. G. 1983. Ecological significance of flight activity in *Colias* butterflies: Implications for reproductive strategy and population structure. *Ecology* 64:546–51.

Kingsolver, J. G., and T. L. Daniel. 1983. Mechanical determinants of nectar feeding strategy in hummingbirds: Energetics, tongue morphology, and licking behavior. *Oecologia* 60:214–26.

Koehl, M.A.R. 1977. Effects of sea anemones on the flow forces they encounter. *J. Exp. Biol.* 69:87–105.

Koehl, M.A.R. 1985. Seaweeds in moving water: Form and mechanical function. In T. J. Givnish, ed., *On the Economy of Plant Form and Function*, pp. 603–34. Cambridge University Press, Cambridge, England.

Krebs, J. R., and N. B. Davies, eds. 1984. *Behavioral Ecology: An Evolutionary Approach*. Sinauer, Sunderland, Mass.

Lehman, J. T. 1984. Grazing, nutrient release, and their importance on the structure of phytoplankton communities. In D. G. Meyers and J. R. Strickler, eds., *Trophic Interactions within Aquatic Ecosystems*, pp. 49–72. AAAS Select Symp. 85.

Lehman, J. T. 1986. The goal of understanding in limnology. *Limnol. Oceanogr.* 31:1160–66.

Lewis, M. R., W. G. Harrison, N. S. Oakey, D. Hebert, and T. Platt. 1986. Vertical nitrate fluxes in the oligotrophic ocean. *Science* 234:870–73.

Loehle, C. 1987. Hypothesis testing in ecology: Psychological aspects and the importance of theory maturation. *Quart. Rev. Biol.* 62:397–409.

May, R. M. 1986. The search for patterns in the balance of nature: Advances and retreats. *Ecology* 67:1115–26.

May, R. M., and R. M. Anderson. 1987. Transmission dynamics of HIV infection. *Nature* 326:137–42.

Mackas, D. L., K. L. Denman, and M. R. Abbott. 1985. Plankton patchiness: Biology in the physical vernacular. *Bull. Mar. Sci.* 37:652–74.

Mangel, M., and C. W. Clark. 1983. Uncertainty, search, and information in fisheries. *J. Cons. Int. Explor. Mer.* 41:93–103.

Mangel, M., and C. W. Clark. 1986. Towards a unified foraging theory. *Ecology* 67:1127–38.

Mangel, M., and C. W. Clark. 1988. *Dynamic Modeling in Behavioral Ecology.* Princeton University Press, Princeton, N.J.

Monteith, J. L. 1973. *Principles of Environmental Physics.* American Elsevier, New York.

Nowell, A.R.M., and P.A. Jumars. 1984. Flow environments of the aquatic benthos. *Ann. Rev. Ecol. Syst.* 15:303–28.

Okubo, A. 1980. *Diffusion and Ecological Problems: Mathematical Models.* Springer-Verlag, New York.

Paine, R. T., and S. A. Levin. 1981. Intertidal landscapes: Disturbance and the dynamics of pattern. *Ecol. Monogr.* 51:145–78.

Peters, R. H. 1986. The role of prediction in limnology. *Limnol. Oceanogr.* 31:1143–59.

Platt, J. R. 1964. Strong inference. *Science* 146:347–53.

Porter, W. P., J. W. Mitchell, W. A. Beckman, and C. R. Tracy. 1975. Environmental constraints on some predator-prey interactions. In Gates and Schmerl (1975), pp. 347–64.

Powell, T. M., and P. J. Richerson. 1985. Temporal variation, spatial heterogeneity, and competition for resources in plankton systems: A theoretical model. *Amer. Natur.* 125:431–64.

Price, M. V., ed. 1986. Symposium: Mechanistic approaches to the study of natural communities. *Amer. Zool.* 26:3–106.

Quinn, J. F., and A. E. Dunham. 1983. On hypothesis testing in ecology and evolution. *Amer. Natur.* 122:602–17.

Rigler, F. H. 1982. Recognition of the possible: An advantage of empiricism in ecology. *Can. J. Fish. Res. Aquat. Sci.* 39:1323–31.

Rosenthal, G., and D. Janzen, eds. 1979. *Herbivores: Their Interaction with Secondary Plant Metabolites.* Academic Press, New York.

Roughgarden, J., and Y. Iwasa. 1986. Dynamics of a metapopulation with space-limited subpopulations. *Theor. Pop. Biol.* 29:235–61.

Rubenstein, D. I., and R. W. Wrangham, eds. 1986. *Ecological Aspects of Social Evolution: Birds and Mammals.* Princeton University Press, Princeton, N.J.

Scheltema, R. S. 1975. Relationship of larval dispersal, gene-flow, and natural selection to geographic variation of benthic invertebrates in estuaries and along coastal regions. *Estuarine Res.* 1:372–91.

Schoener, T. W. 1986. Mechanistic approaches to community ecology: A new reductionism. *Amer. Zool.* 26:81–106.

Shanks, A. L. 1985. Behavioral basis of internal-wave-induced shoreward transport of megalopae of the crap *Pachygrapsis crassipes. Mar. Ecol. Prog. Ser.* 24:289–95.

Sousa, W. P. 1984. The role of disturbance in natural communities. *Ann. Rev. Ecol. Syst.* 15:353–91.

Stephens, D. W., and C. R. Krebs. 1987. *Foraging Theory.* Princeton University Press, Princeton, N.J.

Tilman, D. 1982. *Resource Competition and Community Structure.* Princeton University Press, Princeton, N.J.

Tunnicliffe, V. 1981. Breakage and propagation of the stony coral *Acropora cervicornis. Proc. Natl. Acad. Sci. USA* 78:2427–31.

Tversky, A., and D. Kahneman. 1981. The framing of decisions and the psychology of choice. *Science* 211:453–58.

Via, S., and R. Lande. 1985. Genotype-environment interaction and the evolution of phenotypic plasticity. *Evolution* 39:505–23.

Vogel, S. 1981. *Life in Moving Fluids: The Physical Biology of Flow.* Willard Grant Press, Boston.

Wainwright, S. A., W. D. Biggs, J. D. Currey, and J. M. Gosline. 1976. *Mechanical Design in Organisms.* Wiley, New York.

Webb, P. W. 1984. Body form, locomotion and foraging in aquatic vertebrates. *Amer. Zool.* 24:107–20.

II POPULATION DYNAMICS AND SPECIES INTERACTIONS

Chapter 4
Plant Population Dynamic Theory

STEPHEN W. PACALA

One of the principal factors impeding the further integration of theoretical and empirical ecology is the difficulty of performing density manipulation experiments in the field. These experiments are the most direct way to calibrate and assess the simple density-dependent models that underpin much of population dynamic theory.

Because plants are sedentary and typically lack a widely dispersive juvenile phase, it is straightforward to manipulate densities of internally recruiting plant populations. Such experiments have long demonstrated the importance of density-dependent interactions in plant communities and have led to a rich quantitative literature explaining the effects of competition on plant performance (reviews in Harper 1977; Trenbath 1978). Density manipulations have also been used to assess the importance of interactions other than competition, including herbivory and disease. It seems paradoxical, then, that few attempts have been made by plant ecologists to estimate the parameters of simple density-dependent models and to test the predictions of the theory directly (but see Firbank and Watkinson 1985). Why are measurements of competition coefficients from the Lotka-Volterra competition equations not routine in studies of plant communities? The answer is not that plant ecologists are unaware of the theory or that they consider it irrelevant. For example, plant community ecologists have invested considerable effort in mapping patterns of resource partitioning and in relating these to the results of density manipulation experiments (e.g., Parrish and Bazzaz 1976, 1982; Pickett and Bazzaz 1976, 1978a,b). Similarly, plant demographers commonly rely on demographic theory to summarize plant life histories (e.g., Caswell and Werner 1978; Bierzychudek 1982; Silander 1983).

Rather, I believe that simple density-dependent models are commonly perceived by plant ecologists as, at best, highly abstract descriptors of plant population processes because the models omit several fundamental characteristics of plants. This perception has prompted repeated calls for the development of a theory that includes the following four characteristics of plants (Palmblad 1968; Werner 1976; Antonovics and Levin 1980; Weiner and Conte 1981; Crawley 1986):

1. *Autotrophy.* As autotrophs, most plants require the same few resources (principally light, water, nitrogen). How is it possible that in some cases thousands of species coexist on these (Grubb 1977)? Is there something special about autotrophy?

2. *Sedentary habit and spatially local interactions.* Because interactions among plants are spatially local, population dynamics are inherently spatial; mean population growth rate is a function of spatial distribution, and spatial distribution may change as a consequence of population growth. Also, a plant is at the mercy of its local physical environment, and so plant populations may be affected by spatial heterogeneity at all scales, including fine scale heterogeneity affecting the germination of single seeds (Harper, Williams, and Sagar 1965). Can we adequately account for plant population dynamics without specifying the locations of individuals?

3. *Plastic growth.* Because of the enormous plasticity of plant growth, the fecundities of individuals in the same site may vary by several orders of magnitude. Is mean population size a meaningful metric (Harper 1964)?

4. *Clonal habit.* Many plants, including the dominant species in some communities, propagate vegetatively. Individual ramets within a genet can compete with one another, but in some cases they also share resources (Bazzaz 1985; Pitelka and Ashun 1985). To paraphrase Harper (1985): What is an individual clonal plant?

The past half dozen years have seen the development and empirical assessment of analytically tractable models that include all of the above characteristics except for clonal growth. Below, I briefly review these models and show that the inclusion of each of characteristics 1–3 has provided new and significant insights. Some of the models can be easily calibrated and have proven to be predictive for simple communities in the field. Thus the models can be used in field studies to answer such basic questions as: Will species A and B coexist or will interspecific competition cause the elimination of one? See Roughgarden (1987) for a more comprehensive review of the theoretical ecology of plants.

Like the remainder of models in ecology, plant models are primarily either simple and analytically tractable or complex computer simulations. The simulation models come primarily from the forestry literature and have an impressive record of predictive power. Of the models developed to date, only the computer simulations are spatial models that explicitly include competition for light and nutrients (both attributes 1 and 3 in the same model). I argue that corresponding simplifying models offer the best hope for a theory that is explanatory and at the same time straightforward to apply in empirical studies of the multispecies communities that dominate most locations. I also call for the inclusion of attributes 1–4 into the theory of pollination, herbivory, seed predation, and disease.

This paper's focus on models that include special characteristics of plants should not be taken as an assessment of the relevance of the remainder of population dynamic theory to plant ecology. Throughout, I highlight several important cases in which the special attributes of plants have little effect and the plant models reduce approximately to more standard population models.

MODELS OF AUTOTROPHS: PREDATORS COMPETING FOR ABIOTIC PREY

Models of autotrophs include the dynamics of abiotic resources such as light and water together with the dynamics of plants whose population growth depends on the quantities of resources captured. Simple, analytically tractable models of this sort have been explored primarily by Tilman (1980, 1982, 1985, 1986) whose formulation is an extension of standard predator-prey models. Using the notation in Tilman (1982),

$$dN_i/dt = N_i\,[f_i(\mathbf{R}) - m_i]; \qquad i = 1,2,\ldots,Q$$

$$dR_j/dt = g_j(R_j) - \sum_i N_i f_i(\mathbf{R})\,h_{ij}(\mathbf{R}); \qquad j = 1,2,\ldots,k \qquad (4.1)$$

where N_i is the density of plant species i; R_j is the mean concentration of resource j; $\mathbf{R}$ is the row vector $(R_1,\ldots,R_k)$; $f_i(\mathbf{R})$ and m_i are, respectively, the birth and death rates of species i; $g_j(R_j)$ governs the dynamics of resource renewal and depletion in the absence of plants; and $h_{ij}(\mathbf{R})$ gives the amount of resource j required to produce a new species-i plant.

Tilman (1982) draws important distinctions between the kinds of resources that affect autotrophs. In particular, resources such as light and water are essential. At any one time, the growth of a plant population is set primarily by the concentration of only one of the suite of essential resources. In contrast, resources in the widely studied niche theoretic models developed by MacArthur (1972) are substitutable. An increase in the concentration of either of two substitutable resources results in increased consumer population growth.

The system (4.1) with essential resources has been examined in a variety of studies (e.g., Taylor and Williams 1975; Tilman 1980, 1982) and the principal findings are as follows: If all resources save one are superabundant, then one of the species will exclude all others. The surviving species is the one that can reproduce successfully at the lowest concentration of the limited resource. With k resources in limited supply, up to k species may coexist. Each of these has a different resource on which it alone would survive if all other resources were superabundant.

Now, let S_1 and S_2 be the rates at which two resources are supplied to a site. The values of S_1 and S_2 determine the identities of the (up to two) species that can persist in the site. If a habitat is composed of a loosely coupled assemblage of local populations occupying sites with different values of S_1 and S_2, then the habitat can support many species even though there are only two resources (Tilman 1982). Thus, with spatial heterogeneity in the physical environment, the model can explain the coexistence of many autotrophs on few resources.

Tilman (1982, 1985, 1986) has shown that models such as (4.1) can account for observed species-abundance relations, patterns of species replacement along environmental gradients, collapse of species diversity following enrichment by fertilizer, and primary and secondary succession. Moreover, Tilman (1977) has empirically validated the model for simple algal communities and is currently testing the theory in an extensive study of successional terrestrial vegetation (Tilman, pers. comm.).

SIMPLE MODELS OF INDIVIDUAL PLANTS DISTRIBUTED IN SPACE

Two approaches have been adopted to include spatial locations of individuals and spatially local interactions in simple plant population dynamic models. The first, introduced by Skellam (1951), is to subdivide the habitat into discrete cells. Within-cell spatial locations are not specified and competitive interactions occur only among individuals located in the same cell. Discrete-cell models have been widely studied (e.g., Skellam 1951; Levin 1974; Hamilton and May 1977; Yodzis 1978; DeAngelis, Travis, and Post 1979; Fagerstrom and Agren 1979; Hastings 1980; Chesson and Warner 1981; Hanski 1983; Shmida and Ellner 1984; Crawley and May 1987, and references therein). In the majority of studies focusing on plants, each cell contains a single adult, and competition occurs only among the juveniles within a cell following the death of an adult. Such models may be compactly written as

$$\Delta X_i = -d_i X_{it} + \sum_j d_j X_{jt} P_{ij}; \qquad i = 1,2, \ldots ,Q \tag{4.2}$$

where X_{it} is the fraction of cells occupied by species *i* at time *t*; d_i is the species-*i* death rate; and P_{ij} is the probability that species-*i* captures a cell opened by the death of a species-*j* adult.

In the linear models of Horn (1975, 1976), the P_{ij} are constant. Although this assumption has been criticized (Usher 1979), Horn's linear models have yielded important theoretical insights, are intended primarily as descriptions of systems near equilibrium, are relatively easy to calibrate in the field, and have proven applied value (see the review in Shugart 1984). In the lottery models studied by Fagerstrom and Agren (1979), Chesson and Warner (1981), and Shmida and Ellner (1984), the P_{ij} are determined by lottery competition. Following the death of an adult, a single juvenile is chosen by a weighted random draw from the juveniles present in the cell. The weights describe the relative competitive abilities of the species. Also, the P_{ij} are functions of the abundances and dispersal abilities of each species because these quantities affect the abundances of juveniles within a vacated cell.

Lottery models have yielded a wealth of theoretical insights that include the following ideas: (1) Dispersal ability is a strong determinant of the outcome of competition because it affects the spatial distribution of juveniles and because interjuvenile competition is spatially local (Shmida and Ellner 1984). (2) Random fluctuations in fecundity can, in some cases, overwhelm competitive asymmetries and maintain the diversity of a plant community (Chesson and Warner 1981; Shmida and Ellner 1984). Chesson has labeled this phenomenon the "storage effect" and he is currently testing the idea in communities of desert annuals (pers. comm.). (3) Hubbell and Foster (1986) argue that large guilds of competitively equivalent tree species exist in tropical rain forests and that the relative abundances within each guild are determined by a drift process directly analogous to the genetic drift of selectively neutral alleles. Hubbell and Foster (1986) studied a version of (2) with finite population size and equal demographic parameters for all species. They show that a species' expected waiting time to extinction may approach the geological age of the species itself.

Pacala and Silander (1985) and Pacala (1986a,b, 1987a) adopted an alternative strategy to include sedentary habit and spatially local interactions in plant population models. Suppose that the performance of a plant is a function of the number and identities of the plant's neighbors, here defined as those individuals within a circle of fixed radius about the plant. Thus the fecundity of a species-1 plant in a stand composed of species 1 and species 2 is $f_1(n_1,n_2)$, where n_1 and n_2 are the numbers of neighbors of each species. This function may be estimated by growing plants in a highly heterogeneous spatial pattern, measuring the location and seed set of each plant, and the regressing seed set versus n_1 and n_2. Similar functions may be defined for the effects of local crowding on plant survivorship and for seed dispersal, survivorship, germination, and dormancy.

A population dynamic computer-simulation model is constructed from these performance functions as follows. Individual plants are included as points on a modeled plane. In each generation, the functions are applied separately to each individual to determine if the individual dies, the number of seeds produced by each survivor, and the locations of new seeds and seedlings. In this way, the model projects the abundance and spatial distribution of each species through time.

Pacala and Silander (1985) also developed analytically tractable versions of neighborhood population dynamic models (extended by Crawley and May 1987). For example, consider a two-species community of annual plants lacking dormant seed and let N_{it} be the population density of species i at time t. Further, let $P_{it}(n_1,n_2)$ be the probability that a randomly chosen species-i plant has n_1 and n_2 neighbors, and let $s_i(n_1,n_2)$ be the product of a species-i individual's fecundity, survivorship, and germination probability. Then,

$$N_{it+1} = N_{it} \sum_{n_1} \sum_{n_2} P_{it}(n_1,n_2)\, s_i\,(n_1,n_2); \qquad i = 1,2. \tag{4.3}$$

If dispersal is sufficiently large, then $P_{it}(n_1,n_2)$ is approximately Poisson with means A_iN_{1t} and A_iN_{2t}, where A_i is the area of a species-i neighborhood (Pacala and Silander 1985; Pacala 1986a). Thus in the large dispersal case, (4.3) reduces to a system of first-order recursions on the mean population sizes. Indeed, if the $s_i(n_1,n_2)$ are linear and dispersal is large, then the neighborhood model (4.3) reduces to the Lotka-Volterra competition equations. The implication is that simple nonspatial models such as those employed by Watkinson (1980) and MacDonald and Watkinson (1981) are appropriate in the long dispersal limit.

With short dispersal, however, the $P_{it}(n_1,n_2)$ distributions change form through time due to the aggregation of siblings about the position of their mother and the differential effects of intra- and interspecific interference. Pacala (1986a) demonstrates, for example, that short dispersal may facilitate coexistence by increasing the degree of interspecific spatial segregation. One other general result is that the outcome of competition is in part determined by relations among several spatial scales, including the neighborhood radii, the mean dispersal distances, and the spatial scales of environmental heterogeneity (Pacala 1986a, 1987).

Pacala and Silander (in prep.) have used neighborhood models to investigate the dynamics of two-species weed communities in the field. They estimated all performance functions in each of two years on a series of plots. They then tested the predictions of calibrated population dynamic models constructed from these functions against census data from a second series of naturally fluctuating plots. The results in figure 4.1 clearly demonstrate that the predictions of the models are quantitatively accurate. Analysis of the calibrated models indicates that one of the species (velvet leaf) will exclude the other because of a large asymmetry in the strength of interspecific interference affecting fecundity.

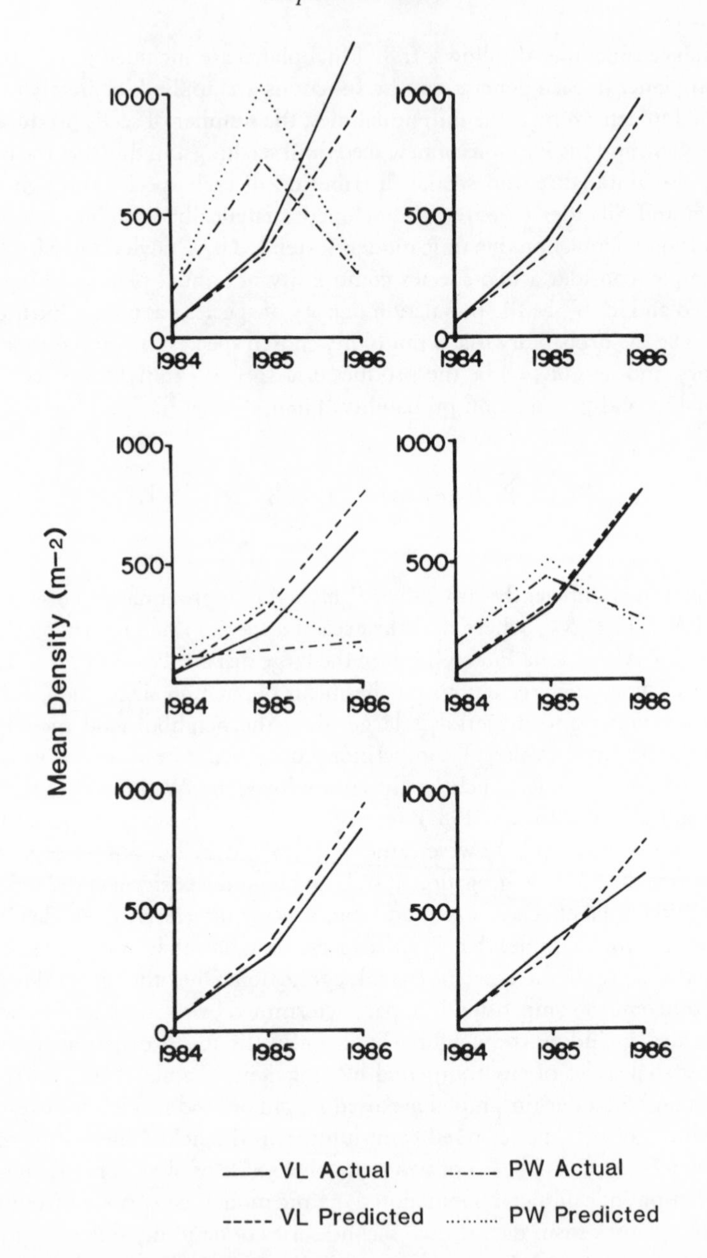

FIGURE 4.1. Plots of mean plot densities of single or two-species populations of velvet leaf (VL) and/or pigweed (PW) at yearly census dates. Actual field census data are shown together with densities predicted by the calibrated neighborhood models.

MODELS INCLUDING WITHIN-POPULATION VARIATION IN PLANT SIZE: THE EFFECTS OF PLASTICITY

Aikman and Watkinson (1980) and Holsinger and Roughgarden (1985) investigated the population dynamic consequences of within-site variation in plant growth. Each of these models incorporates continuous-time within-season plant growth, fecundity that is proportional to plant size and density-dependent growth and mortality. Within-population variability is included by assuming that growth rates vary among individuals according to a normal distribution. The models exhibit several well-known properties of even-aged monocultures such as the constant yield law, the 3/2 thinning law, and the fact that plant size distributions become increasingly skewed during a growing season. In addition, Holsinger and Roughgarden (1985) demonstrate analytically that equilibrium population sizes and between-season population growth rates are increasing functions of the *variance* in plant growth rates. They also show that plasticity can affect the outcome of competition among plant species. All else being equal, the more variable of two species will competitively exclude the less variable.

It is important to note that the high degree of variability in fecundity within a plant population represents demographic stochasticity (May 1973) and so will not generally cause highly stochastic population dynamics. The experimental evidence is that mean plant fecundity is, in fact, highly predictable (see Harper 1977). Thus stochastic fluctuations in population size are likely to be caused by exogenous events in all but small field populations.

SIMULATION MODELS OF FOREST DYNAMICS

An alternative to the simple modeling approach discussed thus far is to include autotrophy, spatial dynamics, and plastic growth in a complex computer simulation. Models of this sort are found primarily in the forestry literature (reviewed in Shugart 1984), and are typically based on submodels of the growth, mortality, and reproduction of individual trees as affected by temperature, light, water, and nutrients. A population model is constructed from many modeled trees. The trees are stratified by height and are located in one or more spatial cells. Other submodels govern the dynamics of resources in each cell, including the effects of shading on the light available to each tree. Some of the models include factors such as fire, storms, and seed predation.

Forest simulators have been developed primarily to predict the dynamics of specific forest systems. For example, the JABOWA model (Botkin, Janak, and Willis

1972) simulates 13 species in New Hampshire; the FORET model (Shugart and West 1977) simulates 33 species in Tennessee; and the KIAMBRAM model (Shugart et al. 1980) simulates 125 species in New South Wales. Owing to their complexity and system specificity, these models have seen relatively little use in theoretical explorations of the factors controlling distribution and abundance in general. By design, forest simulators typically sacrifice simplicity and explanatory power for increased accuracy of system-specific predictions.

Nevertheless, several general insights emerge from forest simulation models. First, it is possible to estimate the many parameters in models of even diverse forest communities. Second, in some cases the models have remarkable predictive power. For example, forest simulators predict altitudinal zonation in New Hampshire (Botkin, Janak, and Willis 1972) and the response of southern Appalachian forests to chestnut blight (Shugart and West 1977). Estimation error does not cause the predictive power of a multispecies forest model to evaporate in a statistical haze. Third, species diversity does not collapse catastrophically in forest models lacking spatial heterogeneity in physical processes (spatial variation in S_1 and S_2 in Tilman 1982), despite the fact that plant species greatly outnumber resources in these models.

DISCUSSION

A variety of plant population dynamic models now exists that accounts for autotrophy, sedentary habit, spatially local interactions, and plastic growth. The models demonstrate that each of these characteristics can have significant population dynamic consequences. Moreover, the models have led to new hypotheses about the structure and dynamics of plant communities. Because plants are sedentary, it is both possible and practical to determine experimentally the merits of alternative models and to test hypotheses that emerge from the theory. Clearly, this empirical endeavor is essential to refine the theory. At the same time, some of the models have been empirically calibrated and have proven to be accurate and informative descriptions of the processes that govern specific communities in the field. These models offer new tools for empirical ecologists to understand the community-level consequences of measured attributes of plants and interplant interactions and to forecast the dynamics of field communities under study.

Still, the theory is in a comparatively early stage of development and new formulations are needed. Of the models developed to date, only complex computer simulations include competition for autotrophic resources, space, sedentary habit, *and* within-population variation in performance. If we are to understand plant communities, as opposed to simply predict their dynamics, then I believe that we must develop analogous simplifying explanatory models.

The advantages of explicitly modeling the dynamics of resource acquisition and renewal are twofold. First, one of the most obvious features of plant communities is that community structure changes markedly in response to temporal and spatial changes in the physical environment. There is simply no obvious way to deduce how the parameters of a phenomenological plant competition model will change in response to changes in the physical environment, unless the competition model is itself based on a submodel of resource acquisition and renewal. Second, it would be prohibitively difficult to estimate the many parameters governing competitive interactions in a phenomenological model of a diverse community. In contrast, mechanistic forest simulators have been developed for communities with over one hundred tree species.

The further development of spatially explicit models is essential because interactions among plants are inherently spatial. In the vertical dimension, competitive asymmetries associated with shading may largely drive secondary succession and amplify within-site variation in plant size. In the horizontal dimensions, spatial heterogeneity in resource availability is caused by plants as well as by spatial variation in physical processes. This "biotic" spatial heterogeneity is responsible for the effects of dispersal on the outcome of competition in discrete-cell and neighborhood models, and may maintain successional diversity in nature (see Crawley and May 1987).

It will not be easy to develop simple, spatially explicit plant models that include resource acquisition and renewal. However, even if analytically tractable approaches prove elusive, the analysis of simple computer models should allow us to identify minimal sets of assumptions that lead to observed phenomena.

New formulations are also needed for the dynamics of vegetatively reproducing species. These species are present in nearly all plant communities and are dominant in some. One particularly intriguing possibility is that the spatial averaging of performance caused by resource sharing among connected ramets serves to buffer clonal plants against spatial heterogeneity in resource availability. Perhaps population dynamic models of clonal plants could be constructed from the clonal growth models developed by Bell (1984).

Finally, plant population dynamics are affected by factors other than competition for abiotic resources, including seed predation, herbivory, disease, pollination, and mycorrhizal associations. Autotrophy, sedentary habit, and plastic growth promise to bring as much to the theory of these interactions as they have to the theory of plant competition.

On balance, I believe that the progress to date warrants optimism. Simplifying models are producing new insights and yet remain sufficiently close to the biology of plant communities to predict the behaviors of simple field systems. Current summarizing models do predict the dynamics of diverse plant communities. Together, these are encouraging signs that an explanatory and predictive multispecies theory is an attainable goal.

ACKNOWLEDGMENTS

I thank J. A. Silander, Jr., for awakening my interest in plants. Comments by J. Silander, K. Holsinger, and M. Crawley on an earlier draft contributed much to the structure and content of this paper. Finally, I gratefully acknowledge the support of the National Science Foundation (BSR-8407844 and BSR-861674).

REFERENCES

Aikman, D. E., and A. R. Watkinson. 1980. A model for growth and self-thinning in even-aged monocultures of plants. *Ann. Bot.* 45:419–27.

Antonovics, J., and D. A. Levin. 1980. The ecological and genetic consequences of density-dependent regulation in plants. *Ann. Rev. Ecol. Syst.* 11:411–52.

Bazzaz, F. A. 1985. Demographic consequences of plant physiological traits: Some case studies. In R. Dirzo and J. Sarukhan, eds., *Perspectives on Plant Population Ecology,* pp. 324–46. Sinauer, Sunderland, Mass.

Bell, A. D. 1984. Dynamic morphology: A contribution to plant population ecology. In R. Dirzo and J. Sarukhan, eds., *Perspectives on Plant Population Ecology,* pp. 48–65. Sinauer, Sunderland, Mass.

Bierzychudek, P. 1982. The demography of jack-in-the-pulpit, a forest perennial that changes sex. *Ecol. Monogr.* 52:335–51.

Botkin, D. B., J. F. Janak, and J. R. Willis. 1972. Some ecological consequences of a computer model of forest growth. *J. Ecol.* 60:849–73.

Caswell, H., and P. A. Werner. 1978. Transient behavior and life history analysis of teasel. *Ecology* 59:53–66.

Chesson, P. L., and R. R. Warner. 1981. Environmental variability promotes coexistence in lottery competitive systems. *Amer. Natur.* 117:923–43.

Crawley, M. J. 1986. The structure of plant communities. In M. J. Crawley, ed., *Plant Ecology,* pp. 1–50. Blackwell Scientific, London.

Crawley, M. J., and R. M. May. 1987. Population dynamics and plant community structure: Competition between annuals and perennials. *J. Theor. Biol.* 125:475–89.

DeAngelis, D. L., C. C. Travis, and W. M. Post. 1979. Persistence and stability of seed-dispersed species in a patchy environment. *Theor. Pop. Biol.* 16:107–25.

Fagerstrom, T., and G. I. Agren. 1979. Theory for coexistence of species differing in regeneration properties. *Oikos* 33:1–10.

Firbank, L. G., and A. R. Watkinson. 1985. On the analysis of competition within two-species mixtures of plants. *J. Appl. Ecol.* 22:503–17.

Grubb, P. J. 1977. The maintenance of species-richness in plant communities: The importance of the regeneration niche. *Biol. Rev.* 52:107–45.

Hamilton, W. D., and R. M. May. 1977. Dispersal in stable habitats. *Nature* 269:578–81.

Hanski, I. 1983. Coexistence of competitors in a patchy environment. *Ecology* 64:493–500.

Harper, J. L. 1964. The individual in the population. *J. Ecol.* 52:149–58.

Harper, J. L. 1977. *Population Biology of Plants.* Academic Press, New York.

Harper, J. L. 1985. Modules, branches and the capture of resources. In J.B.C. Jackson, L. W. Buss, and R. E. Cook, eds., *Population Biology and Evolution of Clonal Organisms,* pp. 1–34. Yale University Press, New Haven.

Harper, J. L., J. T. Williams, and G. R. Sagar. 1965. The behavior of seeds in soil: Part 1, The heterogeneity of soil surfaces and its role in determining the establishment of plants from seed. *J. Ecol.* 53:273–86.

Hastings, A. 1980. Disturbance, coexistence, history and competition for space. *Theor. Pop. Biol.* 18:363–73.

Holsinger, K. E., and J. Roughgarden. 1985. A model for the dynamics of an annual plant population. *Theor. Pop. Biol.* 28:288–313.

Horn, H. S. 1975. Markovian properties of forest succession. In M. Cody and J. Diamond, eds., *Ecology and Evolution of Communities,* pp. 196–211. Belknap Press, Cambridge, Mass.

Horn, H. S. 1976. Succession. In R. M. May, ed., *Theoretical Ecology: Principles and Applications,* pp. 187–204. Saunders, Philadelphia.

Hubbell, S. P., and R. B. Foster. 1986. Biology, chance, and history and the structure of tropical rain forest tree communities. In J. Diamond and T. Case, eds., *Community Ecology.* Harper and Row, New York.

Leon, J. A., and D. B. Tampson. 1975. Competition between two species for two complementary or substitutable resources. *J. Theor. Biol.* 50:185–201.

Levin, S. A. 1974. Dispersion and population interactions. *Amer. Natur.* 114:103–14.

MacArthur, R. H. 1972. *Geographical Ecology.* Harper and Row, New York.

MacDonald, N., and A. R. Watkinson. 1981. Models of an annual plant population with a seedbank. *J. Theor. Biol.* 93:643–53.

May, R. M. 1973. *Stability and Complexity in Model Ecosystems.* Princeton University Press, Princeton, N.J.

Pacala, S. W. 1986a. Neighborhood models of plant population dynamics. II: Multi-species models of annuals. *Theor. Pop. Biol.* 29:262–92.

Pacala, S. W. 1986b. Neighborhood models of plant population dynamics. IV: Single and multi-species models of annuals with dormant seed. *Amer. Natur.* 128:859–78.

Pacala, S. W. 1987. Neighborhood models of plant population dynamics. III: Models with spatial heterogeneity in the physical environment. *Theor. Pop. Biol.* 31:359–92.

Pacala, S. W. 1988. Competitive equivalence: The coevolutionary consequences of sedentary habit. *Amer. Natur.* In press.

Pacala, S. W., and J. A. Silander, Jr. 1985. Neighborhood models of plant population dynamics. I: Single-species models of annuals. *Amer. Natur.* 125:385–411.

Palmblad, I. G. 1968. Competition in experimental populations of weeds with emphasis on the regulation of population size. *Ecology* 49:26–34.

Parrish, J.A.D., and F. A. Bazzaz. 1976. Underground niche separation in successional plants. *Ecology* 57:1281–88.

Parrish, J.A.D., and F. A. Bazzaz. 1982. Responses to plants from three successional communities to a nutrient gradient. *J. Ecol.* 70:233–48.

Pickett, S.T.A., and F. A. Bazzaz. 1976. Divergence of two co-occurring annual species on a soil moisture gradient. *Bull. Torrey Bot. Club* 105:312–16.

Pickett, S.T.A., and F. A. Bazzaz. 1978a. Germination of co-occurring annual species on a soil moisture gradient. *Ecology* 57:169–76.

Pickett, S.T.A., and F. A. Bazzaz. 1978b. Organization of an assemblage of early successionial species on a soil moisture gradient. *Ecology* 59:1248–1255.

Pitelka, L. F., and J. W. Ashun. Physiology and integration of ramets in clonal plants. In J.B.C. Jackson, L. W. Buss, and R. E. Cooke, eds., *Population Biology and Evolution of Clonal Organism,* pp. 399–436. Yale University Press, New Haven.

Roughgarden, J. 1987. The theoretical ecology of plants. In L. J. Gross and R. M. Miura, eds., *Some Mathematical Questions in Biology—Plant Biology,* pp. 235–67. American Mathematical Society, Providence.

Shmida, A., and S. D. Ellner. 1984. Coexistence of plant species with similar niches. *Vegetatio* 58:29–55.

Shugart, H. H. 1984. *The Theory of Forest Dynamics.* Springer-Verlag, New York.

Shugart, H. H., and D. C. West. 1977. Development of an Appalachian deciduous forest succession model and its application to the assessment of the impact of the chestnut blight. *J. Environ. Manag.* 5:161–79.

Shugart, H. H., A. T. Mortlock, M. S. Hopkins, and I. P. Burgess. 1980. A computer simulation model of ecological succession in Australian subtropical rain forest. ORNL-M-7029. Oak Ridge National Laboratory, Oak Ridge, Tenn.

Silander, J. A., Jr. 1983. Demographic variation in the Australian desert cassia under grazing pressure. *Oecologia* 60:227–33.

Skellam, J. G. 1951. Random dispersal in theoretical populations. *Biometrika* 38:196–218.

Taylor, P. A., and O. L. Williams. 1975. Theoretical studies on the coexistence of competing species under continuous-flow conditions. *Can. J. Microbiol.* 21:90–98.

Tilman, D. 1977. Resource competition between planktonic algae: An experimental and theoretical approach. *Ecology* 58:338–48.

Tilman, D. 1980. Resources: A graphical-mechanistic approach to competition and predation. *Amer. Natur.* 116:362–93.

Tilman, D. 1982. *Resource Competition and Community Structure.* Monographs in Population Biology 17. Princeton University Press, Princeton, N.J.

Tilman, D. 1985. The resource ratio hypothesis of plant succession. *Amer. Natur.* 125:827–52.

Tilman, D. 1986. Evolution and differentiation in terrestrial plant communities: The importance of the soil resource:light gradient. In J. Diamond and T. Case, eds., *Community Ecology.* Harper and Row, New York.

Trenbath, B. R. 1978. Models and interpretation of mixture experiments. In J. R. Wilson, *Plant Relations in Pastures,* pp. 145–62. CSRIO, Melbourne.

Usher, M. B. 1979. Markovian approaches to ecological succession. *J. Anim. Ecol.* 48:413–26.

Watkinson, A. R. 1980. Density-dependence in single-species populations of plants. *J. Theor. Biol.* 83:345–57.

Weiner, J., and P. T. Conte. 1981. Dispersal and neighborhood effects in an annual plant competition model. *Ecol. Model.* 13:131–47.

Werner, P. A. 1976. Ecology of plant populations in seasonal environments. *Syst. Bot.* 1:246–68.

Yodzis, P. 1978. *Competition for Space and the Structure of Ecological Communities.* Springer-Verlag, New York.

Chapter 5

Renewing the Dialogue between Theory and Experiments in Population Ecology

PETER KAREIVA

Theoretical population ecology has never been more vigorous. We have models for age-structured populations (Hastings 1986; Nisbet and Gurney 1986), stochastic environments (Chesson 1985), spatial heterogeneity (Levin 1978), and just about every type of species interaction imaginable (see, e.g., Edelstein-Keshet 1986; Addicott 1981; Pacala 1986). Not only is theory tackling the complexity and diversity of nature, but more and more models are being phrased in mechanistic terms (e.g., Tilman 1977, 1982). Although robust general predictions seem out of the question, models are generating numerous hypotheses and insights regarding particular classes of species interactions.

However, the current glories of theoretical population biology remain unknown to most ecologists. Experimental ecology and theoretical ecology are developing as almost separate fields and both pursuits suffer as a result. The main theme of this essay is that the future of theoretical ecology depends not on "better models," but on a constructive dialogue between theoreticians and empiricists. I will try to go beyond this platitude to point out specific avenues that will further the sort of dialogue I think is desirable. I begin with a brief historical tour, because the past provides numerous concrete examples of the rewards gained when mathematical models are intertwined with experiments.

THE BIRTH OF ECOLOGICAL THEORY SHIFTED THE EMPHASIS IN ECOLOGY FROM THE PHYSICAL ENVIRONMENT TO BIOTIC INTERACTIONS

Lotka, Volterra, Gause, Nicholson, and Bailey are men whose names are known to most ecologists. They were the founders of theoretical ecology and it is their models that still appear in most introductory texts. Interestingly, of these five innovators, only Nicholson had much interest in natural populations or experience with them. Gause was a laboratory biologist, Bailey a physicist enlisted by Nicholson to help with mathematics, Volterra an applied mathematician, and Lotka a chemist. In spite of their different backgrounds, these early theorists shared the conviction that biology in general, and ecology in particular, would be more scientific if it could be made more mathematical. Whether, in fact, the mathematical approach did make ecology a "more scientific" discipline is perhaps debatable; but there is no doubt that the early theoretical treatments of species interactions profoundly influenced how ecologists studied the natural world.

Before 1930, ecology was an essentially descriptive compendium of correlations and anecdotes. Fluctuations in the numbers of plants and animals were usually explained in terms of changing physical factors, such as temperature, rainfall, or even sunspots (Kingsland 1985). Models of species interactions introduced a new way of explaining population dynamics—the models made it clear that biotic influences such as competition or predation could reliably cause population densities to change. Furthermore, the models identified predictable patterns of change. For instance, a tendency toward cycling was inextricably linked to predator-prey and host-parasitoid interactions. Interspecific competition was shown to lead to the exclusion of one species unless the interaction satisfied rather special requirements. There was even the hope that simple (yet defensible) assumptions about how an organism moved, fed, and grew could yield precise quantitative predictions of population trajectories (Nicholson and Bailey 1935).

Ecology had suddenly found a theory that was not only predictive but also quantitative, far-reaching, and testable. With hindsight, much of this theory may look either wrong or trivial. Yet, at the time of their conception, the first models of species interactions provided insights unanticipated by field ecologists (e.g., predator-prey cycles).

THE EARLY MODELS PROMPTED ZEALOUS COMPARISONS WITH FIELD OBSERVATIONS AND FIERCE EXPERIMENTAL SCRUTINY IN THE LABORATORY

A dialogue between theory and data was initiated soon after Lotka's (1925) and Volterra's (1931) contributions were published. Biologists did not need to understand the mathematical details of the models in order to appreciate their implications. For example, although its mathematics were a mystery to him, Elton realized that Volterra's model of oscillating predator-prey systems might help explain small-mammal population cycles (Kingsland 1985). Indeed, advocates and disciples of the new mathematical approach tended to find support for Lotka-

Volterra models throughout the natural world. Volterra and D'Ancona (1935) interpreted fluctuations in certain fish populations as verification of Volterra's predicted periodic cycles. Fluctuations in beetle populations that were attacked by a parasitic wasp were also cited as documentation of Volterra cycles (Chapman 1933). Of course, these examples always had alternative explanations and most ecologists remained skeptical of the models (Kingsland 1985). No one attempted to test any of the models experimentally in the field, but at the time field experiments of any kind were rare.

Although Lotka-Volterra and Nicholson-Bailey models were only impressionistically compared to field data, they were rigorously scrutinized in laboratory experiments (e.g., Gause 1934; Crombie 1946; Park 1948; Utida 1953, 1957). The usual approach was to pit one competitor against another, or a predator against its prey. If short-term observations (e.g., of instantaneous rates of population change) could be used to predict long-term population trajectories, then the models were thought to be supported. For practical purposes, these experiments invariably involved small organisms with short generation times, such as protozoans, mites, and insects. Because the organisms were usually cultured in small containers, these investigations have been labeled "bottle experiments." Although such experiments could not mimic the complexity of nature, they were the logical arena in which to first test models of species interactions. Only after Lotka-Volterra equations had proven their merits with *Paramecium* cultures (Gause 1934), did it make much sense to pursue the broader implications of these models.

Of course, the results of bottle experiments were not always in accord with theory. For example, contrary to the predictions of Volterra's predator-prey models, Gause (1934) found that predatory *Paramecium* and their prey (yeast) did not cycle with an amplitude determined by initial densities. In a separate set of experiments he also found that the consumption of *Paramecium* by carnivorous *Didinium* depended more on the predator's than on the prey's density, whereas Volterra's models implied that the consumption of prey depends equally on predator and prey density. Motivated by these experimental results, Gause, Smaragdova, and Witt (1936) altered Volterra's models to include a nonlinear functional response; these revised models produced dynamics more in accord with the experimental data. Lotka-Volterra models were also not always effective at predicting the outcome of interspecific competition in laboratory experiments. For instance, Park (1948) observed that under some circumstances it was impossible to determine which of two flour beetle species would win their competitive struggle. This clash of competition theory and data stimulated the development of stochastic models of interspecific competition (Leslie and Gower 1958). In general, dialogue between theory and experiment was frequent between 1930 and 1960. The two approaches were informing one another in an iterative fashion and, as a result, ecology made great advances.

POPULATION ECOLOGY LANGUISHED WHILE COMMUNITY ECOLOGY BOOMED DURING THE "MACARTHUR ERA"

The impact of Robert MacArthur on ecology was enormous. He made community ecology an exciting and theoretical subject, but in doing so attention was diverted away from population ecology. Because MacArthur's approach often began with the assumption that populations were at a steady state, the study of population dynamics was pushed into the background. MacArthur tended also to shun bottle experiments (MacArthur 1972; Schoener 1972), which had played such an important role in the initial testing and elaboration of Lotka-Volterra and Nicholson-Bailey equations. MacArthur's focus on communities, deemphasis of dynamics in favor of statics, and devaluation of bottle experiments catalyzed major changes in theoretical ecology. An unfortunate side effect was that the link between models and experiments was weakened. This happened because the models that dominated during the MacArthur era (i.e., niche overlap community models) were not subjected to experimental testing (Simberloff 1983). Bottle experiments could have been used, but as already mentioned, they had fallen from favor. The ecologists who were pioneering the use of field experiments (e.g., Connell 1961; Paine 1966) happened not to be in the business of testing niche-overlap models. Temporarily at least, the modeling approach and experimental approach appeared to fall into two distinct camps. The legacy of this peculiar history is that numerous ecologists mistakenly believe that mathematical theory in ecology is divorced from the real world and from experimental scrutiny. If this divorce has occurred, there is no reason that a reconciliation cannot be accomplished.

GENERAL THEORIES ARE REPLACED BY MORE RESTRICTIVE MODELS OF SPECIAL CASES

I suspect very few contemporary mathematical ecologists hold out much hope for a general theory of species interactions. Gone also is the idea that models can be conveniently "tested" by simply reporting a coincidence between observed patterns in nature and a model's predictions. Theoretical population ecology is now concerned largely with models that are tailored to particular systems, or that examine particular complications. Thus, instead of Lotka-Volterra equations, we now have models of plant-insect interactions (Edelstein-Keshet 1986), of competition among annual plants (Pacala 1986), of viruses attacking insects (Anderson and May 1980), and so forth. Also, instead of examining interactions in general, theoreticians tend now to analyze how each additional complicating factor influences the interaction. For example, in the area of predator-prey theory, there have been recent theoretical studies of the effects of age structure on predator-prey systems (e.g., Hastings 1986); the role of variability in predator-prey dynamics (e.g., Chesson 1978); and the importance of spatial heterogeneity in predator-prey interactions (Hastings 1978). The price of this increased sophistication has been a loss of generality (as was anticipated by Levins 1968). Nonetheless, recent theoretical explorations have yielded major new insights about species interactions. Some modern theoretical results of which all ecologists should be aware include the following:

1. *Simple difference equations yield complex dynamics.* Recursion models that project populations at time t forward to time $t + 1$ are capable of generating periodic cycles and chaotic fluctuations under the most minimal assumptions. The key requirement is that the function that transforms a current population into the next time period's population has a "hump" (i.e., it is low at low densities, rises to a peak at some intermediate density, and then falls off at extremely high densities). This assumption is clearly met by numerous fish and insect populations (Ricker 1954). The importance of this result is that many of the seemingly complex fluctuations observed in natural populations may be an inescapable feature of discrete reproductive episodes combined with density-dependent recruitment (May and Oster 1976).

2. *The spatial dimension of environments can facilitate the coexistence of species.* When Huffaker (1958) manipulated the spatial complexity of a simple laboratory environment, he observed dramatic changes in the dynamics of an interaction between predator and prey mites. This single experiment inspired scores of mathematical ecologists to examine the influence of the spatial environment on species interactions. An important finding has been that when the environment is broken into patches, even if all patches are the same, species can coexist that could never coexist were they to interact in one patch alone. Coexistence in such patchy environments depends on the proper ratios of dispersal abilities among the interacting species and on an average level of dispersal that is intermediate (Levin 1974). If there is too much movement among patches, the system behaves as one patch; if there is too little movement, populations will disappear from patches more rapidly than they can be reestablished in empty patches.

3. *Species interactions coupled with movement can generate spatial patterning in the densities of organisms.* Spatial patterns, by which I mean regular inhomogeneities in population densities, are found in a variety of ecological systems (Kareiva and Odell 1987; Levin 1978). Reaction-diffusion models show that such patterns can arise as a result of the interaction and redistribution of species *without any underlying environmental heterogeneity.* Plausible models of predator-prey interactions with dispersal (Levin and Segel 1976) and of competitive in-

teractions with dispersal (provided at least three species are involved; Mimura 1984) can lead to predictable spatial patterns. These spatial patterns, which are sometimes called "diffusive instabilities" (Levin and Segel 1985), occur when inhibiting factors (e.g., predators) spread significantly faster in space than do activating factors (e.g., prey).

4. *Stochastic environments can convert competitive exclusion into competitive coexistence.* Whereas it was once thought that environmental variability acted primarily to eliminate species (e.g., Goh 1975, 1976; Leigh 1975), it has now been shown that environmental variability may act to preserve species diversity. In particular, competing species that cannot coexist in a constant environment may be able to coexist in the presence of environmental variation (Chesson 1985). Recent models by Chesson (1988) allow us to predict under which circumstances variability will preserve species and under which circumstances variability will eliminate species.

5. *The potential role of diseases in population dynamics has been clearly demonstrated.* Diseases were neglected in early considerations of population dynamics. During the last decade, however, a series of papers by Anderson and May (e.g., 1979a, 1979b, 1980) have shown that diseases can stably maintain their host populations at depressed densities or can generate periodic cycles of host abundance. Most importantly, the levels of virulence required for these effects are within the range that has been measured for many viral and bacterial pathogens.

6. *Age structure can have both stabilizing and destabilizing effects on species interactions and population dynamics.* Although age-structure effects have long been known to influence quantitative aspects of population dynamics, their influence on qualitative behavior has only recently been delineated (Nisbet and Gurney 1986). The effects of age structure are complex, but appear to follow two routes: (1) age structure may be destabilizing because it introduces time delays into the negative feedbacks that dampen population growth, or (b) age structure may be stabilizing because it distributes perturbations over several different cohorts in a population (Hastings 1984; Levin and Goodyear 1980). General predictions about when to expect stabilizing effects and when to expect destabilizing effects have not, as yet, been possible.

7. *The extent of clumping among individuals can determine the outcome of species interaction.* Most plants and animals live clustered or "aggregated" together with other individuals of the same species (Taylor 1971). The degree of aggregation can have profound consequences for host-parasite (Anderson and May 1986), predator-prey (Hassell and May 1974), and competitive interactions (Atkinson and Shorrocks 1984; Ives and May 1985). Since aggregation is a product of individual movements, theories that relate aggregation to population dynamics hold the promise of connecting population ecology to behavioral ecology (Hassell and May 1985).

In addition to the above general results, theoreticians have also made progress modeling particular systems. For example, detail-rich models have been able to describe population interactions in rocky intertidal communities (Paine and Levin 1981; Roughgarden, Iwasa, and Baxter 1985), among ladybird beetles and aphids (Kareiva and Odell 1987), among forest defoliators and their host trees (Ludwig, Jones, and Holling 1978), and so forth. Although these species-specific models cannot be easily transferred to other assemblages, they do provide general lessons about how to build a quantitative understanding of population interactions. In all instances the models succeed not because they faithfully reproduce the complexity of nature, but because they identify a biological feature that is key to the process being studied. For instance, Paine and Levin (1981) identify the magnitude and frequency of disturbance as the critical variable influencing competition for space among sessile intertidal organisms; Roughgarden, Iwasa, and Baxter (1985) emphasize larval settlement as a determinant of barnacle population dynamics; Kareiva and Odell (1987) build a predator-prey model upon the critical assumption that predators alter their movement behavior after feeding; Ludwig, Jones, and Holling (1978) explain budworm cycles in terms of inexorably slow changes in forest quality coupled with potentially rapid changes in budworm populations. All of these success stories entailed extensive biological understanding before the appropriate mathematical simplifications could be crystallized.

THE LIST OF CRITICAL EXPERIMENTS THAT NEED DOING IN POPULATION ECOLOGY IS LONG

Theorists have been so busy, and empiricists have been so inattentive to theory, that an enormous backlog of experimental work has accumulated. An obvious agenda would be to start experimentally testing the theoretical results listed in the preceding section. This is not going to be easily accomplished because many theoretically important factors are difficult to manipulate experimentally. For instance, there is no convenient way of altering the age structure of interacting populations. Another problem is that as models have increased in sophistication, fewer and fewer ecologists are inclined to follow their arguments closely. In addition, today's ecologist faces an overwhelming selection of models from which to choose (whereas in the 1940s and 1950s there was only a handful to consider). Even ecologists who are sympathetic and supportive of mathematical theories cannot help but be bewildered by the current state of mathematical ecology. However, rather than be discouraged by the explosion of mathematical modeling, we should be grateful to theory for identifying such a long list of critical experiments:

1. *We need more bottle experiments.* Ecologists gave up bottle experiments too soon. For example, such experiments would be particularly valuable in studying

the chaotic fluctuations associated with nonlinear difference equations. These intrinsically driven cycles are best studied in a laboratory environment, where environmental fluctuations can be eliminated. Although the mathematical significance of chaos is unarguable, there is disagreement as to whether organisms possess the life-history traits required for generating chaos (Prout and McChesney 1985). Not only would bottle experiments reveal if the recruitment functions of real organisms made them candidates for chaotic dynamics, but they may also uncover additional routes to chaos. For instance, in a rare recent series of "bottle experiments," Prout and McChesney (1985) found that a dependence of fruit-fly fertility on egg density introduced a time delay that enhanced the likelihood of chaos. If this delay is ignored, as it has been by many previous researchers, one gets the mistaken impression of simple stable dynamics. Another promising avenue for bottle experiments is the study of host-pathogen coevolution using bacteria-phage interactions (Lenski and Levin 1985). Because of the short generation times and measurable mutation rates of phages and bacteria, it is actually possible to test models of coevolution.

2. *Measure and manipulate dispersal.* The role of dispersal in species interactions is now well established theoretically. The obvious experiment is to manipulate dispersal rates of organisms and observe subsequent population dynamics (Kareiva 1987). If theory is correct in attaching so much importance to dispersal, one should be able fundamentally to alter the outcome of competitive or predator-prey interactions by adjusting rates of movement. Such adjustments are practical for many plant-insect systems (e.g., Kareiva 1984; Bergelson and Kareiva 1987).

3. *Follow populations of competing plant species through several generations.* When competition models were first being examined for animals, multigeneration experiments (Crombie 1945; Gause 1934) helped in the evaluation and refinement of the models. Competition models are just now being formulated for plants (Pacala 1986, 1987), but there are no experimental data concerning population trajectories of competing plants. Instead, experiments on plant competition have always been terminated after a single generation, which is a sensible practice for an agronomist concerned with yields, but is inadequate for a botanist concerned with modeling the dynamics of competition.

4. *Huffaker's experiments warrant repeating and elaboration.* Huffaker's classic experiments with mites and oranges provoked two decades of spatial heterogeneity modeling. These later models now provide a broad menu of predictions concerning the influence of patchiness on species interactions (e.g., Hastings 1978; Caswell 1978; Crowley 1979; Hilborn 1975). It would be useful to return to the sort of laboratory experiments that stimulated these models and manipulate the patchiness of microcosms. In some cases this approach could also be applied to interactions in natural communities (e.g., Kareiva 1987).

5. *Manipulate the degree of aggregation.* Aggregation figures prominently as a stabilizing feature in several predator-prey (Hassell and May 1974; Hassell

1978) and competition (Ives and May 1985) models. In theory at least, the tendency of organisms to cluster together could be a primary factor permitting coexistence of species. This idea has been evaluated to date only by showing that observations of unmanipulated field systems are either consistent or inconsistent with phenomenological models of aggregation and interaction. A direct test would be to manipulate the degree of aggregation in selected species and look for predicted changes in predator-prey or competitive contests. Such an approach would be practical with insects, where eggs or early instar individuals can be arranged experimentally in a variety of dispersions.

6. *Test the influence of environmental variability on competing species.* Chesson's models (1985, 1986, 1988) concerning variability and competition are widely discussed and among the most interesting theoretical advances in modern ecology. Yet these models have never been tested. Since Chesson (1988) is able to relate the influence of variability on coexistence to its influence on recruitment rates, one could select species whose recruitment functions respond differently to variability and then ask whether the competitive interactions respond to variability as predicted by theory. Alternatively, one could select a particular interaction and manipulate the magnitude of variability. These will be challenging experiments to perform, but the importance of the theory makes them worth the investment.

7. *Manipulate habitat geometry.* Several models indicate that the size, shape, and spatial arrangement of habitat patches should influence an organism's population density (Okubo 1980), its likelihood of extinction or establishment (Lefkovitch and Fahrig 1985; Quinn and Hastings 1988), and its interactions with other species (Okubo 1980). At the same time, it is widely recognized that one of mankind's major impacts is to fragment and alter the geometry of habitats. The combination of ready-made theory and a practical need for information on the effects of habitat shape or fragmentation should be a strong impetus for experimental manipulations of habitat geometry. Some of the manipulations suggested by theory are (a) manipulate the size of habitat patches and determine whether there is some minimum size below which populations go extinct (Okubo 1980) or explode (Ludwig, Aronson, and Weinberger 1979); (b) manipulate the ratio of a patch's perimeter to its area and look for changes in the density of residents (Kareiva 1984); (c) manipulate the spatial arrangement of a fixed number of patches (Lefkovitch and Fahrig 1985); and so forth. Unfortunately, unless direct observations are made of movements into, out of, or among habitat patches, there will be no way of ascribing the results of such experiments to particular mechanisms. Studies of habitat geometry are becoming fashionable under the rubric of "landscape ecology," but in the absence of guiding models, key data are neglected and the value of the experiments is diminished.

8. *Build population models out of microscale observations of individual behavior.* It should be possible to predict population dynamics from observations of indi-

vidual feeding behavior and of birth and death schedules. This approach does not yield general models but it does allow one to determine how well a particular system is understood. The experiments that underlie this research stragegy can be laborious, but they sometimes lead one to discover important consequences of "minor" changes in behavior (see Hassell and May 1974 or Kareiva and Odell 1987 for examples). I think the key to successful behavior-based models is parsimony in the number of parameters included in the models and cleverness about quantitatively summarizing complex behaviors. However, other theoreticians have argued that we need to take advantage of new computer software (e.g., artificial intelligence programs) and build increasingly complicated models if we are to meld behavioral ecology and population ecology (see, for example, Franklin and Taylor 1988).

9. *Test three-trophic-level models.* Somewhere between classical population ecology and community ecology lie three-trophic-level models. These models are amenable to the same sorts of experiments that have proven useful in studying pairwise competitive interactions and predator-prey interactions (see, e.g., Morin 1983). However, although three-trophic-level interactions are much discussed by ecologists (e.g., Price et al. 1980), they have yet to be attacked by a joint modeling and experimental approach.

10. *Explore the population-level consequences of phenotypic and genetic variability.* Organisms in any population usually differ markedly because of phenotypic plasticity or genetic variability. Regardless of its causes, ecologists tend to ignore variability among organisms, even though they have no assurances that such a simplification is justified. More attention (theoretical and experimental) needs to be given to the consequences of variability for population interactions. Here, I think the mathematical investigations will be more difficult than the experiments. It is a simple matter to manipulate the magnitude of variation within interacting populations (e.g., Antonovics and Ellstrand 1984), but it is not a simple matter to include variability among individuals in equations of population change. Thus, for their own sake as well as to direct theory, we need experimental studies in which the effects of variability on predator-prey, plant-herbivore, and competitive interactions are examined. In addition to its basic importance, the issue of variability also has an applied dimension. It is well known that genetic variability is being eroded in many species. While evolutionary biologists have been quick to point out the evolutionary consequences of this reduced variability, ecologists have said little about the implication of diminished variability for population dynamics.

I am sure there are many other critical experiments than those in the above list. In general, the influence of heterogeneity, variability, and movement demand the most experimental attention. These are the influences that have received the greatest and most original theoretical attention over the last few decades. It is time to inspect this theory in the light of data. The most satisfying experiments will involve the clean manipulation of a single factor targeted by theory (e.g., movement rate or spatial variability) and then comparison of results to predictions. Another satisfying approach involves independently estimating model parameters and quantitatively predicting population dynamics in a novel situation (see, e.g., Kareiva and Odell 1987). Where manipulation or independent estimation of parameters is impossible, models can still be tested by asking whether they are sufficient to explain observed dynamics given plausible parameters (e.g., Ludwig, Jones, and Holling 1978).

Rather than wanting to test models, most field ecologists are interested in particular biological questions or systems. Even in these situations models can guide an experimental program in profitable ways. For example, William Murdoch and colleagues have been studying biocontrol systems for the last decade. Their central question has been: What factors contribute to the success of biological control programs (Murdoch, Chesson, and Chesson 1985)? Although there has never been a specific model proposed, each step of the inquiry has addressed the assumptions or predictions of general phenomenological models concerning host-parasitoid interactions. The result is that weaknesses in the theory have been exposed (e.g., in the treatment of aggregation) and alternative models developed. At the same time, the models, although imperfect, have informed Murdoch and colleagues as to which data to collect. The concrete results of this interaction between theory and experiment are considerable: (1) we know that aggregation of the sort required for effective biological control is either hard to detect or unimportant (Chesson and Murdoch 1986; Murdoch, Chesson, and Chesson 1985; Reeve and Murdoch 1985); (2) a stable biocontrol system has been found in which the existence of a refuge appears to be key to stability (Reeve and Murdoch 1986); (3) the attention of theorists is turning toward more mechanistic models of aggregation (Kareiva and Odell 1987). This progress did not require an explicit model to be tested, but simply that the experimentalist was well aware of the pertinent models.

WHY EMPIRICISTS NEED THEORIES

It is well known that empiricists are needed to occasionally constrain the imaginations of theorists. Less obvious is the need empiricists have for theory, especially since so often they seem to be quite productive without worrying about models. Several authors have eloquently described the contributions mathematical models make to experimental ecology (Levins 1968; Levin 1981; May 1981). These contributions include clarifying hypotheses and chains of argument, identifying key components in systems, suggesting critical experiments, and introducing new ideas. Models may also give an empiricist confidence in a particular idea and the motivation to collect data that otherwise would have been ignored (see discussion of Dobzhansky and theoretical population genetics models in Mayr and Provine 1980). In addition, models are needed to determine whether specific processes can account for observed phenomena. For instance, if mark-recapture studies indicate that butterflies move on average 10 meters from their release point within one day, how far should the butterflies have moved after one week? Or at what rate should the range of these butterflies expand into a new habitat, assuming logistic population growth and only local movement? These questions can be addressed only by examining models.

Simultaneous consideration of spatial and temporal population dynamics provides another reason for models. Verbal reasoning simply does not work well when one is trying to determine the influences of both movement and local interactions. Counterintuitive results such as the "diffusive instabilities" mentioned previously cannot be understood without mathematics. It is hard even to present spatiotemporal data, much less interpret it, without some modeling framework.

Ecologists have also become concerned with processes operating at large spatial scales (i.e., landscapes) and over long time periods. Experiments in ecology typically collect data from a few square meters over a couple of field seasons (Kareiva and Andersen 1988). A central question is, to what extent does the understanding we obtain from our brief, small experiments extrapolate to large-scale phenomena? This question can be economically answered by using models to make the extrapolation, and then comparing predictions to observed processes. In general, the study of large-scale processes will increasingly demand models since experiments at large scales are prohibitively expensive or unreplicable. In these situations, one value of models is the guidance they can provide as to which experiments can provide the most information. When only a few experiments can be done, one cannot allow empiricism to proceed in an ad hoc fashion.

One final use of models is as a framework for data standardization. As it is currently practiced, ecology involves too little standardization. For instance, hundreds of mark-recapture studies have been performed with the goal of describing dispersal in insects. Yet the data from these studies have been reported in such haphazard and idiosyncratic ways that it is impossible to synthesize the results. A model might dictate a way of reporting such data; for example, numerous different movement models indicate that a diffusion coefficient (in terms of distance^2/time) would be a useful parameter to calculate from mark-release experiments (Okubo 1980). If theoreticians could convince researchers doing mark-release experiments to report diffusion coefficients, the study of dispersal would be greatly facilitated. It does not matter that diffusion models will not describe the movement of most organisms—the point is they identify a key parameter to measure. The functional response is a good example of models promoting useful data standardization. Because of Holling's (1965) and Hassell's

(1978) models, most experimentalists who study predator-prey interactions report a functional response; often they even do so according to standard equations. As a result of this practice, the effectiveness of different predators, studied by different researchers, can usually be compared (e.g., Lentern 1986; Olszak (1986). Naturalists often make the point that each species is unique and thus general models are probably impossible; but precisely because each species is different, we need to report data in some consistent fashion that readily allows comparisons. Models can indicate which standardizations will be most useful.

Models are more necessary for experimental ecology now than ever before, because ecology in general has become more sophisticated. The "easy questions," which yield to straightforward manipulations such as competitor removals or predator removals, have mostly been answered. Attention has now turned to processes that can produce counterintuitive results. Spatial patterning in homogeneous environments, coexistence due to stochastic variation, and complex dynamics from recursion equations are all examples of phenomena that cannot be understood by intuition and verbal reasoning. They are also phenomena likely to play a major role in natural populations. Experimentalists are interested in interactions between competition and predation, rather than just predation or competition alone. The effect of such interactions is a quantitative question, again a question best addressed with the aid of models. Environmental problems demand quantitative answers from ecologists, not just platitudes about diversity and productivity, or about stability. Experimentalists have done a good job of documenting the sorts of effects species have on one another, but it is now time for models to predict the magnitude of these effects given various initial conditions.

HOW SHOULD THEORETICIANS BE INVESTING THEIR ENERGIES?

I often wonder if theoreticians realize the extent to which their work is ignored by field ecologists. It would be educational for all involved if each theoretician tried convincing at least one empiricist to test some favorite model. Not only might useful experiments get performed, but models would be improved through the process. The current standard for linking models to nature seems to be phrases of the form, "this model could apply to ______, which shares the features ______," or "this model is a metaphor for ______," etc. Such phrases are necessary and informative, but are not enough to stir the typical empiricist to action. Of course many models are not intended to make testable predictions, but are only indications of what might be. The messages these metaphorical models carry for field ecologists could nonetheless be made more transparent.

There are, however, more than communication problems to be overcome before theory becomes better integrated into ecology. Theoreticians need to put greater effort into devising mechanistic models that include concrete, directly measurable parameters. Because they are often less general and more resistant to elegant analysis, such mechanistic models are less interesting to a theoretician. The theoretician's reward for developing special-case mechanistic models will be the attention of biologists. A good example is Tilman's recent model of resource competition among plants (Tilman 1982). Although many botanists disagree with Tilman's ideas, his model is extensively discussed and evaluated because it is mechanistic, and it suggests straightforward experiments (i.e., manipulating nutrient ratios).

Theoreticians also need to provide guides to the diversity of models now available. To some extent this will require original theoretical work. For example, virtually all populations are age-structured and live in heterogeneous environments; experimentalists need to know under what circumstances age structure is likely to be of overriding importance, and under what circumstances spatial heterogeneity is likely to have the greatest impact. To answer these questions alternative complications should be simultaneously analyzed in models, with the goal of identifying where in the parameter space different influences are most dramatic. Here, ratios of parameters can be especially useful—for instance, the ratio of a predator's dispersal rate to its prey's dispersal rate determines whether pattern formation is expected (Okubo 1980). The fact that ecological theory is pluralistic will not be a weakness if some form of "dichotomous key" to theory can be devised.

Studies in which models are applied to simulated noisy data have begun to appear in the literature (Hassell 1985; Ludwig and Walters 1985). These will be extremely useful both to theoreticians and to empiricists. Theoreticians can learn what level of model complexity is commensurate with the quality of data available. In some cases, estimation errors override the advantages of using "more realistic" comprehensive models (Ludwig and Walters 1985). Empiricists might learn that when a certain model holds, standard practices of collecting data gloss over key features of the dynamics. For example, Hassell (1985) found that k-factor analysis (which is supposed to quantify regulating factors) often fails to unmask population regulation in data sets contaminated by stochastic variation. Hassell is able to use this analysis to recommend alternative approaches to collecting data on host-parasite interactions.

In addition to the above general platitudes, certain specific problems in population ecology deserve more theoretical attention. It would be useful to build models of population dynamics at the landscape level from an understanding of local dynamics plus long-range transport. The framework for such models is well established (e.g., Levin 1978), but has never been applied to a large-scale system. Theoretical studies of species interactions that include dispersal need to examine the effects of taxis, aggregation, and other forms of transport that are more complex than simple diffusion. Models of plant-herbivore interactions have not yet dealt with the rich variety of plant responses to herbivory (e.g., induced defenses, compensatory regrowth, etc.). Theoretical description of plant population dynamics and plant competition pose some of the greatest problems (Schaeffer and Leigh 1975). Existing models tend to be built from curve-fitting routines rather than any underlying mechanism. To deal with the spatial dimension, potentially undesirable assumptions about the dispersion of seeds are often made. Frequency dependence, which several recent experiments suggest may play a major role in plant competition (Antonovics and Kareiva 1988), is not adequately treated in current models. Finally, important work also needs to be done with basic models of predator-prey interactions. In particular, the significance of differing foraging behaviors for population dynamics has not been adequately explored (Hassell and May 1985). While it is clear that animals forage in predictable ways, theoreticians have at best only phenomenologically described the population-level consequences of assorted foraging behaviors (the exception is Hassell and May 1974). It would be especially interesting to learn whether behavior that approximates optimal foraging has any tendency to stabilize or destabilize predator-prey interactions.

My recommendations in this area have been concerned largely with style and communication, because I think theoreticians are doing an excellent job of addressing the important questions in ecology. Getting the models noticed and tested by field biologists has been the weak link.

THE ROAD TO RENEWED DIALOGUE BETWEEN THEORETICAL AND EXPERIMENTAL ECOLOGY IS DANGEROUS

Because there are so many ways for empiricists and theoreticians to misunderstand one another, dialogue between their two approaches can be treacherous. Although I have made a strong plea for more testing of models, I worry that this appeal might be misconstrued into a demand that theoreticians conduct experiments. That would be a waste of talent and would probably produce some absolutely disastrous experiments. Empiricists need to realize that models and theories go through a development process from abstract to specific—it would be folly to demand that models be immediately couched in testable form. It is also folly to demand that models faithfully adhere to reality (although they should touch it in some tangible way); one of the virtues of theory is that the imagination is freed from details so that new connections are revealed.

On the other hand, I also worry that my plea for improved communication on the part of theorists might be misunderstood. There is always the danger that theoreticians will be overzealous about "selling their models" and thereby mislead empiricists. To avoid this, theoreticians need to be ever skeptical of their own models and restrained in the claims they make for theory. Perhaps if empiricists were more motivated to understand models, they would not need hyperbole to prick their interest.

Many ecologists are completely ignorant of mathematical ecology. This ignorance is initially tolerated in undergraduate ecology courses, in which mathematical models are usually avoided or discussed in a defensive manner. Graduate programs further nurture this ignorance by allowing candidates to receive Ph.D.s in ecology while knowing so little theory. The generally poor training in mathematical ecology that most ecologists receive is inexcusable. There is no escaping the fact that ecology is largely a science of births, deaths, numbers, derivatives, and fluxes. But ecologists have been timid about insisting on mathematical literacy. Progress in ecology will be slow as long as experimentalists are handicapped by a fear or ignorance of basic mathematical reasoning.

A MILITANT SUMMARY

In summary, population ecology is rich with theory. Where theories have been combined with experimental research, substantial advances have been made (e.g., Hassell and May 1974; Reeve and Murdoch 1986; Anderson and May 1979a,b; Paine and Levin 1981; Lenski and Levin 1985). But the sad truth is that ecological theory exists largely in a world of its own, unnoticed by mainstream ecology. If ecology is ever to become more than a compendium of analysis of variance tables, there must be a strong dialogue between theoreticians and empiricists. Militant and immediate action is needed along two fronts: (1) theoreticians must become more effective at making their models accessible and testable, and they must produce some agreed-upon guidelines to help the empiricist through the maze of models; and (2) empiricists must become more educated in the application of mathematical reasoning to ecological questions. Past mistakes and the fact that no grand unified theory is on the horizon should not detract from the contributions models are likely to make to ecology.

ACKNOWLEDGMENTS

The ideas of P. McEvoy, G. Odell, and D. Wiernasz have influenced me greatly as I wrote this essay. I am grateful for their provocation.

D. Doak, G. Dwyer, W. Morris, and P. Turchin provided references and championed theory in a variety of useful ways. I have used their suggestions freely. Pat Doak and Mark Andersen helped get the manuscript presentable. Most importantly, R. T. Paine has been a “model” colleague—harassing, cajoling, and finally bludgeoning as much common sense as he could into this discourse on mathematical models; he may not have saved my soul but he has shown me the path.

REFERENCES

Addicott, J. F. 1981. Stability properties of two-species models of mutualism: Simulation studies. *Oecologia* 49:42–49.

Anderson, R., and R. May. 1979a. Population biology of infectious diseases, I. *Nature* 280:361–67.

Anderson, R., and R. May. 1979b. Population biology of infectious diseases, II. *Nature* 280:455–61.

Anderson, R., and R. May. 1980. Infectious diseases and population cycles of forest insects. *Science* 210:658–61.

Anderson, R., and R. May. 1986. Helminth infections of humans: Mathematical models, population dynamics and control. *Adv. Parasitol.* 24:1–101.

Antonovics, J., and N. Ellstrand. 1984. Experimental studies of the evolutionary significance of sexual reproduction. I. A test of the frequency-dependent selection hypothesis. *Evolution* 38:103–15.

Antonovics, J., and P. Kareiva. 1988. Frequency-dependent selection and competition: Empirical approaches. *Proc. Roy. Soc. Lond. (B).* In press.

Atkinson, W. D., and B. Shorrocks. 1984. Aggregation of larval Diptera over discrete and ephemeral breeding sites: The implications for coexistence. *Amer. Natur.* 124:336–51.

Bergelson, J., and P. Kareiva. 1987. Barriers to movement and the response of herbivores to alternative cropping patterns. *Oecologia* 71:457–60.

Caswell, H. 1978. Predator mediated coexistence: A nonequilibrium model. *Amer. Natur.* 112:127–54.

Chapman, R. 1933. The causes of fluctuations of populations of insects. *Proc. Haw. Ent. Soc.* 8:279–92.

Chesson, P. 1978. Predator-prey theory and variability. *Ann. Rev. Ecol. Syst.* 9:323–47.

Chesson, P. 1985. Coexistence of competitors in spatially and temporally varying environments: A look at the combined effects of different sorts of variability. *Theor. Pop. Biol.* 28:263–87.

Chesson, P. 1986. Environmental variation and the coexistence of species. In J. Diamond and T. Case, eds., *Community Ecology,* pp. 240–56. Harper and Row, New York.

Chesson, P. 1988. A general model of the role of environmental variability in communities of competing species. In A. Hastings, ed., *Community Ecology,* pp. 68–83. Springer Verlag, New York.

Chesson, P., and W. Murdoch. 1986. Aggregation of risk: Relationships among host-parasitoid models. *Amer. Natur.* 127:696–715.

Connell, J. 1961. Effects of competition, predation by *Thais lapillus,* and other factors on natural populations of the barnacle *Balanus balanoides*. *Ecol. Monogr.* 31:61–104.

Crombie, A. 1945. On competition between different species of graminivorous insects. *Proc. Roy. Soc. Lond. (B)* 132:362–95.

Crombie, A. 1946. Further experiments on competition. *Proc. Roy. Soc. Lond. (B)* 133:76–109.

Crowley, P. 1979. Predator-mediated coexistence: On equilibrium interpretation. *J. Theor. Biol.* 80:129–44.

Edelstein-Keshet, L. 1986. Mathematical theory for plant-herbivore systems. *J. Math. Biol.* 24:25–58.

Elton, C. 1935. *Animal Ecology,* 2d ed. Macmillan, New York.

Franklin, R., and R. Taylor. 1988. Artificial intelligence simulation modeling for behavioral ecology. *Ecol. Model.* In press.

Gause, G. F. 1934. *The Struggle for Existence.* Williams & Wilkins, Baltimore.

Gause, G. F., N. P. Smaragdova, and A. A. Witt. 1936. Further studies of interaction between predators and prey. *J. Anim. Ecol.* 5:1–18.

Goh, B. 1975. Stability, vulnerability and persistence of complex ecosystems. *Ecol. Model.* 1:105–16.

Goh, B. 1976. Nonvulnerability of ecosystems in unpredictable environments. *Theor. Pop. Biol* 10:83–95.

Hassell, M. 1978. *The Dynamics of Arthropod Predator-Prey Systems.* Princeton University Press, Princeton, N.J.

Hassell, M. 1985. Insect natural enemies as regulating factors. *J. Animal Ecol.* 54:323–34.

Hassell, M., and R. May. 1974. Aggregation in predators and insect parasites and its effects on stability. *J. Anim. Ecol.* 43:567–94.

Hassell, M., and R. May. 1985. From individual behaviour to population dynamics. In R. Sibly and R. Smith, eds., *Behavioral Ecology,* pp. 3–32. Blackwell, Oxford.

Hastings, A. 1978. Spatial heterogeneity and the stability of predator-prey systems: Predator-mediated coexistence. *Theor. Pop. Biol.* 14:380–95.

Hastings, A. 1984. Simple models for age-dependent predation. *Lect. Notes in Bio-Math.* 54:114–21.

Hastings, A. 1986. Interacting age-structure populations. In G. Hallam and S. Levin, eds., *Mathematical Ecology,* pp. 287–94. Springer-Verlag, New York.

Hilborn, R. 1975. The effect of spatial heterogeneity on the persistence of predator-prey interactions. *Theor. Pop. Biol.* 8:346–55.

Holling, C. S. 1965. The functional response of predators to prey and its role in mimicry and population regulation. *Mem. Entomol. Soc. Can.* 45:1–60.

Huffaker, C. 1958. Experimental studies on predation: Dispersion factors and predator-prey oscillations. *Hilgardia* 27:343–83.

Hutchinson, G. E. 1978. *An Introduction to Population Ecology.* Yale University Press, New Haven.

Ives, A. R., and R. May. 1985. Competition within and between species in a patchy environment: Relations between microscopic and macroscopic models. *J. Theor. Biol.* 115:65–92.

Kareiva, P. 1984. Predator-prey dynamics in spatially-structured populations: Manipulating dispersal in a coccinellid-aphid interaction. *Lect. Notes in BioMath.* 54:368–89.

Kareiva, P. 1987. Habitat fragmentation and the stability of predator-prey interactions. *Nature* 321:388–91.

Kareiva, P., and M. Andersen. 1988. Spatial aspects of species interactions: The wedding of models and experiments. In A. Hastings, ed., *Community Ecology,* pp. 38–54. Springer Verlag, New York.

Kareiva, P., and G. Odell. 1987. Swarms of predators exhibit "preytaxis" if individual predators use area restricted search. *Amer. Natur.* 130:233–70.

Kingsland, S. 1985. *Modeling Nature.* University of Chicago Press, Chicago.

Lefkovitch, L., and L. Fahrig. 1985. Spatial characteristics of habitat patches and population survival. *Ecol. Model.* 30:297–308.

Leigh, E. G. 1975. Population fluctuations, community stability and environmental variability. In M. Cody and J. Diamond, eds., *Ecology and Evolution of Communities,* pp. 51–73. Harvard University Press, Cambridge, Mass.

Lenski, R., and B. Levin. 1985. Constraints on the coevolution of bacteria and virulent phage: A model, some experiments, and predictions for natural communities. *Amer. Natur.* 125:585–602.

Lentern, J. 1986. Evaluation of effectiveness and utilization. In Hodek, ed., *Ecology of Aphidophaga,* pp. 505–10. Dr. W. Junk, Prague.

Leslie, P. H., and J. Gower. 1958. The properties of a stochastic model for two competing species. *Biometrika* 45:316–30.

Levin, S. 1974. Dispersion and population interactions. *Amer. Natur.* 108:207–28.

Levin, S. 1978. Population models and community structure in heterogeneous environments. In S. Levin, ed., *Studies in Mathematical Biology,* part 2, pp. 439–76. American Mathematical Association, Washington, D.C.

Levin, S. 1981. The role of theoretical ecology in the description and understanding of populations in heterogeneous environments. *Amer. Zool.* 21:865–75.

Levin, S., and C. Goodyear. 1980. Analysis of an age-structured fishery model. *J. Math. Biol.* 9:245–74.

Levin, S., and L. Segel. 1976. Hypothesis for the origin of plankton patchiness. *Nature* 259:659.

Levin, S., and L. Segel. 1985. Pattern generation in space and aspect. *SIAM Review* 27:45–67.

Levins, R. 1968. *Evolution in Changing Environments.* Princeton University Press, Princeton, N.J.

Lotka, A. 1925. *Elements of Physical Biology.* Williams & Wilkins, Baltimore. (Reprinted 1956 by Dover Press, New York.)

Ludwig, D., D. Aronson, and H. Weinberger. 1979. Spatial patterning of the spruce budworm. *J. Math. Biol.* 8:217–58.

Ludwig, D., and C. Walters. 1985. Are age-structured models appropriate for catch-effort data? *Can. J. Fish. Aquat. Sci.* 42:1066–72.

Ludwig, D., D. Jones, and C. S. Holling. 1978. Qualitative analysis of insect outbreak systems: The spruce budworm and the forest. *J. Anim. Ecol.* 47:315–32.

MacArthur, R. 1972. *Geographical Ecology.* Harper and Row, New York.

May, R. 1981. The role of theory in ecology. *Amer. Zool.* 21:903–10.

May, R., and G. Oster. 1976. Bifurcations and dynamic complexity in simple ecological models. *Amer. Natur.* 110:573–99.

Mayr, E., and W. Provine. 1980. *The Evolutionary Synthesis.* Harvard University Press, Cambridge, Mass.

Mimura, M. 1984. Spatial distribution of competing species. *Lect. Notes BioMath.* 54:492–501.

Morin, P. 1983. Predation, competition, and the composition of larval anuran guilds. *Ecol. Monogr.* 53:119–38.

Murdoch, W., Chesson, J., and P. Chesson. 1985. Biological control in theory and practice. *Amer. Natur.* 125:344–66.

Nicholson, A., and V. Bailey. 1935. The balance of animal populations. *Proc. Zool. Soc. Lond.* 3:551–98.

Nisbet, R., and W. Gurney. 1986. The formulation of age-structure models. In G. Hallam and S. Levin, eds., *Mathematical Ecology,* pp. 95–116. Springer-Verlag, New York.

Okubo, A. 1980. *Diffusion and Ecological Problems: Mathematical Models.* Springer-Verlag, New York.

Olszack, R. 1986. The effectiveness of three species of coccinellid adults in controlling small colonies of the green apple aphid. In Hodek, ed., *Ecology of Aphidophaga,* pp. 381–84. Dr. W. Junk, Prague.

Pacala, S. 1986. Neighborhood models of plant population dynamics. II: Multi-species models of annuals. *Theor. Pop. Biol.* 29:262–92.

Pacala, S. 1987. Neighborhood models of plant population dynamics. III: Models with spatial heterogeneity in the physical environment. *Theor. Pop. Biol.* 31:359–92.

Paine, R. 1966. Food web complexity and species diversity. *Amer. Natur.* 100:65–75.

Paine, R., and S. Levin. 1981. Intertidal landscapes: Disturbance and the dynamics of pattern. *Ecol. Monogr.* 51:145–78.

Park, T. 1936. Studies in population physiology. VI: The effect of differentially conditioned flour upon the fecundity and fertility of *Tribolium confusum. J. Exp. Zool.* 73:394–404.

Park, T. 1948. Experimental studies of interspecies competition: I, Competition between populations of the flour beetles *Tribolium confusum* and *Tribolium castaneum. Ecol. Monogr.* 18:265–307.

Park, T., E. Gregg, and C. Lutherman. 1941. Studies in population physiology. x, Interspecific competition in populations of granary beetles. *Physiol. Zool.* 14:395–430.

Price, P., et al. 1980. Interactions among three trophic levels. *Ann. Rev. Ecol. Syst.* 11:41–65.

Prout, T., and F. McChesney. 1985. Competition among immatures affects their adult fertility: Population dynamics. *Amer. Natur.* 126:521–58.

Quinn, J., and A. Hastings. 1988. Extinction in subdivided habitats. *Conserv. Biol.* 1:198–208.

Reeve, J., and W. Murdoch. 1985. Aggregation by parasitoids in the successful control of the California red scale: A test of theory. *J. Anim. Ecol.* 54:797–816.

Reeve, J., and W. Murdoch. 1986. Biological control by the parasitoid *Aphytis melinus,* and population stability of the California red scale. *J. Anim. Ecol.* 55:1069–82.

Ricker, W. 1954. Stock and recruitment. *J. Fish. Res. Board Can.* 11:559–623.

Roughgarden, J., Y. Iwasa, and C. Baxter. 1985. Demographic theory for an open marine population with space-limited recruitment. *Ecology* 66:54–67.

Schaeffer, W., and E. Leigh. 1975. The prospective role of mathematical theory in plant ecology. *Syst. Bot.* 1:109–232.

Schoener, T. W. 1972. Mathematical ecology and its place among the sciences. *Science* 178:389–91.

Simberloff, D. 1983. Competition theory, hypothesis testing, and other community-ecological buzzwords. *Amer. Natur.* 122:626–35.

Taylor, L. R. 1971. Aggregation as a species characteristic. In G. Patil, E. Pielou, and W. Waters, eds., *Statistical Ecology,* vol. 1, pp. 357–77. Pennsylvania State University Press, Philadelphia.

Tilman, D. 1977. Resource competition between planktonic algae: An experimental and theoretical approach. *Ecology* 58:338–48.

Tilman, D. 1982. *Resource Competition and Community Structure.* Princeton University Press, Princeton, N.J.

Utida, S. 1953. Interspecific competition between two species of bean weevil. *Ecology* 34:301–307.

Utida, S. 1957. Population fluctuation: An experimental and theoretical approach. In *Cold Spring Harbor Symp. Quant. Biol.* 22:139–51.

Volterra, V. 1931. *Leçons sur la théorie mathématique de la lutte pour la vie.* Marcel Brelot, Paris.

Volterra, V., and U. D'Ancona. 1935. Les associations biologiques au point de vue mathématique. Hermann, Paris.

Chapter 6

Discussion: Population Dynamics and Species Interactions

DAVID TILMAN

Ecology is the scientific discipline that attempts to determine the causes of patterns in the distribution, abundance, and dynamics of the earth's biota. The earth's ecosystems are complex. In any given habitat, there are tens to hundreds or even thousands of different species. These influence each other both through direct pairwise interactions and through indirect interactions mediated by intermediate species, processes, or substances (Levine 1976; Holt 1977; Vandemeer 1980; Schaffer 1981). Because it is impractical, if not impossible, to observe all the potential interactions among all species and processes, ecological research involves the simplifying assumption that much of the complexity of nature is either unimportant or can be subsumed within a few summary variables. Schaffer (1981, p. 383) defined such simplification as the process of ecological abstraction: "Accordingly, when the empiricist fits data to equations describing the growth rates of particular species, he has, in a sense, 'abstracted' these species from a more complex matrix of interactions in which they are embedded. Nevertheless, because the species studied, as opposed to the variables in the abstracted equations, continue to interact with the remaining, unspecified components of the ecosystem, the parameter values obtained perforce reflect, in part, the species and interactions omitted from the model."

The study of population dynamics and population interactions is, of necessity, a process of ecological abstraction. Ecologists attempt to find, through empirical observation, experimentation, and theory, the critical subset of parameters and interspecific interactions that are needed to describe and predict ecological patterns. How might the process of ecological abstraction best proceed? This paper presents five somewhat related points that may increase the efficiency of ecological research. These musings are based, in part, on discussions at the Asilomar

Conference, but have been colored by my predispositions. Although some points may seem obvious, the approach they encompass has not been used as frequently as such a categorization would imply.

STUDY MAJOR, BROAD, REPEATABLE PATTERNS

Because the purpose of ecology is to understand the causes of patterns in nature, we should start by studying the largest, most general, and most repeatable patterns. The wealth of natural-history information that has already been collected is an invaluable but all too often overlooked guide in choosing the organisms, habitats, and questions for study. All ecological research, of necessity, is performed in a particular locality on a particular group of organisms and environmental processes, at a particular time. This tends to make any particular study narrow, but can be overcome if the study is used as a test of the potential causes of broader patterns. Although ecologically unique and unusual events have a certain appeal, studies of the unique and unusual are done to the unavoidable exclusion of studies of broad, major, general patterns. With well over two million species of plants and animals on the earth, there are an almost unlimited number of unique events caused by the peculiarities of the life history, morphology, physiology, or behavior of a particular species, and by chance environmental events. Although all studies of natural history have some value, we must choose research questions based on their generality if we are to avoid having too great a proportion of the resources available for ecological research spent studying rare events and unusual patterns. For instance, oscillations in the density of a small mammal species in a particular locality would be of interest if the pattern were repeatable through time. The oscillations would be of greater interest if this same species exhibited similar oscillations in many other localities. If other species of small mammals also exhibited similar oscillations across many habitats, the phenomenon would be highly general and repeatable, and thus of great interest. Patterns that are global in extent should be studied before those that are purely local. However, both local and global studies are needed to determine which patterns are general and which are unique. This seeming contradiction is resolved later in this chapter.

Some of the classic papers of our discipline have been those that pointed out major patterns (e.g., Cowles 1899; Tansley 1949; Elton and Nicholson 1942; Hutchinson 1959, 1961; MacArthur and Wilson 1967). May (1986) took this approach in his MacArthur Award lecture, highlighting some intriguing, general patterns in food-web structure, in relative abundance patterns among species, in the relationship between the physical size of species and their species richness, and the relationship between the size of individuals and their abundance in a region. Some of these patterns are relatively well studied. Others are unstudied. For instance, we do not yet have a single terrestrial food web that has been studied in sufficient depth that all species on all trophic levels, including the decomposer level, have been quantified. Until such work is accomplished, patterns in food webs will be inadequately described, and it will be impossible to determine their causes.

There are many ways to judge ecological generality. A few years ago, a paper was presented at the annual meeting of the Ecological Society of America in which it was suggested, as a tongue-in-cheek response to Hurlbert (1984), that all ecological experiments had an element of pseudoreplication unless they were replicated on different planets on which life had independently evolved. After all, all organisms on earth are derived from a single ancestor, and thus their responses are potentially constrained by this common heritage! This raises an important point. Should we not, as ecologists, first seek those general features of life that are likely to be repeatedly observed given the general constraints of a planet such as earth? Although we cannot explore other solar systems, we can make comparisons among continents whose biota have been geographically isolated for eons. There are many patterns in the morphology, physiology, life history, behavior, and dynamics of species, and in the structure of communities and ecosystems, that reoccur from continent to continent even though the taxa are often unrelated (e.g., Beard 1944, 1955, 1983; Cody 1969, 1973; Whittaker 1975; Mooney 1977; Cody and Mooney 1978; Orians and Paine 1983; Walter 1985). In allocating the limited time each of us has in our careers, and the funds that are available to support research in our discipline, should we not seek the causes of such similarities before we study less general processes? General patterns provide a framework within which less general patterns can be more effectively studied.

Because all habitats and all species have unique features, patterns that are repeatable across many habitats or from one species to another are unlikely to be explained by the unique features of each habitat or species. Rather, the existence of broad, general patterns suggests the existence of broad, general causative forces. If this is so, theory based on these forces should be able to predict the major patterns we see in nature. Ecology is conceptually young. Our highest priority should be to understand these forces and their ramifications for patterns in nature. Every such advance will allow us to explain some of the variance we see in nature and to determine the patterns that are not explained by these forces. It is the unexplained patterns—the unexplained variance—that then guide our future research.

LOOK FOR ENVIRONMENTAL CONSTRAINTS AND ORGANISMAL TRADE-OFFS

The forces that cause pattern in ecology result from the constraints of the abiotic and biotic environment and from the unavoidable trade-offs that organisms encounter in dealing with these environmental constraints. An environmental constraint is any factor that acts to reduce the reproductive rate or increase the mortality rate of a population. There are many such constraints. As has long been recognized, the most general constraint comes from the universal requirement of all living organisms for energy and matter. Hutchinson (1959) expressed this well when he said, "In any study of evolutionary ecology, food relations appear as one of the most important aspects of the system of animate nature. There is quite obviously much more to living communities than the raw dictum 'eat or be eaten,' but in order to understand the higher intricacies of any ecological system, it is most easy to start from this crudely simple point of view."

Each individual organism exists within a web of consumer-resource relations. Its reproductive rate is constrained by the availabilities of the items it consumes—its resources. Its survivorship is constrained by the organisms that attempt to consume it. The universality of consumer-resource interactions has motivated both theory and experiments (e.g., MacArthur 1972; Schoener 1971, 1976, 1986; and Tilman 1977, 1980, 1982, 1988), but has not yet become as central a concept in ecology as its universality demands. In addition to consumer-resource interactions, populations are also constrained by physical factors such as temperature, humidity, and pH, and by mortality from nonbiological agents of disturbance, such as fire, windstorms, waves, or climatic variability. These physical factors can be included in a consumer-resource approach by explicitly stating how they influence resource availability, resource-dependent growth, and mortality (Tilman 1982, 1988).

In dealing with environmental constraints, individual organisms face trade-offs. If there were no trade-offs, there would be no forces favoring biotic diversity, genetic polymorphisms, or individual plasticity (e.g., Tilman 1982:234–65). The reproductive rate and survivorship that an organism attains when it experiences a particular suite of environmental constraints is determined by its morphology, physiology, life history, and behavior. Any beneficial traits that can be attained without giving up other beneficial traits should rapidly become fixed in a population. The differences among individuals, then, should be based on unavoidable trade-offs. (An unavoidable trade-off is a type of "constraint," but I prefer to call it a trade-off because its exists only if the cost of gaining one beneficial trait is the loss of some other beneficial trait or traits.)

What types of trade-offs are likely to be unavoidable? All traits that are based on allocation of some limiting item are subject to an unavoidable trade-off. Thus, a plant that allocates a carbon or nitrogen atom toward the production of a leaf cannot allocate that same atom toward the production of a root or stem or seed (Tilman 1988). For a plant to gain the ability to grow on nutrient-poor soils, it must necessarily give up some of the leaf area or stem biomass that would allow it to grow better on a rich soil. Similarly, an animal that allocates some protein to locomotor musculature cannot also allocate that protein to digestive functions. Protein allocated to one physiological process cannot also be allocated to a different physiological process. Time allocated to one method of foraging cannot be allocated to a different method. Time allocated to courtship and mating cannot be used for foraging or the raising of young. The process of allocation of any limiting item—energy, protein, a trace metal, time, and so on—necessarily effects not just the process to which it is allocated, but also the processes to which it is not allocated. Different patterns of allocation lead to different morphologies, physiologies, life histories, and/or behaviors. A potentially beneficial trait is gained, but at the unavoidable expense of a different potentially beneficial trait. Thus, the suite of traits that leads to maximal fitness for an individual in response to one set of environmental constraints is unlikely to do so in response to a different set of constraints.

Ecological patterns result from the interplay between environmental constraints and the trade-offs that organisms face in dealing with them. General and repeatable constraints will lead to general and repeatable patterns if there are unavoidable trade-offs that organisms face in dealing with these constraints. Many components of an organism's morphology, physiology, life history, and behavior are determined by the pattern of allocation to these components, and thus represent unavoidable trade-offs. Individual organisms overcome some of these trade-offs, to some extent, through plasticity. However, there are limits to plasticity because plasticity has costs. An individual, though plastic, rarely can assume the full range of variation observed within an entire species. Species, though variable both from individual plasticity and from genetic differences among individuals, do not have the full range of variation that is observed among species. This, in its own right, is a general pattern that merits further consideration.

If the existence of general patterns implies that organisms face similar constraints, then an important step in studying such patterns is to determine what the environmental constraints are and what the patterning of each constraint is with respect to the others. This is most easily done experimentally. Based on Hutchinson's assertion, it might be best to consider initially the position of a species in its food web, and thus to determine experimentally which resources limit it. It is also necessary to manipulate the densities of the organisms that consume it, to determine their effect on the species. Observational studies of the relationships between the density of a species, the availability of its limiting resource(s), and the densities of species that consume it would help determine the generality of the patterns suggested by the manipulative experiments. Other experimental studies might reveal other environmental constraints.

THEORY SHOULD EXPLICITLY INCLUDE CONSTRAINTS AND TRADE-OFFS

Whatever the environmental constraints might be, theory should be developed to deal explicitly with those constraints. The constraints may be separate and distinct, or they may be interdependent. Such interdependence, if it exists, is important to include, because it can greatly influence the ecological pattern that a theory predicts. If a large, general, repeatable pattern is being studied, theory should start by considering only the most universal constraints. It is also necessary to determine which traits are important in species' responses to these constraints, and the unavoidable trade-offs that individual organisms face with respect to these traits. A theory that explicitly dealt with the major constraints and trade-offs could then explore their logical implications. It could determine the extent to which broad, general patterns and deviations away from these patterns could be explained by these constraints and trade-offs.

Environmental constraints and organismal trade-offs represent the underlying mechanisms whereby organisms interact with each other and with their environment. Most ecological theory has been more phenomenological than that called for above. Rather than explicitly dealing with environmental constraints and organismal trade-offs, it has sought greater generality by using summary variables that are not explicitly tied to the mechanisms of interaction between environmental constraints and organismal trade-offs (Schoener 1986). This has been valuable in establishing a conceptual framework for ecology. Although further theory of this sort will be useful, there is an even greater need for theory that more explicitly deals with mechanism (Tilman 1987).

Clearly, phenomenology and mechanism are not absolute entities but idealized ends of a spectrum. Any theory that explicitly includes environmental constraints and organismal trade-offs will be more mechanistic than most current theory. It is likely that, along the spectrum from phenomenological to mechanistic theory, there will be a point that is optimal for explaining any given ecological pattern. It is very possible to produce theory that is too mechanistic, that loses generality without gaining significant predictive power. The optimal point will be found only through the usual trial-and-error process of science. It is always possible to produce a theory that is either more or less mechanistic than a given theory. I do not present mechanistic theory as an absolute good, but rather suggest that many present theories may lack predictive power because they are not sufficiently mechanistic, that is, they do not explicitly deal with environmental constraints and organismal trade-offs.

THEORY, OBSERVATION, AND EXPERIMENT MUST INTERACT

Let me start this section by offering a simplified view of the scientific process, which is often described as having three parts: observation, hypothesis formation, and experimentation. Scientific advances come from the repetitive application of these three activities. Observations lead to hypotheses that are often formalized in mathematical theory. Theory is used to make predictions that are tested with experiments or additional observations. This leads to the rejection, modification, or extension of the theory, and thus to new predictions that are again testable through observation or experimentation. Thus observation, experimentation, and theory are interdependent. Each gains its value from the others. Theory is valuable if it explains patterns that have been observed and makes testable predictions. Observations are valuable as tests of theoretical predictions and as they suggest new hypotheses. Experiments are valuable as tests of theoretical predictions.

Theory, observation, and experimentation advance together and cannot stand alone. The more each relates to the other, the more powerful each becomes. Even the most ardent empiricist is, of necessity, a theoretician, for it is impossible to measure everything in an ecosystem. The act of choosing some items for measurement and ignoring others represents the formation of a hypothesis as to the possible causes of pattern. Theory that is based on easily observed or easily measured items is more easily tested than theory based on abstract summary variables. Environmental constraints, by their nature, are often easily observed and measured, for they are the entities to which organisms are actually responding. Theory based on them is more easily related to observation and experimentation than theory based on more abstract summarizations.

Although many models of predator-prey or host-parasite interactions have been based on directly measurable items and processes, many models of the growth of a single population and of interspecific competition have not. Rather, models of intraspecific and interspecific competition have too often been based on the inclusion of density dependence, even though there is little evidence of direct density dependence in such interactions. Direct density dependence implies that an increase in the density of a species is directly responsible for a decrease in per capita growth rate. With the exception of territorial animals whose density is regulated in response to aggressive interactions, most population regulation is mediated through some intermediate entity, usually one or more limiting resources. Woodpecker reproduction can be limited by the number of suitable nesting holes; *Daphnia* reproduction by the density of algae. A change in *Daphnia* density influences algal density, and algal density, in turn, influences growth rate. As such, intraspecific and interspecific competition is often a consumer-resource interaction. Even for territorial animals, it is difficult to imagine how territoriality would evolve except in response to consumer-resource interactions.

Theory that describes the phenomenon of competition by using density-dependent summary variables may be broadly applicable to all types of competition, but the density-dependent summary variables cannot be directly observed. Lotka-Volterra models of competition use an abstract concept, the competition coefficient alpha, that cannot be directly observed but must be measured via field experiments. This has meant that many studies of competition have served merely to measure alpha (e.g., see reviews in Schoener 1983; Connell 1983).

Alpha, though, is just a measure of the effect of the density of one species on the growth rate of another. As soon as it is admitted that species are imbedded in a matrix of many other species, an alpha measured in the field is an ecological abstraction that depends on the densities of all other species at the time it as measured (Schaffer 1981; Tilman 1987), and is thus highly specific to those conditions. Work motivated by a "general" theory becomes highly specific and is of limited utility in seeking the causes of broad ecological patterns.

Alternatively, studies of competition based on a more mechanistic approach might gather information on the foraging behavior of species and on the availabilities of the limiting resources for which they were competing. Like alphas, information on foraging is obtained through experiments. However, unlike alphas, foraging information is potentially applicable to habitats other than that in which it is obtained and can be used to predict the dynamics and outcome of competition in other habitats. As such, a seemingly less general study of consumer-resource interactions may be generalizable to many more habitats than a study based on the "more general" Lotka-Volterra model. Indeed, this has been the case for studies of nutrient competition among freshwater algae (e.g., Tilman 1977, 1982; Tilman, Kilham, and Kilham 1982; Sommer 1983; Tilman et al. 1986). The inclusion of a few simple mechanisms of interaction related to environmental constraints (limitation by phosphorus and silicon) and species tradeoffs (superior competitive ability for silicon being gained at the expense of competitive ability for phosphate) has allowed prediction of phytoplankton patterns in a wide range of lakes around the world.

If ecology is to develop a general theory, it will probably be based on consumer-resource interactions because of the universality of consumer-resource interactions. What other approach is so easily extended to whole food webs and ecosystems, as well as to the evolution of morphology, foraging behavior, physiology, and life histories? Although future research may be able to identify universal aspects to consumer-resource interactions, there are many aspects that are likely to be specific to the constraints of the physical environment and the patterning among these constraints. Such a consumer-resource-based theory may well prove to be the foundation for MacArthur's (1972) vision of ecology, as quoted by May (1986): "I predict that there will be erected a two- or three-way classification of organisms and their geometrical and temporal environments, this classification consuming most of the creative energy of ecologists. The future principles of the ecology of coexistence will then be of the form 'for organisms of type A, in environments of structure B, such and such relations will hold.' "

TREAT ECOLOGY AS A SINGLE DISCIPLINE

Our graduate students are trained, our papers are reviewed, and our grants are funded as if ecology consists of subdisciplines: evolutionary ecology, population ecology, community ecology, and ecosystem ecology. Although there are pragmatic reasons for the National Science Foundation to use such categories, the lack of communication and synthesis among subdisciplines limits our training and vision. The natural world is not divided into evolutionary, population, community, and ecosystem spheres. Organisms have evolved in and live in a single world. Explanations for patterns in this world must be consistent not only within a particular subdiscipline, but across all of these. An explanation of an ecosystem-level phenomenon that is inconsistent with evolutionary theory is either incorrect, or evolutionary theory is incorrect, and vice versa. If evolutionary theory had been rigorously applied to many of the ideas suggested by Clements (1916), we could have avoided decades of misdirected research that treated communities and ecosystems as superorganisms. Conversely, if population ecologists had started, in 1916, to seek the causes of the broad, general patterns Clements described, that subdiscipline could have advanced much more quickly. There is much about the evolution of organismal traits that can be best understood in terms of ecosystem-level constraints, just as there are many ecosystem-level patterns that are best explained in terms of constraints on the evolution of individual organisms. The vastness of the ecological literature makes it difficult for anyone's knowledge to bridge these subdisciplines. However, major advances are likely to come from those who attempt syntheses across these subdisciplinary boundaries, for this will allow them to test their hypotheses against the accumulated knowledge of all of ecology, not just that of a single subdiscipline.

The study of consumer-resource interactions is a natural starting place for such a synthesis. The life histories of organisms, their morphology, physiology, and behavior are already studied in an evolutionary context. It is these traits that are directly relevant to the interactions between a species and its resources and between a species and other species for which it is a resource. Constraints of optimal foraging, for instance, have ramifications for community and ecosystem structure and dynamics that can be determined as soon as they are applied to a complete food web. Although I am arguing, contrary to the Clementsian view, that ecosystem structure should be predictable, in principle, from knowledge of lower-level processes, I am also arguing that evolutionary, population, and community ecology must be done in the context of the whole ecosystem.

Recent discussions of indirect effects (e.g., Levine 1976; Holt 1977; Lawlor 1979; Vandermeer 1980) have emphasized the problems inherent in interpreting interactions in multispecies communities as if only the species of interest existed. Of all the food-web linkages that exist, probably the most overlooked link is that controlled by decomposers. An indirect result of the life of a decomposer is the resupply of the mineral nutrients that are required by and often constrain plants. Because the amounts and proportions of these nutrients can have a large impact on competitive interactions among plants (Tilman 1982, 1988), all processes that influence the decomposer species are potentially important. For instance, if the litter of different plant species differed in its suitability as food for decomposers, this could lead to changes in nutrient mineralization rates (Pastor, Naiman, and Dewey 1987). In theory, this could form a positive feedback loop that would magnify initial differences in local species composition, leading to multiple stable equilibria. Differential grazing by herbivores on plants that differed in litter quality could have similar feedback effects (Pastor, Naimon, and Dewey 1988). Only further observation, experimentation, and theory will determine if such feedback effects are important, and which aspects of food-web structure need to be considered to predict particular types of patterns.

SUMMARY

Ecological research is the process of seeking simplifying assumptions that allow us to abstract much of the complexity of nature into a few variables. There are many reasons to be optimistic that this is a viable process. The very existence of broad, general, repeatable patterns suggests that many of the unique aspects of organisms and habitat are of minor importance. Schaffer (1981) showed that there is a conceptual foundation for ecological abstraction. Much of the dynamics of complex models can often be predicted by a few equations (Tilman 1988). However, we do not yet know the level of mechanism and detail at which this abstraction might be most effectively pursued. A comparison of the mechanistic studies reviewed in Schoener (1986) with studies that are more traditional suggests that our field is still in a phase for which inclusion of additional mechanistic detail can lead to major gains in predictive ability.

In this paper, I have suggested that we should study broad, general patterns. In studying such patterns, we should pursue ecological abstraction by using the simplest possible approach that explicitly includes the most universal constraints of the environment and the unavoidable trade-offs that organisms face in dealing with these constraints. The most universal constraints may come from consumer-resource interactions because all species are, of necessity, parts of food webs. A major advantage of studies at this level of mechanism is that the critical variables (availabilities of various resources, foraging behavior, morphological or physiological processes relevant to resource acquisition and use) can be directly observed. Theories based on these observations are less likely to be habitat-specific than those based on more phenomenological summary variables. Moreover, such an approach is easily expanded to additional trophic levels, as needed. The process of seeking ecological abstraction may be accelerated if there is closer interaction among empiricists, experimentalists, and theoreticians. Such an interaction would be aided if theory were based on variables that could be directly observed. Finally, such a process should cut across the current subdisciplinary lines in ecology: an observation of a general pattern on any conceptual level is of potential importance in testing concepts that had initially been developed on other levels.

REFERENCES

Beard, J. S. 1944. Climax vegetation in tropical America. *Ecology* 25:127–59.

Beard, J. S. 1955. The classification of tropical America vegetation-types. *Ecology* 36:89–100.

Beard, J. S. 1983. Ecological control of the vegetation of Southwestern Australia: Moisture versus nutrients. In F. J. Kruger, D. T. Mitchell, and J.U.M. Jarvis, eds., *Mediterranean-Type Ecosystems,* pp. 66–73. Springer-Verlag, New York.

Clements, F. E. 1916. Plant succession: An analysis of the development of vegetation. Carnegie Inst. Washington Pub. No. 242.

Cody, M. L. 1969. Convergent characteristics in sympatric populations: A possible relation to interspecific territoriality. *Condor* 71:222–35.

Cody, M. L. 1973. Coexistence, coevolution in seabird communities. *Ecology* 54:31–44.

Cody, M. L., and H. A. Mooney. 1978. Convergence versus nonconvergence in Mediterranean-climate ecosystems. *Ann. Rev. Ecol. Syst.* 9:265–321.

Connell, J. H. 1983. On the prevalence and relative importance of interspecific competition: Evidence from field experiments. *Amer. Natur.* 122:661–96.

Cowles, H. C. 1899. The ecological relations of the vegetation on the sand dunes of Lake Michigan. *Bot. Gaz.* 27:95–117, 167–202, 281–308, 361–91.

Elton, C., and M. Nicholson. 1942. The ten-year cycle in numbers of the lynx in Canada. *J. Anim. Ecol.* 11:215–44.

Holt, R. D. 1977. Predation, apparent competition and the structure of prey communities. *Theor. Pop. Biol.* 12:197–229.

Hurlbert, S. H. 1984. Pseudoreplication and the design of ecological field experiments. *Ecol. Monogr.* 54:187–211.

Hutchinson, G. E. 1959. Homage to Santa Rosalia, or why are there so many kinds of animals? *Amer. Natur.* 93:145–59.

Hutchinson, G. E. 1961. The paradox of the plankton. *Amer. Natur.* 95:137–45.

Lawlor, L. R. 1979. Direct and indirect effects of n-species competition. *Oecologia* 43:355–64.

Levine, S. H. 1976. Competitive interactions in ecosystems. *Amer. Natur.* 110:903–10.

MacArthur, R. H. 1972. *Geographic Ecology.* Harper and Row, New York.

MacArthur, R. H., and E. O. Wilson. 1967. *The Theory of Island Biogeography.* Princeton University Press, Princeton, N.J.

May, R. M. 1986. The search for patterns in the balance of nature: Advances and retreats. *Ecology* 67(5):1115–26.

Mooney, H. A. 1977. *Convergent Evolution in Chile and California.* Dowden, Hutchinson & Ross, Stroudsburg, Penn.

Orians, G. H., and R. T. Paine. 1983. Convergent evolution at the community level. In D.

J. Futuyma and M. Slatkin, eds., *Coevolution,* pp. 431–58. Sinauer, Sutherland, Mass.

Pastor, J., R. J. Naimon, and B. Dewey. 1988. A hypothesis of the effects of moose and beaver foraging on soil nitrogen and carbon dynamics, Isle Royale. *Alces* 23. In press.

Schaffer, W. M. 1981. Ecological abstraction: The consequences of reduced dimensionality in ecological models. *Ecol. Monogr.* 51:383–401.

Schoener, T. W. 1971. Theory of feeding strategies. *Ann. Rev. Ecol. Syst.* 2:369–404.

Schoener, T. W. 1976. Alternatives to Lotka-Volterra competition: Models of intermediate complexity. *Theor. Pop. Biol.* 10:309–33.

Schoener, T. W. 1983. Field experiments on interspecific competition. *Amer. Natur.* 122:240–85.

Schoener, T. W. 1986. Mechanistic approaches to community ecology: A new reductionism? *Amer. Zool.* 26:81–106.

Sommer, U. 1983. Nutrient competition between phytoplankton species in multispecies chemostat experiments. *Arch. Hydrobiol.* 96(4):399–416.

Tansley, A. G. 1949. *The British Isles and Their Vegetation.* Cambridge University Press, Cambridge, England.

Tilman, D. 1977. Resource competition between planktonic algae: An experimental and theoretical approach. *Ecology* 58:338–48.

Tilman, D. 1980. Resources: A graphical-mechanistic approach to competition and predation. *Amer. Natur.* 116:362–93.

Tilman, D. 1982. *Resource Competition and Community Structure.* Princeton University Press, Princeton, N.J.

Tilman, D. 1987. The importance of the mechanisms of interspecific competition. *Amer. Natur.* 129(5):769–74.

Tilman, D. 1988. *Plant Strategies and the Dynamics and Structure of Plant Communities.* Princeton University Press, Princeton, N.J.

Tilman, D., S. Kilham, and P. Kilham. 1982. Phytoplankton community ecology: The role of limiting nutrients. *Ann. Rev. Ecol. Syst.* 13:349–72.

Tilman, D., R. Kiesling, R. Sterner, S. Kilham, and F. A. Johnson. 1986. Green, bluegreen and diatom algae: Taxonomic differences in competitive ability for phosphorus, silicon and nitrogen. *Arch. Hydrobiol.* 106(4):473–85.

Vandermeer, J. H. 1980. Indirect mutualism: Variations on a theme by Stephen Levine. *Amer. Natur.* 116:441–48.

Walter, H. 1985. *Vegetation of the Earth and Ecological Systems of the Geobiosphere,* 3d ed. Springer-Verlag, New York.

Whittaker, R. H. 1975. *Communities and Ecosystems,* 2d ed. Macmillan, London.

ECOLOGY AND EVOLUTION III

Chapter 7

Blending Ecology and Genetics: Progress toward a Unified Population Biology

JOSEPH TRAVIS AND LAURENCE D. MUELLER

THEORY AND DATA IN POPULATION BIOLOGY

The first step in examining the interplay of theoretical and empirical investigations in population biology is to bring the goal of these investigations into focus. That goal is to understand the genetic and phenotypic diversity within and among populations in terms of the microevolutionary forces that regulate that diversity. Those forces are functions ultimately of the numbers of individuals, the ecological factors that impose risks of mortality on those individuals and constrain their reproductive abilities, and the temporal and spatial patterns of population dynamics. Investigations in population biology revolve around the attempt to understand the joint dynamics of numbers of individuals and genetic and phenotypic diversity.

In this essay we review three topical areas in which there has been considerable progress in examining the reciprocal influences of ecological and genetic processes. Our choice of areas is based on a subjective assessment of how well empirical and theoretical approaches have combined to facilitate that progress. In other words, we have chosen to focus on three successful areas of interplay and complementarity between theoretical and empirical work.

Our subjective selection does not include several topics that could conceivably be discussed in this context. The impact of heterogeneous environments and multiple niches has been reviewed recently by Hedrick (1986). The general connection of heterozygosity with measures of organism performance and demographic parameters has also been reviewed relatively recently (Mitton and Grant 1984), and some theoretical aspects of that connection have been explored subsequently by Turelli and Ginzburg (1983) and Smouse (1986) for viability selection and by Feldman and Liberman (1985) for fertility selection. There have been some exciting advances in the statistical methodology for describing and analyzing various demographic and genetic processes: recent work in this vein includes Lenski and Service (1982), Swallow and Monahan (1984), Manly (1985a), de Kroon et al. (1986), Meyer et al. (1986), Muenchow (1986), and Westcott (1986). More general papers that contain some methods that may be new to empirical workers and that deserve attention include those by Mueller and Ayala (1981a), O'Donald and Majerus (1985), and Anderson et al. (1986).

ECOLOGICAL PROCESSES AND NATURAL SELECTION

Overview

Two recent reviews of selection in natural populations (Manly 1985b; Endler 1986) suggested that we know little about the ecological genesis of selective mortality or selective fertility. Without such knowledge it is impossible to predict how consistent selection pressures will be across generations or how variable such pressures will be among spatial locations. Most of our "ecological" knowledge of selection is based on two types of cases: effects of anthropogenous alterations in the environment and survival patterns in the face of catastrophic events. Neither type of information is necessarily indicative of the selective pressures that have molded the genetic diversity in populations through their normal cycles of expansion and contraction in numbers.

Recent theoretical work on the measurement of selection and on the evolutionary dynamics of continuous traits (O'Donald 1968, 1970, 1971; Lande 1976, 1979, 1980, 1984; Bulmer 1980; Lande and Arnold 1983; Arnold and Wade 1984; Turelli 1984, 1985; Barton and Turelli 1987) has inspired a growing number of empirical studies of phenotypic selection and quantitative genetics. In many cases these studies include the dissection of the ecological genesis of selection pressures. In the subsequent sections we attempt to draw some lessons from a number of recent studies and to use those lessons to point to some new or unsolved problems in ecological genetics.

The Ecological Genesis of Selection Pressures

A plethora of selective pressures act simultaneously on a variety of traits in a natural population. Two reviews by Bell (1976, 1984) illustrate this point well in describing the manifold selective pressures faced by populations of sticklebacks.

In order to make any progress, one must choose one or a small number of characters for careful study. Yet even a single trait is usually subject to many selective agents that may act antagonistically and may vary spatially or temporally in their relative importance (Endler 1978, 1980, 1983). These facts can make the detection and measurement of selection difficult, such that successful studies of phenotypic selection are likely to have focused on traits under relatively strong selection. This is a possible source of bias in our conclusions.

Differential fertility appears at least as important as differential viability in contributing to total fitness differences (Endler 1986). In some cases the differential fertility arises through differences in the age at first reproduction that give some phenotypes shorter generation times (eye color polymorphism in skuas, O'Donald and Davis 1976; P-gene polymorphism for body size in some poeciliid fishes, Borowsky 1984). In other cases the differences appear to result from different fecundity rates (Schmidt and Levin 1985; Antonovics and Ellstrand 1984).

In several cases the ecological genesis of a selective pressure has been uncovered through detailed observational ecology or through a combination of observation and experiment. The extensive studies of Grant and his colleagues (Grant 1986) on Darwin's finches provide a notable case study. Strong viability selection in the population of *Geospiza fortis* on the island of Daphne Major in the Galapagos in 1977 favored larger birds with stouter bills. This selection pressure was generated by a severe drought during what was normally the wet season that produced a dramatic change from other years in the numbers and types of seeds available (Boag and Grant 1981; Price et al. 1984). As the pool of available seeds was depleted during the drought, the birds were forced to turn to increasingly larger, less preferred seeds, which were harder and more difficult to crack. The population numbers continued to fall throughout the drought as smaller birds with bills that were less stout were unable to find sufficient numbers of seeds that they could eat. By the drought's end, the larger birds with stouter bills were disproportionately represented among the survivors. The heritability of those traits allowed Grant and his colleagues to observe a short-term response to selection in the increased sizes and altered bill morphologies of the offspring (Grant 1986).

An excellent example of a careful observational study of selection and its ecological genesis is that of Hairston and Walton (1986) on the timing of production of diapause eggs in the copepod *Diaptomus sanguineus*. During early spring, as predatory fish increase their foraging activity, the copepods switch from producing subitaneous eggs (which hatch immediately but do not survive passage through the guts of fish) to producing diapause eggs (which hatch after a long dormant period but can survive passage through fish guts). Hairston and Walton observed one of two adjacent ponds to dry completely and lose its fish population. The absence of fish released the copepods from the pressure to switch to diapause eggs. This change altered the balance of selection pressures in favor of directional selection for longer production of subitaneous eggs, because longer production of subitaneous eggs confers a net reproductive advantage over a switch to diapause eggs. That selection pressure was strong enough to shift the mean switch date by twenty-six days over the succeeding two generations. No parallel shift occurred in the adjacent pond that had not dried.

An experimental approach to understanding the genesis of selective mortality was undertaken by Travis and his colleagues in their studies of larval growth rate in the anurans that breed in temporary ponds. These populations are regulated by the interaction of predation and competition (Wilbur 1980, 1984; Morin 1981, 1983). The critical predators of larvae in these ponds are gape-limited (larval salamanders and adult newts) or limited by the size of the feeding appendages (insects) such that predation rates are decreasing functions of larval body size (Travis, Keen, and Juilianna 1985a; Cronin and Travis 1986). For each predator there is also an upper limit to the body size of larvae that can be handled (Caldwell, Thorp, and Jervey 1980). As a result of these interactions, larvae that grow more rapidly experience lower cumulative risks of predation. Enclosure experiments on full-sib families that grew at different rates revealed family-level selective mortality due to predation by insects alone at high prey densities (Travis 1983) and by a combination of insect and salamander predators at even low prey densities (Travis, Keen, and Juilianna 1985b).

Studies on the ecological genesis of selection have shown how variable the action of selection can be. The ecological milieu in which an array of populations exist changes over space and time, and this changing ecological context produces a pattern of fluctuating modes of selection and variable selection intensities. For *G. fortis,* although selection for larger birds with stouter bills recurs whenever drought conditions recur, its intensity varies among years (Price et al. 1984). In some years smaller birds have a net advantage through either a viability difference or a lower age at first reproduction (Grant 1986).

Selection intensities also fluctuate in larval anurans. *Hyla crucifer* larvae experienced no selective mortality with respect to growth rate when exposed only to newts, despite the fact that newts are voracious gape-limited predators (Morin 1986). Travis, Keen, and Juilianna (1985b) observed no selection on growth rate of *H. gratiosa* larvae when only predatory dragonfly naiads or only larval salamanders were present, although the combination of predators generated strong selective mortality. Natural history suggests that these effects reflect the importance of predation refuges: when only salamander predators are present, larvae escape by hiding in litter (Morin 1986), an option precluded by the presence of litter-dwelling dragonfly naiads.

Similar conclusions about the nature of phenotypic selection are suggested by several short-term studies or more strictly phenomenological studies. In Kalisz's (1986) three-year study of selection on seed germination times in *Collinsia verna,* the net selection pressure varied widely in direction and intensity among spatial locations and generations. The relative contributions of viability and fertility selection to the total fitness differences also varied. Studies by Schmitt and Antonovics (1986) on *Anthoxanthum odoratum* also found selection pressures to vary dramatically over short spatial distances, and Maddox and Cappuccino (1986) found that genotypic differences in resistance to herbivores in *Solidago altissima* changed drastically with changes in the plants' moisture regime. Travis and Henrich (1986) showed that the intensity of fertility selection on female body size in the poeciliid fish *Heterandria formosa* was highly sensitive to variation in food levels.

Problems in Ecological Genetics of Continuous Traits

The patterns seen in ecologically oriented studies of selection raise several issues in ecological genetics. The first issue arises when selective pressures act on a suite of traits in a consistently directional fashion. Theoretical studies (cited by Travis, Emerson, and Blouin 1987) have suggested that any remaining genetic variation in the individual traits should be due largely to dominance variance, and many studies of *Drosophila* have been consistent with that prediction (Breese and Mather 1960; Mather and Cooke 1962; Keller and Mitchell 1964; Kearsey and Kojima 1978; Kidwell 1969; for *Tribolium* see Dawson 1966). Sulzbach and Lunch (1984) offered a verbal model that long-term directional selection on a suite of traits should result in a dominance genetic correlation matrix whose principal axes describe a gradient from favorable trait combinations to deleterious ones. This is a multivariate modification of Fisher's (1958) notion that present-day directional dominance should indicate the directional history of selection. Their studies of thermoregulatory and nesting behavior in *M. musculus* and Travis, Emerson, and Blouin's (1987) examination of growth rate and larval period in *H. crucifer* were consistent with this prediction. The generality of these observations is unknown, and a mathematical version of this suggestion has not been developed.

A second problem revolves around the empirical prominence of strong fertility selection. The evolutionary response to strong fertility selection can be more complicated than it would be for viability selection. With certain types of fertility selection, a wider range of dynamic behaviors is possible than with viability selection, with historical effects playing a prominent role in determining the particular genetic outcome (see Feldman, Christiansen, and Liberman 1983; Feldman and Liberman 1985). General principles derived from weak fertility selection (Nagylaki 1987) may not apply to situations of strong selection, and it appears that more theoretical work on strong fertility selection is needed.

The third issue is raised by the inconsistent nature of many local selection pressures. Even with high selection intensities on continuous traits, the actual selection coefficients at each constituent locus may be quite small, with alleles at many loci behaving as though under weak selection and alleles at other loci behaving as though completely neutral (Milkman 1978, 1982; Kimura and Crow 1978; Lynch 1984; Campbell 1984). The fluctuating nature of phenotypic selection may increase the likelihood that patterns of allele frequency change will be indistinguishable from genetic drift (Gillespie 1979). If major evolutionary changes in metric phenotypes only occur in response to sporadic bouts of sustained directional selection, which are precipitated only by radical ecological changes (such as that seen by Hairston and Walton), then the genetic dynamics may be characterized by occasional bursts of directional selection amid long interludes of near-neutrality. Gillespie (1984) has developed models of molecular evolution in which evolution proceeds by a series of bursts of natural selection. Such models appear to be consistent with observed patterns of nucleotide substitution in nuclear and mitochondrial DNA sequences (Gillespie 1986). Thus, at the genetic level, selection on some types of continuous traits may act similarly to selection on molecular variants.

Fluctuating selection pressures at the phenotypic level have been suggested as an important mechanism for reducing the rate of loss of genetic variation (Grant 1986). There is some theoretical support for this suggestion from studies of single-locus dynamics (Karlin and Levikson 1974; Karlin and Liberman 1974). The effects of such a selective regime on the mean and variance of a continuous, polygenic trait deserve further exploration.

The proper interpretation of what would appear to be fluctuating phenotypic selection could become more complicated if the fitness functions acting on continuous traits generate significant levels of epistasis in fitness at the genetic level. Purely additive gene effects on the phenotype can be readily translated into epistatic effects at the genetic level through the ways in which phenotypic variation is translated into fitness variation (Nei 1963; Jain and Allard 1966; Lewontin 1974; Rose 1982). When epistatic effects are present, part of the overall response to selection will be caused by the generation of linkage disequilibrium. If selection is relaxed, the linkage disequilibrium will decay, and its associated component of the response to selection will disappear, causing the mean value of the trait to "slip" backwards (Kojima 1961; Gill 1965a,b; Bulmer 1980). This effect, if observed in isolation, could easily appear to be a reversal of the direction of phenotypic selection. In the absence of ecological knowledge, changes of sign in selection differentials should be interpreted cautiously.

It is tempting to speculate that the presence of significant epistasis will enhance the role of fluctuating selective regimes in slowing the rate of loss of genetic variation by producing a "genetic drag" on allele frequency responses. The empirical problem with this speculation is that we have too little knowledge of what fitness functions actually look like or how components of fitness actually combine to generate overall fitness to say how general such epistatic effects might be. There has been no specific mathematical investigation of this speculation that we know of.

A final issue is more of an empirical caveat. The development of tractable models of the dynamics of quantitative traits under natural selection may require the assumption that large numbers of genes contribute to the trait (Barton and Turelli 1987). Empirical studies are clearly necessary to evaluate the validity of these assumptions whenever possible. Luckinbill et al.'s (1987) genetic analysis of lines of *Drosophila melanogaster* selected in different directions for life-history variation suggested that only a few genes were responsible for the divergence between lines. O'Donald and Majerus (1985) found a similar result for the divergence of female mating preference lines in a ladybird beetle. These may not be isolated cases. The classic polygenic model of phenotypic selection and response in large populations is remarkably robust (Hill 1969), but it does generate inaccurate predictions when its genetic assumptions are violated. For genes of large effect or extensive nonadditive effects, the classic model can grossly overestimate the response to selection after only a few generations (Kojima 1961; Gill 1965a,b; Latter 1965a,b).

LIFE-HISTORY EVOLUTION

Overview

A major goal of evolutionary biology has been the development of a general theory that will allow the prediction of those life-history traits most likely to evolve in different ecological settings. Three major lines of theoretical development have ensued (Stearns 1976, 1977). One line of inquiry, beginning with Cole (1954) and Murphy (1968), has concentrated on the response to different selection regimes generated by different levels of juvenile and adult mortality. The second approach, beginning with MacArthur (1962; MacArthur and Wilson 1967), concentrated on the different selection pressures operating at different extremes of population density. A third approach, two variants of which are attributable to Medawar (1957) and Williams (1957), has focused on the age-specificity of survivorship and fertility and has concentrated on the genetic causes of such effects. This third approach differs from the first two in that it was not undertaken to address large-scale differences among populations or species. However, observations of widespread genetic variation for life-history traits within populations (Istock 1983) have prompted the development of explicit genetic models that have begun to unify the first and third approaches (Lande 1982; Rose 1982, 1985). Three sets of developments, intellectual descendants of the three original approaches, illustrate recent progress in understanding life-history evolution.

Demographic Models of Life Histories

One line of empirical research attempted to match differences among populations in age-specific reproductive effort to the different age-specific mortality regimes experienced by those populations. The problem was complicated by the extensive phenotypic plasticity displayed by many organisms. This plasticity often served to obscure the resolution of just what the expressed "life history" really was. In *Plantago lanceolata,* different populations displayed different patterns of reproductive allocation that appeared to match the theoretical predictions for their respective mortality regimes. However, those differences were not maintained in a common environment (Primack and Antonovics 1982), suggesting that plasticity in allocation itself might be the major adaptation (Bradshaw 1965; Schlichting 1987). Similar observation in many systems (e.g., facultative viviparity in mosquitoes, Bradshaw 1986; flexible maturation patterns in fishes, Policansky 1983) prompted the development of models for "plastic life histories." These models explore the evolution of rules for phenotypic expression that optimize fitness over a range of commonly encountered demographic conditions (Caswell 1983; Kaplan and Cooper 1984; Werner and Gilliam 1984; Stearns and Koella 1986). Explicit tests of these models have yet to be performed.

Although much attention has been devoted to the evolution of age-specific reproductive effort, less attention has been devoted to examining how different demographic regimes might select for different ways of "packaging" that effort. Smith and Fretwell (1974) and Brockelman (1975) focused on how offspring survival probabilities would guide the compromise between offspring size and offspring number. Gillespie (1975), Downhower and Brown (1975), and Wilbur (1977), inspired by completely different empirical problems, independently showed that as the uncertainty in offspring survival probabilities rose, there would be an increased selection pressure to divide the offspring in a single bout of reproduction into more clutches or nests. Winkler and Wallin (1987) have unified some of these ideas and developed a general model for the simultaneous evolution of reproductive effort and the specific allocation of that effort in response to specific demographic conditions.

This topic has received even less empirical attention. Only Travis, Farr, et al. (1987) have attempted to link overall allocation and "packaging" strategy to a demographic context in a specific attempt to evaluate theoretical predictions. They found that morphological constraints played a major role in addition to the selective regime in governing the manner in which reproductive effort was "packaged." The generality of such constraints is unknown. The linkage of patterns of reproductive effort with patterns of "packaging" is obvious in broad taxonomic comparisons (Pianka 1970), but the quality of the match between the predicted patterns and the particular demographic regimes that are supposed to generate selection for those patterns remains to be discovered.

Density-Dependent Selection and Life-History Responses

The initial ideas of *r*- and *K*-selection (MacArthur and Wilson 1967) were developed along two lines. The verbal theory (Pianka 1970; Stearns 1976, 1977;

Parry 1981) attributed a large array of life-history evolution to r- and K-selection, whereas the mathematical theory (Anderson 1971; Charlesworth 1971, 1980; Roughgarden 1971; Clarke 1972; Anderson and Arnold 1983; Asmussen 1983) made more modest predictions. Initial tests of the theory relied on observations from field populations to which different regimes of density-dependent population regulation were attributed. Despite some early reports that confirmed predictions from the verbal models (Gadgil and Solbrig 1972; McNaughton 1975), an almost equal number of studies with contradictory results have appeared (Tinkle and Hadley 1975; Wilbur 1976).

Stearns (1977) has discussed the difficulties with many field studies, and recently it has become clear that carefully controlled laboratory studies will be the most useful for testing some of the more specific aspects of the mathematical theories. There are presently a small number of such studies that have utilized *Escherichia coli* (Luckinbill 1978, 1984) and *Drosophila* (Taylor and Condra 1980; Barclay and Gregory 1981, 1982; Mueller and Ayala 1981b). Only the studies of Luckinbill and Mueller and Ayala have dealt with phenotypes that are components of the mathematical theories. The other studies have dealt largely with the predictions of the verbal theory, and the results have been mixed (see Mueller 1985 for a review of these studies).

The results of Luckinbill and Mueller and Ayala are contradictory. At the present time it does not seem possible to explain these differences other than to note the very different life histories of the two organisms and to admit that this theory may not be valid for all life forms. We next describe in some detail what is known about density-dependent natural selection in laboratory populations of *Drosophila melanogaster*. From this discussion it will be apparent that population densities can have a significant impact on the evolution of life histories. Furthermore, these studies have verified key predictions of the theory of density-dependent natural selection and thereby justify the central role this theory has assumed in the field of evolutionary ecology (Roughgarden 1979).

The results summarized below are from a long-term study of six independent populations of *D. melanogaster*. The origin and maintenance of these populations are described by Mueller and Ayala (1981b).

DENSITY-DEPENDENT RATES OF POPULATION GROWTH

After just eight generations of selection, rates of population growth were determined for the six experimental populations at one low and two high densities (Mueller and Ayala 1981b). The r-selected population had the higher rate of growth at the low density, whereas the K-selected population did better at the high densities. This experiment was important because the basic mathematical theories assume that density-dependent rates of population growth are equivalent to fitness and that natural selection should cause the increase in these quantities. It is known that this assumption is not strictly true (Prout 1980; Mueller and Ayala

1981c), but this experiment demonstrated that rates of population growth respond to selection in a manner consistent with theoretical expectations.

DENSITY-DEPENDENT VIABILITY AND SIZE

In an attempt to understand which life-history characteristics gave rise to the previously noted growth rate differences, a number of additional experiments were conducted (Bierbaum, Mueller, and Ayala 1988). Survival of first-instar larvae to adulthood was determined at one low and two high densities. Although the r-selected larvae survived better at low density, this difference was not significant. At the high densities the K-selected larvae had substantially higher survivorship. In addition, the K-selected males and females emerging from the high-density cultures were heavier by 15% than their r-selected counterparts. Larger adults will generally have greater fitness due to sexual selection in males (Wilkinson 1987) and fecundity in females (Robertson 1957; Mueller 1987).

Clearly these differences in survivorship and female size could account for the large differences in rates of population growth observed at high densities. The increased size and survivorship of the high-density populations under crowded conditions does not appear to be due to increased digging behavior, which would provide access to unused food, or to an ability to pupate on less food (Mueller, unpublished). However, pupation site choice may affect viability in crowded cultures.

Under standardized conditions Mueller and Sweet (1986) have measured the height above the food medium at which larvae pupate in vial cultures. These experiments have revealed tremendous differences between the r and K populations; members of the K populations always pupate higher than their r counterparts. Because pupation on the surface of the medium or close to it in crowded environments may substantially increase the probability of death, Mueller and Sweet conjectured that increased pupation height in crowded cultures is adaptive.

COMPETITIVE ABILITY

In crowded cultures many resources are in short supply, including food. The ecological and evolutionary consequences of population growth in food-limited environments has recently been explored by Mueller (1988a). Using previous empirical data on larval competition (Bakker 1961) and a model of viability (Nunney 1983), Mueller (1988a) has demonstrated that natural selection will favor an increase in competitive ability for food at high densities, but increasing competitive ability will *not* have any long-term effect on viability or population-carrying capacity.

In an experimental system that is consistent with the models discussed above, the competition coefficients for food have been estimated for the r- and K-selected populations (Mueller 1988b). This study showed that the K populations have competitive abilities 62% greater than those of the r populations.

Evolution of Senescence

The two most prominent theories of the evolution of senescence are the mutation-accumulation theory and the pleiotropy theory. The mutation-accumulation theory (Medawar 1957; Edney and Gill 1968) states that senescence is due to the accumulation of deleterious alleles, in mutation-selection balance, whose effects are expressed late in life. Such alleles will have only small effects on fitness and thus be only weakly acted upon by natural selection. The pleiotropy theory (Williams 1957; Rose 1982, 1985) suggests that alleles with beneficial effects for survival or fecundity early on in life will have deleterious pleiotropic effects late in life. The early beneficial effects of such alleles substantially outweigh these deleterious late effects, and consequently such alleles are favored by natural selection.

A corollary of the pleiotrophy hypothesis is that many early and late components of fitness, e.g., early fecundity and survivorship, ought to have negative genetic correlations. The evidence for such correlations or costs to reproduction have been reviewed by Reznick (1985) and Clark (1987). These studies have shown essentially all possible results. However, the methods used vary dramatically. It now seems clear that two widely used techniques are biased and more likely to produce positive correlations. If the populations used have been inbred, then positive correlations in life-history traits appear even when the parental populations show negative correlation (Rose 1984a). Second, if measurements of life-history traits are made in environments that differ from those the populations evolved in, positive correlations appear (Service and Rose 1985).

In *Drosophila* the current evidence (Rose and Charlesworth 1980) indicates that the normal pattern of senescence seen in this insect is not due to mutation accumulation. In addition, populations selected for reproduction late in life exhibit increased survivorship but reduced fecundity early in life (Rose 1984b; Luckinbill et al. 1984; Clare and Luckinbill 1985). However, in small populations of *Drosophila* in which only young adults reproduce, senescence, as measured by age-specific female fecundity, may be accelerated by the accumulation of deleterious recessive alleles (Mueller 1987). Likewise, in *D. melanogaster* there appear to be higher frequencies of deleterious late-acting alleles affecting virility than early-acting ones (Kosuda 1985).

A secondary phenomenon of relevance to these discussions is the possibility that evolution can compensate for deleterious pleiotropic effects of alleles. For instance, mutations in *Escherichia coli* that confer resistance to T4 virus also reduce competitive ability. In the absence of virus, resistant bacteria may evolve higher competitive ability by genetic changes that compensate for the deleterious effects of the resistance mutations (Lenski 1988).

EVOLUTIONARY BIOLOGY AND POPULATION DYNAMICS

Overview

The ways in which changing population sizes affect levels of genetic variation have received attention in several contexts. The initial survival probabilities of rare mutants are higher in expanding populations than in stationary populations (Ewens 1969). Fluctuations in population size can increase the rate of loss of genetic variability dramatically (Nei et al. 1975; Motro and Thomson 1982) and alter multilocus genotype frequencies to produce significant changes in the components of genetic variance (Goodnight 1987). In the next section we highlight some ecologically oriented ways of examining the effect of changing population sizes on levels of genetic variation. In the subsequent sections we examine how the problem can be turned around to show how genetic variation affects population dynamics.

Demographic Variability and Population Structure

When populations fluctuate there are often dramatic changes in their internal demographic mechanics. Wade (1980, 1984) showed how the ratio of the variance effective population size to the census number of adults decreased with increasing population density of *Tribolium castaneum* through changes in the distributions of family size (among other factors). This relationship could allow the effective size to fluctuate much less from one generation to the next than the census numbers and would vitiate the expected effects of large fluctuations in adult numbers on levels of genetic variation. The generality of this interesting result is unknown.

The heterogeneity of family sizes and population fluctuations play key causal roles in determining the variance effective size and the inbreeding effective size (Ewens 1982). Examination of the joint behavior of population fluctuations and demographic heterogeneity can thus offer a critical link between population ecology and population genetics. Mueller et al. (1985), Heywood (1986), and Wood (1987) have developed ecological approaches to estimating effective population sizes in a single generation directly from levels of demographic heterogeneity. These initial models deserve wider use and further elaboration. The effective population size is a fundamental parameter in population genetics, yet existing extimation procedures that use genetic data are indirect, imprecise (Pollak 1983; Tajima and Nei 1984), and often biased (Mueller et al. 1985).

The Interplay of Phenotypic Distributions and Vital Rates: Genetic Variation Directly Affecting Population Dynamics

In many populations of plants and animals, vital rates vary functionally with a phenotypic trait, usually individual size. In such situations the phenotype distribution determines the overall natality and mortality rates of the population. Density-dependent regulatory processes may work either through a direct feedback of density on vital rates or an indirect feedback through a change in the phenotype distribution (e.g., smaller body sizes at higher densities and consequently lower fecundity rates). Plant population ecologists have been exploring specific interrelationships of this type for a long time. Recently a number of general phenotype models have been developed to explore the joint dynamics of the phenotype distribution and the numbers of individuals (Streifer 1974; Oster 1976; Caswell 1982a,b; Kirkpatrick 1984; Roughgarden, Iwasa, and Baxter 1985; Pacala 1986; Mueller 1988a).

These models allow us to see directly how phenotypic diversity (and, with the addition of heritability, genetic diversity) can affect the actual numbers of individuals present at a given time. More traditional population genetics models have made the connection between population numbers and genotypic diversity through somewhat imaginative interpretations of the significance of the value of "mean fitness" (see Smouse 1976 and Prout 1980 for a discussion of these issues). The more explicit operational connection provided by these models can clarify our understanding of a number of postulates about the mode of microevolutionary change, not the least of which is the operation of the shifting balance model (Wright 1982).

Two critical problems must be addressed before such models can be used to explore particular systems. First, the actual feedback mechanism from density to phenotype to vital rate must be described. Some mechanisms stabilize population dynamics and others destabilize them (Gurney and Nisbet 1985). Second, the extent of genetic variation in the susceptibility to the effects of density must be determined. Density-dependent selective forces change the interrelation of population dynamics and genetic variation substantially (Prout 1980; Desharnais and Constantino 1983). Too few studies have been done under natural conditions for any conclusions to be drawn about the general strength of such density-dependent effects (Shaw 1986). In one system, however, knowledge of the feedback mechanisms and the extent of genetic variation in density response have been combined with careful modeling to illuminate some ways in which stable population dynamics can actually evolve.

Evolution of Stable Population Dynamics

Even simple systems can in principle display immensely complex and chaotic population dynamics (Guckenheimer, Oster, and Ipaktchi 1977; Nisbet and

Gurney 1982). Yet laboratory populations of *Drosophila melanogaster* kept in serial transfer systems always exhibit stable dynamics (Mueller and Ayala 1981d). This fact raises the issue of whether population stability itself can evolve and, if so, by what mechanism.

Initial work by Thomas, Pomerantz, and Gilpin (1980) suggested that group selection would be necessary for the evolution of stability, but models of Heckel and Roughgarden (1980) and Turelli and Petry (1980) showed that individual selection could mold population growth rates at higher densities so that stability would evolve. Mueller and Ayala (1981a) showed that experimental populations with different genetic compositions displayed different parameters of population growth. This result showed phenomenologically that genetic variation existed that could drive the evolution of stability without a group-selection process.

Subsequent theoretical work by Mueller and Ayala (1981d) and Mueller (1988a) has clarified an actual mechanism of individual selection through which stability can evolve. At higher densities, increased tolerance to density stress is favored, and larval survivorship increases. However, the consequence of increased survivorship is lowered per-capita fecundity rates, which could occur through reduced adult size or reduced reproductive effort (Mueller 1988a). Some compromise between increased larval survival and decreased adult fecundity is necessary for stable dynamics to evolve. It is noteworthy that an analysis of Nicholson's famous experimental blowfly population has uncovered just this sort of change during the period of time over which that the population's demography was monitored (Stokes et al. 1988).

ACKNOWLEDGMENTS

We thank M. W. Feldman, R. M. May, and J. Roughgarden for discussions and suggestions. Our research has been supported by NSF grants DEB 81-02782, BSR 83-05823, BSR 84-13233, and BSR 84-15529 (JT), and by NIH grant GM34303 (LM).

REFERENCES

Anderson, W. W. 1971. Genetic equilibrium and population growth under density-regulated selection. *Amer. Natur.* 105:489–98.

Anderson, W. W., and J. Arnold. 1983. Density-regulated selection with genotypic interactions. *Amer. Natur.* 121:649–55.

Anderson, W. W., J. Arnold, S. A. Sammons, and D. G. Yardley. 1986. Frequency-dependent viabilities of *Drosophila pseudoobscura* karyotypes. *Heredity* 56:7–17.

Antonovics, J., and N. C. Ellstrand. 1984. Experimental studies of the evolutionary significance of sexual reproduction. I: A test of the frequency-dependent selection hypothesis. *Evolution* 38:103–15.

Arnold, S. J., and M. J. Wade. 1984. On the measurement of natural and sexual selection: Theory. *Evolution* 38:709–19.

Asmussen, M. A. 1983. Density-dependent selection incorporating intraspecific competition. II: A diploid model. *Genetics* 103:335–50.

Bakker, K. 1961. An analysis of factors which determine success in competition for food among larvae of *Drosophila melanogaster. Arch. Neerl. Zool.* 14:200–81.

Barclay, H. J., and P. T. Gregory. 1981. An experimental test of models predicting life-history characteristics. *Amer. Natur.* 117:944–61.

Barclay, H. J., and P. T. Gregory. 1982. An experimental test of life history evolution using *Drosophila melanogaster* and *Hyla regilla. Amer. Natur.* 120:26–40.

Barton, N. H., and M. Turelli. 1987. Adaptive landscapes, genetic distance and the evolution of quantitative characters. *Genet. Res. Camb.* 49:157–73.

Bell, M. A. 1976. Evolution of phenotypic diversity in *Gasterosteus aculeatus* superspecies on the Pacific coast of North America. *Syst. Zool* 25:211–27.

Bell, M. A. 1984. Evolutionary phenetics and genetics: The threespine stickleback, *Gasterosteus aculeatus,* and related species. In B. J. Turner, ed., *Evolutionary Genetics of Fishes,* pp. 431–528. Plenum Press, New York.

Bierbaum, J. T., L. D. Mueller, and F. J. Ayala. 1988. Density-dependent evolution of life history characteristics in *Drosophila melanogaster. Evolution.* In press.

Boag, P. T., and P. R. Grant. 1981. Intense natural selection on a population of Darwin's Finches (Geospizinae) in the Galápagos. *Science* 214:82–85.

Borowsky, R. 1984. The evolutionary genetics of *Xiphophorus.* In B. J. Turner, ed., *Evolutionary Genetics of Fishes.* pp. 235–310. Plenum Press, New York.

Bradshaw, A. D. 1965. Evolutionary significance of phenotypic plasticity in plants. *Adv. in Genetics* 13:115–55.

Bradshaw, W. E. 1986. Variable iteroparity as a life-history tactic in the pitcher-plant mosquito *Wyeomyia smithii. Evolution* 40:471–78.

Breese, E. L., and K. Mather. 1960. The organization of polygenic activity within a chromosome in *Drosophila.* II: Viability. *Heredity* 14:375–99.

Brockelman, W. Y. 1975. Competition, the fitness of offspring, and optimal clutch size. *Amer. Natur.* 109:677–99.

Bulmer, M. G. 1980. *The Mathematical Theory of Quantitative Genetics.* Clarendon Press, Oxford.

Caldwell, J. P., J. H. Thorp, and T. O. Jervey. 1980. Predator-prey relationships among larval dragonflies, salamanders, and frogs. *Oecologia* 46:285–89.

Campbell, R. B. 1984. The manifestation of phenotypic selection at constituent loci. I: Stabilizing selection. *Evolution* 38:1033–38.

Caswell, H. 1982a. Optimal life histories and the maximization of reproductive value: A general theorem for complex life cycles. *Ecology* 63:1218–22.

Caswell, H. 1982b. Stable population structure and reproductive value for populations with complex life cycles. *Ecology* 63:1223–31.

Caswell, H. 1983. Phenotypic plasticity in life history traits: Demographic effects and evolutionary consequences. *Amer. Zool.* 23:35–46.

Charlesworth, B. 1971. Selection in density-regulated populations. *Ecology* 52:469–74.

Charlesworth, B. 1980. *Evolution in Age Structured Populations.* Cambridge University Press, Cambridge, England.

Clare, M. J., and L. S. Luckinbill. 1985. The effects of gene-environment interaction on the expression of longevity. *Heredity* 55:19–29.

Clark, A. G. 1987. Senescence and the genetic correlation hang-up. *Amer. Natur.* 129:932–40.

Clarke, B. 1972. Density-dependent selection. *Amer. Natur.* 106:1–13.

Cole, L. C. 1954. The population consequences of life history phenomena. *Quart. Rev. Biol.* 29:103–37.

Cronin, J. T., and J. Travis. 1986. Size-limited predation on larval *Rana areolata* (Anura: Ranidae) by two species of backswimmer (Hemiptera: Notonectidae). *Herpetologica* 42:171–74.

Dawson, P. S. 1966. Developmental rate and competitive ability in *Tribolium*. *Evolution* 20:104–16.

de Kroon, H., A. Plaisier, J. van Groenendael, and H. Caswell. 1986. Elasticity: The relative contribution of demographic parameters to population growth rate. *Ecology* 67:1427–31.

Desharnais, R. A., and R. F. Constantino. 1983. Natural selection and density-dependent population growth. *Genetics* 105:1029–40.

Downhower, J. F., and L. Brown. 1975. Superfoetation in fishes and the cost of reproduction. *Nature* 256:345.

Edney, E. B., and R. W. Gill. 1968. Evolution of senescence and specific longevity. *Nature* 220:281–82.

Endler, J. A. 1978. A predator's view of animal color patterns. *Evol. Biol.* 11:319–64.

Endler, J. A. 1980. Natural selection on color patterns in *Poecilia reticulata*. *Evolution* 34:76–91.

Endler, J. A. 1983. Natural and sexual selection on color patterns in poeciiid fishes. *Env. Biol. Fishes* 9:173–90.

Endler, J. A. 1986. *Natural Selection in the Wild.* Princeton University Press, Princeton, N.J.

Ewens, W. J. 1969. *Population Genetics.* Methuen, London.

Ewens, W. J. 1982. On the concept of the effective population size. *Theor. Pop. Biol.* 21:373–78.

Feldman, M. W., and U. Liberman. 1985. A symmetric two-locus fertility model. *Genetics* 109:229–53.

Feldman, M. W., F. B. Christiansen, and U. Liberman. 1983. On some models of fertility selection. *Genetics* 105:1003–10.

Fisher, R. A. 1958. *The Genetical Theory of Natural Selection,* 2d ed. Dover, New York.

Gadgil, M., and O. T. Solbrig. 1972. The concept of r- and K-selection: Evidence from wild flowers and some theoretical considerations. *Amer. Natur.* 106:14–31.

Gill, J. L. 1965a. A Monte Carlo evaluation of predicted selection response. *Aust. J. Biol. Sci.* 18:999–1007.

Gill, J. L. 1965b. Effects of finite size on selection advance in simulated genetic populations. *Aust. J. Biol. Sci.* 18:599–617.

Gillespie, J. H. 1975. Natural selection for within-generation variance in offspring number II: Discrete haploid models. *Genetics* 81:403–13.

Gillespie, J. H. 1979. Molecular evolution and polymorphism in a random environment. *Genetics* 93:737–54.

Gillespie, J. H. 1984. Molecular evolution over the mutational landscape. *Evolution* 38:1116–29.

Gillespie, J. H. 1986. Variability of evolutionary rates of DNA. *Genetics* 113:1077–91.

Goodnight, C. J. 1987. On the effect of founder events on epistatic genetic variance. *Evolution* 41:80–91.

Grant, P. R. 1986. *Ecology and Evolution of Darwin's Finches.* Princeton University Press, Princeton, N.J.

Guckenheimer, J., G. Oster, and A. Ipaktchi. 1977. The dynamics of density-dependent population models. *J. Math. Biol.* 4:101–47.

Gurney, W.S.C., and R. M. Nisbet. 1985. Fluctuation periodicity, generation separation, and the expression of larval competition. *Theor. Pop. Biol.* 28:150–80.

Hairston, N. G., Jr., and W. E. Walton. 1986. Rapid evolution of a life history trait. *Proc. Natl. Acad. Sci. USA* 83:4831–33.

Heckel, D. G., and J. Roughgarden. 1980. A species near its equilibrium in a fluctuating environment can evolve a lower intrinsic rate of increase. *Proc. Natl. Acad. Sci. USA* 77:7497–7500.

Hedrick, P. W. 1986. Genetic polymorphism in heterogeneous environments: A decade later. *Ann. Rev. Ecol. Syst.* 17:535–66.

Heywood, J. S. 1986. The effect of plant size variation on genetic drift in populations of annuals. *Amer. Natur.* 127:851–61.

Hill, W. G. 1969. On the theory of artificial selection in finite populations. *Genet. Res.* 13:143–63.

Istock, C. A. 1983. The extent and consequences of heritable variation for fitness characters. In C. E. King and P. S. Dawson, eds., *Population Biology: Retrospect and Prospect,* pp. 61–96. Columbia University Press, New York.

Jain, S. K., and R. W. Allard. 1966. Effects of linkage, epistasis, and interbreeding on population changes under selection. *Genetics* 53:633–59.

Kalisz, S. 1986. Variable selection on the timing of germination in *Collinsia verna* (Scrophulariaceae). *Evolution* 40:479–91.

Kaplan, R. H., and W. S. Cooper. 1984. The evolution of developmental plasticity in reproductive characters: An application of the "adaptive coin-flipping" principle. *Amer. Natur.* 123:393–410.

Karlin, S., and B. Levikson. 1974. Temporal fluctuations in selection intensities: Case of small population size. *Theor. Pop. Biol.* 6:383–412.

Karlin, S., and Liberman, U. 1974. Random temporal variation in selection intensities: Case of large population size. *Theor. Pop. Biol.* 6:355–82.

Kearsey, M. S., and K. Kojima. 1967. The genetic architecture of body weight and egg hatchability in *Drosophila melanogaster. Genetics* 56:23–37.

Keller, E. C., Jr., and D. F. Mitchell. 1964. Interchromosomal genotypic interactions in *Drosophila.* II: An analysis of viability characters. *Genetics* 49:293–307.

Kidwell, J. F. 1969. A chromosomal analysis of egg production and abdominal chaeta number in *Drosophila melanogaster. Can. J. Genet. Cytol.* 11:547–57.

Kimura, M., and J. F. Crow. 1978. Effect of overall phenotypic selection on genetic change at individual loci. *Proc. Natl. Acad. Sci. USA* 75:6168–71.

Kirkpatrick, M. 1984. Demographic models based on size, not age, for organisms with indeterminate growth. *Ecology* 65:1874–84.

Kojima, K. 1961. Effects of dominance and size of population on response to mass selection. *Genet. Res.* 2:177–88.

Kosuda, K. 1985. The aging effect on male mating activity in *Drosophila melanogaster. Behav. Genet.* 15:297–303.

Lande, R. 1976. The maintenance of genetic variability by mutation in a polygenic character with linked loci. *Genet. Res.* 26:221–35.

Lande, R. 1979. Quantitative genetic analyses of multivariate evolution, applied to brain : body size allometry. *Evolution* 33:402–16.

Lande, R. 1980. The genetic covariance between characters maintained by pleiotropic mutations. *Genetics* 94:203–15.

Lande, R. 1982. A quantitative genetic theory of life history evolution. *Ecology* 63:607–15.

Lande, R. 1984. The genetic correlation between characters maintained by selection, linkage, and inbreeding. *Genet. Res.* 44:309–20.

Lande, R., and S. J. Arnold. 1983. The measurement of selection on correlated characters. *Evolution* 37:1210–26.

Latter, B.D.H. 1965a. The response to artificial selection due to autosomal genes of large effect. II: The effects of linkage on limits to selection in finite populations. *Aust. J. Biol. Sci.* 18:1009–23.

Latter, B.D.H. 1965b. The response to artificial selection due to autosomal genes of large effect. I: Changes in gene frequency at an additive locus. *Aust. J. Biol. Sci.* 18:585–98.

Lenski, R. E. 1988. Experimental studies of pleiotropy and epistasis in *Escherichia coli,* II: Compensation for maladaptive effects associated with resistance to virus T4. *Evolution* 42:433–40.

Lenski, R. E., and P. M. Service. 1982. The statistical analysis of population growth rates calculated from schedules for survivorship and fecundity. *Ecology* 63:655–62.

Lewontin, R. C. 1974. *The Genetic Basis of Evolutionary Change.* Columbia University Press, New York.

Luckinbill, L. S. 1978. *r*- and *K*-selection in experimental populations of *Escherichia coli. Science* 202:1201–1203.

Luckinbill, L. S. 1984. An experimental analysis of a life history theory. *Ecology* 65:1170–84.

Luckinbill, L. S., R. Arking, M. G. Clare, W. C. Cirocco, and S. A. Buck. 1984. Selection for delayed senescence in *Drosophila melanogaster. Evolution* 38:996–1003.

Luckinbill, L. S., M. J., Clare, W. L. Krell, W. C. Cirocco, and P. A. Richards. 1987. Estimating the number of genetic elements that defer senesence in *Drosophila. Evol. Ecol.* 1:37–46.

Lynch, M. 1984. The selective value of alleles underlying polygenic traits. *Genetics* 108:1021–33.

MacArthur, R. H. 1962. Some generalized theorems of natural selection. *Proc. Natl. Acad. Sci. USA* 48:1893–97.

MacArthur, R. H., and E. O. Wilson. 1967. *The Theory of Island Biogeography.* Princeton University Press, Princeton, N.J.

McNaughton, S. J. 1975. *r*- and *K*-selection in *Typha. Amer. Natur.* 109:251–61.

Maddox, G. D., and N. Cappuccino. 1986. Genetic determination of plant susceptibility to an herbivorous insect depends on environmental context. *Evolution* 40:863–66.

Manly, B.F.J. 1985a. *The Statistics of Natural Selection on Animal Populations.* Chapman and Hall, New York.

Manly, B.F.J. 1985b. Tests of the theory of natural selection: An overview. *J. Roy. Soc. New Zealand* 15:411–32.

Mather, K., and Cooke, P. 1962. Differences in competitive ability between genotypes of *Drosophila. Heredity* 17:381–407.

Medawar, P. D. 1957. *The Uniqueness of the Individual.* Methuen, London.

Meyer, J. S., C. G. Ingersoll, L. L. McDonald, and M. S. Boyce. 1986. Estimating uncertainty in population growth rates: Jackknife vs. bootstrap techniques. *Ecology* 67:1156–66.

Milkman, R. 1978. Selection differentials and selection coefficients. *Genetics* 88:391–403.

Milkman, R. 1982. Toward a unified selection theory. In R. Milkman, ed., *Perspectives on Evolution,* pp. 105–18. Sinauer, Sunderland, Mass.

Mitton, J. B., and M. C. Grant. 1984. Associations among protein heterozygosity, growth rate, and developmental homeostasis. *Ann. Rev. Ecol. Syst.* 15:479–500.

Morin, P. J. 1981. Predator salamanders reverse the outcome of competition among three species of larval anurans. *Science* 212:1284–86.

Morin, P. J. 1983. Predation, competition, and the composition of larval anuran guilds. *Ecol. Monogr.* 53:119–38.

Morin, P. J. 1986. Interactions between intraspecific competition and predation in an amphibian predator-prey system. *Ecology* 67:713–20.

Motro, U., and G. Thomson. 1982. On heterozygosity and the effective size of populations subject to size changes. *Evolution* 36:1059–66.

Mueller, L. D. 1985. The evolutionary ecology of *Drosophila.* In M. K. Hecht, W. C. Sterre, and B. Wallace, eds., *Evolutionary Biology,* vol. 19, pp. 37–98. Plenum Press, New York.

Mueller, L. D. 1987. Evolution of accelerated senescence in laboratory populations of *Drosophila. Proc. Natl. Acad. Sci. USA* 84:1974–77.

Mueller, L. D. 1988a. Density-dependent population growth and natural selection in food-limited environments: The *Drosophila* model. *Amer. Natur.* In press.

Mueller, L. D. 1988b. Evolution of competitive ability in *Drosophila* due to density-dependent natural selection. *PNAS* 85:4383–386.

Mueller, L. D., and F. J. Ayala. 1981a. Dynamics of single-species population growth: Experimental and statistical analysis. *Theor. Pop. Biol.* 20:101–17.

Mueller, L. D., and F. J. Ayala. 1981b. Trade-off between *r*-selection and *K*-selection in *Drosophila* populations. *Proc. Natl. Acad. Sci. USA* 78:1303–1305.

Mueller, L. D., and F. J. Ayala. 1981c. Fitness and density-dependent population growth in *Drosophila melanogaster. Genetics* 97:667–77.

Mueller, L. D., and F. J. Ayala. 1981d. Dynamics of single-species population growth: Stability or chaos? *Ecology* 62:1148–54.

Mueller, L. D., and V. F. Sweet. 1986. Density-dependent natural selection in *Drosophila:* Evolution of pupation height. *Evolution* 40:1354–56.

Mueller, L. D., B. A. Wilcox, P. R. Ehrlich, D. G. Heckel, and D. D. Murphy. 1985. A direct assessment of the role of genetic drift in determining allele frequency variation in populations of *Euphydryas editha. Genetics* 110:495–511.

Muenchow, G. 1986. Ecological use of failure time analysis. *Ecology* 67:246–50.

Murphy, G. S. 1968. Pattern in life history and the environment. *Amer. Natur.* 102:391–403.

Nagylaki, T. 1987. Evolution under fertility and viability selection. *Genetics* 115:367–75.

Nei, M. 1963. Effect of selection on the components of genetic variances. In W. D.

Hanson and H. F. Robinson, eds., *Statistical Genetics and Plant Breeding,* pp. 501–15. Natl. Acad. Sci., Natl. Res. Counc. Publ. 982, Washington, D.C.

Nei, M., T. Maruyama, and R. Chakraborty. 1975. The bottleneck effect and genetic variability in populations. *Evolution* 29:1–10.

Nisbet, R. M., and W.S.C. Gurney. 1982. *Modelling Fluctuating Populations.* Wiley-Interscience, New York.

Nunney, L. 1983. Sex differences in larvae competition in *Drosophila melanogaster:* The testing of a competition model and its relevance to frequency dependent selection. *Amer. Natur.* 121:67–93.

O'Donald, P. 1968. Measuring the intensity of natural selection. *Nature* 220:197–98.

O'Donald, P. 1970. Change of fitness by selection for a quantitative character. *Theor. Pop. Biol.* 1:219–32.

O'Donald, P. 1971. Natural selection for quantitative characters. *Heredity* 27:137–53.

O'Donald, P., and J.W.F. Davis. 1976. A demographic analysis of the components of selection in a population of Arctic skuas. *Heredity* 36:343–350.

O'Donald, P., and M.E.N. Majerus. 1985. Sexual selection and the evolution of preferential mating in ladybirds. I: Selection for high and low lines of female preference. *Heredity* 55:401–12.

Oster, G. 1976. Internal variables in population dynamics. In S. A. Levin, ed., *Lectures on Mathematics in the Life Sciences,* vol. 8, pp. 37–68. American Mathematical Society, Providence.

Pacala, S. W. 1986. Neighborhood models of plant population dynamics. 4: Single-species and multispecies models of annuals with dormant seeds. *Amer. Natur.* 128:859–78.

Parry, G. D. 1981. The meanings of *r*- and *K*-selection. *Oecologia* 48:260–64.

Pianka, E. R. 1970. On *r*- and *K*-selection. *Amer. Natur.* 104:592–96.

Policansky, D. 1983. Size, age, and demography of metamorphosis and sexual maturation in fishes. *Amer. Zool.* 23:57–63.

Pollak, E. 1983. A new method for estimating the effective population size from allele frequency changes. *Genetics* 104:531–48.

Price, T. D., P. R. Grant, H. L. Gibbs, and P. T. Boag. 1984. Recurrent patterns of natural selection in a population of Darwin's finches. *Nature* 309:787–89.

Primack, R. B., and J. Antonovics. 1982. Experimental ecological genetics in *Plantago.* VII: Reproductive effort in populations of *P. lanceolata* L. *Evolution* 36:742–52.

Prout, T. 1980. Some relationships between density-independent and density-dependent population growth. In M. K. Hecht, W. C. Steere, and B. Wallace, eds., *Evolutionary Biology,* vol. 13, pp. 1–68. Plenum Press, New York.

Reznick, D. 1985. Costs of reproduction: An evaluation of the empirical evidence. *Oikos* 44:257–67.

Robertson, F. W. 1957. Studies in quantitative inheritance. XI: Genetic and environmental correlation between body size and egg production in *Drosophila melanogaster. J. Genet.* 55:428–43.

Rose, M. R. 1982. Antagonistic pleiotropy, dominance, and genetic variation. *Heredity* 48:63–78.

Rose, M. R. 1984a. Genetic covariation in *Drosophila* life history: Untangling the data. *Amer. Natur.* 123:565–69.

Rose, M. R. 1984b. Laboratory evolution of postponed senescence in *Drosophila melanogaster. Evolution* 38:1004–10.

Rose, M. R. 1985. Life history evolution with antagonistic pleiotropy and overlapping generations. *Theor. Pop. Biol.* 28:342–58.

Rose, M. R., and B. Charlesworth. 1980. A test of evolutionary theories of senescence. *Nature* 287:141–42.

Roughgarden, J. 1971. Density-dependent natural selection. *Ecology* 52:453–68.

Roughgarden, J. 1979. *Theory of Population Genetics and Evolutionary Ecology: An Introduction.* Macmillan, New York.

Roughgarden, J., Y. Iwasa, and C. Baxter. 1985. Demographic theory for an open marine population with space-limited recruitment. *Ecology* 66:54–67.

Schlichting, C. D. 1987. The evolution of phenotypic plasticity in plants. *Ann. Rev. Ecol. Syst.* 17:667–93.

Schmidt, K. P., and D. A. Levin. 1985. The comparative demography of reciprocally sown populations of *Phlox drummondii* Hook. I: Survivorships, fecundities, and finite rates of increase. *Evolution* 39:396–404.

Schmitt, J., and J. Antonovics. 1986. Experimental studies of the evolutionary significance of sexual reproduction. III: Maternal and paternal effects during seedling establishment. *Evolution* 40:817–29.

Service, P. M., and M. R. Rose. 1985. Genetic covariation among life-history components: The effect of novel environments. *Evolution* 39:943–45.

Shaw, R. 1986. Response to density in a wild population of the perennial herb *Salvia lyrata:* Variation among families. *Evolution* 40:492–505.

Smith, C. C., and S. D. Fretwell. 1974. The optimal balance between size and number of offspring. *Amer. Natur.* 108:499–506.

Smouse, P. E. 1976. The implications of density-dependent population growth for frequency- and density-dependent selection. *Amer. Natur.* 110:849–60.

Smouse, P. E. 1986. The fitness consequences of multiple-locus heterozygosity under the multiplicative overdominance and inbreeding depression models. *Evolution* 40:946–58.

Stearns, S. C. 1976. Life-history tactics: A review of the ideas. *Quart. Rev. Biol.* 51:3–47.

Stearns, S. C. 1977. The evolution of life history traits: A critique of the theory and a review of the data. *Ann. Rev. Ecol. Syst.* 8:145–71.

Stearns, S. C., and J. C. Koella. 1986. The evolution of phenotypic plasticity in life-history traits: Predictions of reaction norms for age and size at maturity. *Evolution* 40:893–913.

Stokes, T. K., W.S.C. Gurney, R. M. Nisbet, and S. P. Blythe. 1988. Parameter evolution in a laboratory insect population. *Theoret. Pop. Biol.* In press.

Streifer, W. 1974. Realistic models in population ecology. *Adv. Ecol. Res.* 8:199–266.

Sulzbach, D. S., and C. B. Lynch. 1984. Quantitative genetic analysis of temperature regulation in *Mus musculus*. III: Diallel analysis of correlations between traits. *Evolution* 38:541–52.

Swallow, W. H., and M. F. Monahan. 1984. Monte Carlo comparison of ANOVA, MIVQUE, REML, and ML estimates of variance components. *Technometrics* 26:47–57.

Tajima, F., and N. Nei. 1984. Note on genetic drift and estimation of effective population size. *Genetics* 106:569–74.

Taylor, C. E., and C. Condra. 1980. *r*- and *K*-selection in *Drosophila pseudoobscura*. *Evolution* 34:1183–93.

Thomas, W. R., M. J. Pomerantz, and M. E. Gilpin. 1980. Chaos, asymmetric growth and group selection for dynamical stability. *Ecology* 61:1312–20.

Tinkle, D. W., and N. F. Hadley. 1975. Lizard reproductive effort: Calorie estimates and comments on its evolution. *Ecology* 56:427–34.

Travis, J. 1983. Variation in growth and survival of *Hyla gratiosa* larvae in experimental enclosures. *Copeia* 1983:232–37.

Travis, J., and S. Henrich. 1986. Some problems in estimating the intensity of selection through fertility differences in natural and experimental populations. *Evolution* 40:786–90.

Travis, J., S. B. Emerson, and M. Blouin. 1987. A quantitative genetic analysis of larval life history traits in *Hyla crucifer*. *Evolution* 41:145–156.

Travis, J., W. H. Keen, and J. Juilianna. 1985a. The role of relative body size in a predator-prey relationship between dragonfly naiads and larval anurans. *Oikos* 45:59–65.

Travis, J., W. H. Keen, and J. Juilianna. 1985b. The effects of multiple factors on viability selection in *Hyla gratiosa* tadpoles. *Evolution* 39:1087–99.

Travis, J., J. A. Farr, S. Henrich, and R. T. Cheong. 1987. Testing theories of clutch overlap with the reproductive ecology of *Heterandria formosa*. *Ecology* 68:611–23.

Turelli, M. 1984. Heritable genetic variation via mutation-selection balance: Lerch's zeta meets the abdominal bristle. *Theor. Pop. Biol.* 25:138–93.

Turelli, M. 1985. Effects of pleiotropy on predictions concerning mutation-selection balance for polygenic traits. *Genetics* 111:165–95.

Turelli, M., and L. Ginzburg. 1983. Should individual fitness increase with heterozygosity? *Genetics* 104:191–209.

Turelli, M., and D. Petry. 1980. Density-dependent selection in a random environment: An evolutionary process that can maintain stable population dynamics. *Proc. Natl. Acad. Sci. USA* 77:7501–7505.

Wade, M. J. 1980. Effective population size: The effects of sex, genotype, and density on the mean and variance of offspring numbers in the flour beetle, *Tribolium castaneum*. *Genet. Res.* 36:1–10.

Wade, M. J. 1984. Variance-effective population number: The effects of sex-ratio and density on the mean and variance of offspring numbers in the flour beetle, *Tribolium castaneum*. *Genet. Res.* 43:249–56.

Werner, E. E., and J. F. Gilliam. 1984. The ontogenetic niche and species interactions in size-structured populations. *Ann. Rev. Ecol. Syst.* 15:393–425.

Westcott, B. 1986. Some methods of analysing genotype-environment interaction. *Heredity* 56:243–53.

Wilbur, H. M. 1976. Life history evolution of seven milkweeds of the genus *Asclepias*. *J. Ecol.* 64:223–40.

Wilbur, H. M. 1977. Propagule size, number, and dispersion pattern in *Ambystoma* and *Asclepias*. *Amer. Natur.* 111:43–68.

Wilbur, H. M. 1980. Complex life cycles. *Ann. Rev. Ecol. Syst.* 11:67–93.

Wilbur, H. M. 1984. Complex life cycles and community organization in amphibians. In P. W. Price, C. N. Slobodchikoff, and W. S. Gaud, eds., *A New Ecology: Novel Approaches to Interactive Systems*, pp. 196–224. Wiley, New York.

Wilkinson, G. S. 1987. Equilibrium analysis of sexual selection in *Drosophila melanogaster*. *Evolution* 41:11–21.

Williams, G. C. 1957. Pleiotropy, natural selection and the evolution of senescence. *Evolution* 11:398–411.

Winkler, D. W., and K. Wallin. 1987. Offspring size and number: A life history model linking effort per offspring and total effort. *Amer. Natur.* 129:708–20.

Wood, J. W. 1987. The genetic demography of the Gainj of Papua New Guinea. 2: Determinants of effective population size. *Amer. Natur.* 129:165–87.

Wright, S. 1982. The shifting balance of theory and macroevolution. *Ann. Rev. Gen.* 16:1–19.

Chapter 8

Fossils, Macroevolution, and Theoretical Ecology

STEVEN M. STANLEY

Paleobiology generates theory, and it also contributes information that must constrain theoretical advances in neontology. The fossil record provides the only documentation of the long-term evolution of species and of evolution above the species level. Although this record as a whole is highly incomplete, in places it is of high enough quality to resolve important evolutionary issues. To employ fossils fruitfully in addressing particular problems, we must single out appropriate segments of the record and use their data judiciously.

Certainly the most important paleobiological issue now impinging on theoretical ecology is the controversy surrounding the punctuational model of evolution (Eldredge 1971; Eldredge and Gould 1972; Stanley 1975, 1979). This model can be defined as the assertion that most evolutionary change in the history of life is associated with brief periods of rapid transformation rather than accomplished by the more gradual modification of species; in its more restricted form, the punctuational model associates most evolution with speciation. The alternative, gradualistic model holds that most evolutionary change occurs by the gradual transformation of established species (phyletic evolution).

I have always believed that the most profound implications of the punctuational model are for ecological theory. Looking backward to Darwin, we can see that the prevailing themes of *On the Origin of Species* were gradualistic. Darwin also accepted the concept of ecological plenitude, which formed part of his intellectual heritage. He envisioned a highly competitive natural world, so tightly packed with species that it was difficult for evolution to take place—hence the wedging simile that, by his own account, triggered his formulation of natural selection (if one species could change just a bit, then another could change, and all species of a community could evolve by a kind of subtle domino effect). Theoretical ecology has moved far beyond its primitive state during Darwin's lifetime, yet many of its tenets are still based on the idea that evolution readily moves species toward adaptive optima. If the punctuational model is valid—or even if stasis is moderately common—we must reassess theoretical formulations based on optimization.

It is not only by testifying as to the pattern of evolution that the fossil record can influence theoretical ecology. Fossil data also shed light on rates and patterns of speciation and extinction—and on the factors that control these processes, which in turn determine what assemblages of species coexist in the course of evolutionary time.

EMPIRICAL TESTS OF THE PUNCTUATIONAL MODEL

Many of the entities recognized as species in the fossil record may actually include more than one very similar species; but for the many higher taxa whose fossil remains reveal numerous characters of primary value in species recognition, the following can fairly be claimed: when two or more populations separated in geologic time are united within a single species by a competent taxonomist, then these populations encompass very little total evolutionary change, in comparison to the degree of difference separating two genera, for example. Thus the great longevity of many species recognized in the fossil record must be taken to indicate that approximate evolutionary stasis is very common in nature. Mean duration of species recognized in the fossil record is measured in millions of years for such groups as bivalve mollusks, benthic and planktonic foraminifera, echinoids, corals, freshwater fishes, terrestrial mammals, and flowering plants. In population genetic terms, this means that in these groups evolution spanning 10^6–10^7 generations has usually accomplished very little.

Certainly some examples of gradual change in form, usually following a zigzag course, have been compellingly documented in the fossil record. Perhaps the most convincing case is for the *Globorotalia inflata* lineage of planktonic foraminiferans (Malmgren and Kennett 1981). Many reported studies of phyletic trends have never been shown to represent true lineages, however; these studies may be ignoring speciation events between the ancestral and descendant populations. In addition, most such studies have been undertaken for examples that investigators thought, in advance, illustrated substantial phyletic evolution; stasis has been widely ignored (Gould and Eldredge 1977). Furthermore, the vast majority of published examples of alleged trends represent only change in body size, which is far more labile in evolution than is shape (Stanley and Yang 1987).

To date, only two studies have been designed to measure phyletic change impartially and comprehensively in terms of both morphology and taxonomy (Cheetham 1986; Stanley and Yang 1987). Each of these studies examined what happened over the course of millions of years to a large, arbitrarily assembled group of species, as indicated by a large set of morphologic characters. Cheetham employed forty-six characters to study the fates of the thirteen species constituting a single clade of marine bryozoans (the genus *Metrarabdotos*). His samples came from a relatively continuous stratigraphic sequence in the Dominican Republic spanning the interval from about 8 to 3.5 million years ago. The characters, though all skeletal, reflected a wide variety of adaptive traits. As it turned out, the species, most of which originated during the time interval considered, were separated by great distances in multivariate space. On the other hand, no individual species exhibited a rate of phyletic evolution significantly different from zero. In other words, the pattern that this thorough analysis revealed was strikingly punctuational. Furthermore, most ancestral and descendant species overlapped in time, so that rapid evolutionary steps had to have been associated with speciation (branching) events; they were not sudden transformations of established species.

Stanley and Yang (1987) uncovered a similar pattern of stasis in a set of nineteen arbitrarily chosen lineages of marine bivalve mollusks for which morphology was assessed via twenty-four characters. These variables, like those employed by Cheetham, depicted a wide range of adaptive features. The primary goal of the analysis was to determine to what degree 4 million-year-old populations of these lineages differed from living populations. Geographic variability of Recent populations served as a yardstick. As it turned out, multivariate morphometrics revealed that the ancient populations did not differ from Recent populations substantially more than conspecific Recent populations differed from each other. In addition, three of the lineages were traced back to 17 million-year-old populations that were not appreciably different in form from Recent populations.

Both of the above studies focused on shape rather than size. Size, it turns out, is far more labile in evolution than shape, although changes in size can be difficult to assess because genetic and ecophenotypic controls are difficult to disentangle from one another (Stanley 1985; Stanley and Yang 1987). The punctuational patterns uncovered in the two studies vindicate previous analyses that have concluded from the great longevities of fossil species that stasis has prevailed in the history of life (Stanley 1979, 1985). The morphological characters used by Cheetham included those that taxonomists have traditionally used to define species, and Stanley and Yang (1987) found no evidence of undersplitting by taxonomists, but rather, in two cases, an unjustified splitting of two virtually static lineages into chronospecies.

I would argue that the comprehensive analyses of morphologies and of species longevities constitute a strong case for the prevalence of stasis. At the very least, a skeptic must acknowledge that stasis is common. This means that its cause requires an explanation and also that stasis must be incorporated into ecological theory. The degree to which ecologists adjust their theories to accommodate the punctuational model will presumably depend upon the degree to which they come to believe that stasis prevails.

OPTIMIZATION

It seems fair to state that a substantial fraction of ecological theory is based on the notion that evolution is effective at optimizing ecological strategies. This is especially evident in applications of game theory (e.g., Lewontin 1961; Slobodkin and Rapoport 1974; Maynard Smith 1982). Evolutionary trade-offs have long been recognized in ecological theory, especially in considerations of energy allocation, but acceptance of the idea that a punctuational pattern of evolution prevails in nature would further complicate theoretical treatments.

One issue is the relative lability of various kinds of adaptations. I have already noted that body size for animals is far more easily altered than are most aspects of shape. Since Hutchinson (1959) paid homage to Santa Rosalia, body size has been widely recognized to be an important factor in niche partitioning, but is there any reason to assume that evolution alters behavioral features any more readily for established species than it alters anatomical features? Are most behavioral transformations associated with speciation events? Perhaps change in reproductive output, which, like change in body size, is a simple quantitative matter, not a qualitative one, is also relatively easy for populous, well-established species.

Critical to these issues is the question of what constraints operate to stymie phyletic evolution. Here scientists have become somewhat polarized. One emphasis has been on morphogenetic constraints, the idea being that, given a certain set of traits that develops in a certain way under the influence of a particular set of genes, it can be highly improbable that evolution will yield another, more highly adaptive set of traits of a particular type (Gould and Lewontin 1979). Of course, Darwin himself stressed the limitations imposed by phylogenetic background. At issue is just how severely evolutionary history constrains future evolution. Population geneticists have tended to focus instead upon stabilizing selection as the cause of approximate stasis (Charlesworth, Lande, and Slatkin 1982; Lande 1986). Perhaps a compromise will be reached; pleiotropy and developmental canalization may well make many variants maladaptive—even freakish—so that they are selected against.

Newman, Cohen, and Kipnis (1985) and Lande (1986) have recently developed population genetic models based on diffusion theory and Wright's adaptive landscape, to attempt to explain stepwise evolutionary change; shifts of adaptive peak result primarily from drift and are therefore most likely to occur in small populations. Whether such models can account for a prevalence of stasis is open to question, however, if the adaptive landscape is constantly shifting at moderate rates. Why a species' evolution should not track such changes in the landscape remains unclear. Lande (1986) argues that potentially useful genetic variability is normally available for adjustment to environmental change, but if stasis prevails, one must wonder to what extent pleiotropic factors and gene interactions yield deleterious side effects for variability that appears to have value.

Also important is the point that theoretical population genetic models have been focusing on stepwise evolution within established species, with the implication that these models explain punctuational patterns in general. In fact, the empirical study of Cheetham (1986) clearly associates most change with speciation, not with the transformation of established species.

SPECIES SELECTION

The concept of species selection builds on the notion of Mayr (1963) that species are evolutionary experiments. It has two aspects, analogous to the reproductive and survival components of selection at the individual level. These are differences among species in rates of speciation and extinction (Stanley 1975). It is important to understand that adherence to the concept of species selection in no way denies a dominant role in evolution for selection at the individual level; in a punctuational scheme, it simply requires that most evolution (presumably accomplished primarily by natural selection) be associated with speciation events. The fact that the directions taken by speciation events within a clade are varied and unpredictable means that speciation generates species of varying macroevolutionary potential; these are the raw material for trends. Species selection is not the only source of large-scale trends. Two other macroevolutionary mechanisms with analogues at the population level must also contribute to trends (Stanley 1979). One is phylogenetic drift, which is analogous to genetic drift. The other is directed speciation, which is crudely analogous to mutation pressure; this is the tendency for morphogenetic or environmental constraints to favor speciation events that move in a particular adaptive direction. When ecological constraints are at work here, they may often constitute selection at the level of isolated populations, most of which are unable to expand into new species.

The arguments of Vrba (1980, 1984), Vrba and Eldredge (1984), and Vrba and Gould (1986) notwithstanding, I believe that selection can be considered to occur at a given level of biological organization, such as the level of the species, even if the trait being selected for is emergent at a lower level of biological organization. One of the problems here is that the sorting process among species must usually involve the simultaneous effects of traits that are emergent below the level of the species and of traits that are not. For example, a hypothetical group of species in a higher taxon might suffer preferential extinction, so that there would be a shift in the composition of the higher taxon, because these species shared a particular trait of mobility that had two deleterious effects: (1) it made them more vulnerable to attack by predators (a trait emergent at the level of the individual that appeared also at the level of the species), and (2) it reduced their geographic range (a trait relevant only at the level of the species if it had no bearing on the fates of individuals). Let us assume further that the deleterious trait of mobility was present in all individuals of these species and in no individual of any other species. It would then be not only reasonable but also parsimonious to recognize species as units of selection.

Some population biologists have been comfortable in applying their mathematical tools to the analysis of species selection in the broad sense (Slatkin 1981), while others, maintaining a reductionist view, have insisted on so restrictive a definition for species selection as to render it of limited importance (Maynard Smith 1983). I personally find the broader concept to be of great heuristic value in the study of the fossil record.

As a final note on this subject, I want to stress that species selection is often a diffuse process. When its agents are biological, they seldom represent one-on-one interactions between species. Rather, the appearance of a large set of predators or competitors may work to the disadvantage of another large set of species in survival or speciation. Such changes can alter the probability of extinction or speciation without serving as the sole cause of every extinction or speciation event.

CONTROLS OF RATES OF SPECIATION AND EXTINCTION

A variety of estimates reveals that animal taxa that experience relatively high rates of speciation during adaptive radiation are also characterized by relatively high rates of extinction. More generally, the two kinds of rate are correlated in the animal world (Stanley 1979). Obviously, a taxon characterized by a high rate of extinction would not survive for long. On the other hand, species selection in a higher taxon should tend to increase rate of speciation and decrease rate of extinction: this process must constantly strain the relationship between rate of speciation and rate of extinction. The fact that the correlation remains suggests that common factors govern the two kinds of rate. One of these appears to be the level of stereotypical behavior; advanced behavior probably increases rate of speciation by fostering reproductive isolation, while as a form of ecological specialization it increases rate of extinction. Another factor appears to be dispersal ability; ineffective dispersal tends to promote speciation, but it also restricts geographic range and increases probability of extinction. To these factors, Vrba (1980) has added the notion that stenotopy with respect to limiting factors may favor both extinction and speciation. Finally, a taxon's susceptibility to moderately heavy predation or other forms of disturbance may diminish, destabilize, and fragment populations, accelerating both extinction and speciation (the fission effect, Stanley 1986). It is notable that all of these suggested controls are ecological in nature. Their investigation represents a potentially fruitful area for cross-fertilization between the fields of macroevolution and theoretical ecology. One interesting point here relates to the fact that extinction represents the reduction of both geographic range and population size to zero; fossil data suggest that population size may be the more important of the two variables in affecting probability of extinction for many taxa (Stanley 1986).

COEVOLUTION AND SPECIES SELECTION

Nowhere is there a more obvious conflict between the punctuational model of evolution and the traditional theories of population biology than in the interpretation of coevolution. Actually, coevolution was initially defined to embrace all kinds of evolution resulting from biological interactions (Ehrlich and Raven 1964), but it has since taken on a strongly gradualistic connotation, the idea being that two species track each other's evolution over a long interval of geologic time. Bakker (1983) and Stanley, van Valkenburgh, and Steneck (1983) have challenged the generality of the gradualistic pattern, noting that the fossil record reveals that (1) in terms of ecological efficiency, taxa have often lagged far behind others with which they have been associated ecologically, and (2) taxa have often failed to respond to the presence of gaps in the ecosystem: for long intervals, they have failed to seize evolutionary opportunities. The implication is that coevolution may operate primarily at the level of species selection: the response of one higher taxon to the presence of another may come about chiefly through a change in composition resulting from differential rates of speciation and/or extinction. This, however, is an imprecise, coarse-grained process, not the gradual, finely tuned process commonly envisioned as coevolution. Especially important is the fact that a large adaptive step can be taken only when the appropriate kind of speciation event takes place, and this is unpredictable in time and space. In other words, the nature of optimization is again at issue.

THE RED QUEEN HYPOTHESIS

Fossil data are sometimes taken to offer stronger support for patterns than they really do, and we must be careful that dubious patterns do not gain acceptance too readily. It seems to me that the Red Queen hypothesis (Van Valen 1973) illustrates this danger. This hypothesis, which states that an evolutionary improvement for one species operates to the detriment of others or that evolution is a zero-sum game, has been widely acclaimed by population biologists and has engendered much theoretical work (e.g., Maynard Smith 1976; Schaffer and Rosenzweig 1978; Stenseth 1979; Stenseth and Maynard Smith 1984). Meanwhile, the empirical basis of the Red Queen hypothesis has been seriously undercut. The hypothesis was devised to explain apparent evidence that probability of extinction for a taxon is independent of the taxon's age. Early criticisms of this evidence noted that survivorship for taxa had not actually been demonstrated to be log-linear and, in fact, appeared to be nonlinear; furthermore, if the curves plotted for survivorship of higher taxa had been log-linear, then curves for species would have been concave; yet the theory devised to explain the apparent linearity was for interactions between species (Raup 1975; Sepkoski 1975). It has also been noted that a constant probability of extinction for taxa within a group with respect to the duration of that group does not necessarily imply that each taxon's probability of extinction is constant through real time (McCune 1982). Finally, we can note that an increment in the fitness of one species may be accompanied by decrements in the fitness of others, and yet these others may belong to unrelated higher taxa; survivorship within single higher taxa cannot be used to indicate the operation of a zero-sum game unless species of other taxa are arbitrarily excluded from the game.

My aim here is not to ridicule the Red Queen hypothesis or its creator, but to suggest that theoreticians may sometimes be too easily drawn to a pattern that offers special appeal for its mathematical convenience, even though it lacks empirical strength. Our efforts to bond paleobiology to ecological theory must be rooted as firmly as possible in sound data.

REFERENCES

Bakker, R. T. 1983. The deer flees, the wolf pursues: Incongruencies in predator-prey coevolution. In D. J. Futuyma and M. Slatkin, eds., *Coevolution,* pp. 350–82. Sinauer, Sunderland, Mass.

Charlesworth, B., R. Lande, and M. Slatkin, M. 1982. A neo-Darwinian commentary on macroevolution. *Evolution* 36:474–98.

Cheetham, A. H. 1986. Tempo of evolution in a Neogene bryozoan: Rates of morphologic change within and across species boundaries. *Paleobiology* 12:190–202.

Ehrlich, P. R., and P. H. Raven. 1964. Butterflies and plants: A study in coevolution. *Evolution* 18:586–608.

Eldredge, N. 1971. The allopatric model and phylogeny in Paleozoic invertebrates. *Evolution* 25:156–67.

Eldredge, N., and S. J. Gould. 1972. Punctuated equilibria: An alternative to phyletic gradualism. In T.J.M. Schopf, ed., *Models in Paleobiology,* pp. 82–115. Freeman, Cooper, San Francisco.

Gould, S. J., and N. Eldredge. 1977. Punctuated equilibria: The tempo and mode of evolution reconsidered. *Paleobiology* 3:115–51.

Gould, S. J., and R. C. Lewontin. 1979. The spandrels of San Marco and the Panglossian paradigm: A critique of the adaptationist programme. *Proc. Roy. Soc. Lond. (B)* 205:581–98.

Hutchinson, G. E. 1959. Homage to Santa Rosalia, or why are there so many kinds of animals? *Amer. Natur.* 93:145–59.

Lande, R. 1986. The dynamics of peak shifts and the pattern of morphological evolution. *Paleobiology* 12:343–54.

Lewontin, R. C. 1961. Evolution and the theory of games. *J. Theor. Biol.* 1:382–403.

McCune, A. R. 1982. On the fallacy of constant extinction rates. *Evolution* 36:610–14.

Malmgren, B. A., and J. B. Kennett. 1981. Phyletic gradualism in a late Cenozoic planktonic foraminiferal likeage; DSDP Site 284, southwest Pacific. *Paleobiology* 7:230–40.

Maynard Smith, J. 1976. A comment on the Red Queen. *Amer. Natur.* 110:325–30.

Maynard Smith, J. 1982. *Evolution and the Theory of Games.* Cambridge University Press, Cambridge, England.

Maynard Smith, J. 1983. Current controversies in evolutionary biology. In M. Grene, ed., *Dimensions of Darwinism,* pp. 273–86. Cambridge University Press, Cambridge, England.

Mayr, E. 1963. *Animal Species and Evolution.* Harvard University Press, Cambridge, Mass.

Newman, C. M., J. E. Cohen, and C. Kipnis. 1985. Neo-darwinian evolution implies punctuated equilibria. *Nature* 315:400–401.

Raup, D. M. 1975. Taxonomic survivorship curves and Van Valen's law. *Paleobiology* 1:82–96.

Schaffer, W. M., and M. L. Rosenzweig. 1978. Homage of the Red Queen. I: Coevolution of predators and their victims. *Theor. Pop. Biol.* 14:135–57.

Sepkoski, J. J. 1975. Stratigraphic biases in the analysis of taxonomic survivorship curves. *Paleobiology* 1:343–55.

Slatkin, M. 1981. A diffusion model of species selection. *Paleobiology* 7:421–25.

Slobodkin, L. B., and A. Rapoport. 1974. An optimal strategy of evolution. *Quart. Rev. Biol.* 49:181–200.

Stanley, S. M. 1975. A theory of evolution above the species level. *Proc. Natl. Acad. Sci. USA.* 72:646–50.

Stanley, S. M. 1979. *Macroevolution: Pattern and Process.* Freeman, San Francisco.

Stanley, S. M. 1985. Rates of evolution. *Paleobiology* 11:13–26.

Stanley, S. M. 1986. Population size, extinction, and speciation: The fission effect in Neogene Bivalvia. *Paleobiology* 12:89–110.

Stanley, S. M., and X. Yang. 1987. Approximate evolutionary stasis for bivalve morphology over millions of years: A multivariate, multilineage study. *Paleobiology* 13:113–39.

Stanley, S. M., B. van Valkenburgh, and R. S. Steneck. 1983. Coevolution and the Fossil record. In D. J. Futuyma and M. Slatkin, eds., *Coevolution,* pp. 238–349. Sinauer, Sunderland, Mass.

Stenseth, N. C. 1979. Where have all the species gone? On the nature of extinction and the Red Queen hypothesis. *Oikos* 33:196–227.

Stenseth, N. C., and Maynard Smith, J. 1984. Coevolution in ecosystems: Red Queen evolution or stasis? *Evolution* 38:870–80.

Van Valen, L. 1973. A new evolutionary law. *Evol. Theory* 1:1–30.

Vrba, E. S. 1980. Evolution, species and fossils: How does life evolve? *South African J. Sci.* 76:61–84.

Vrba, E. S. 1984. What is species selection? *Syst. Zool.* 33:318–28.

Vrba, E. S., and N. Eldredge. 1984. Individuals, hierarchies and processes: Towards a more complete evolutionary theory. *Paleobiology* 10:146–71.

Vrba, E. S., and S. J. Gould. 1986. The hierarchical expansion of sorting and selection: Sorting and selection cannot be equated. *Paleobiology* 12:217–28.

Chapter 9

Discussion: Ecology and Evolution

MARCUS W. FELDMAN

The two papers that introduce the interaction between evolutionary and ecological theory do so on vastly different time scales. Stanley's paleobiological discussion presumes that morphological change exhibited in the fossil record is, for most of the time, extremely slow. Those events characterized by him as speciation are extreme discontinuities of stasis. The time scale here involves millions of generations. Travis and Mueller, on the other hand, address data and theory relevant to evolution within species and, mostly, within populations. The interaction between ecology and evolution is broached entirely differently in the two papers. In Stanley's, the questions are: (1) Can ecological theory yield scenarios that are likely to produce the kind of fossil record described by Stanley? and (2) Does the existence of stasis, as described by Stanley, suggest ecological interactions that materially alter current ecological theory? Travis and Mueller, on the other hand, ask questions that address species differences but not the process of speciation. They discuss interactions between population genetics and ecology, and their particular focus is on the roles that traditionally ecological variables, such as population density, age structure, and competition, should play in determining fitness. The comments that follow are informed by the discussion that took place at the conference but generally represent my personal biases.

I begin with some rather basic questions about stasis. The observation of stasis requires an extremely complete fossil record for a given taxon, and for most organisms, even if stasis were a fact, the fossil record is too incomplete to infer its existence. This is an empirical question: How general is the phenomenon? The second point concerns the nature of the traits for which stasis and punctuation are inferred. These are morphological traits such as size and shape, although Stanley suggests that size is more evolutionarily labile than shape. Many of the traits referred to are skeletal characters that are assumed to reflect "a wide variety of adaptive" traits. The problem here is the difficulty of validating that a trait in the fossil record was adaptive.

This last point gave rise to considerable discussion on whether such morphological traits were likely to have been the "stuff" of speciation or whether behavioral and physiological traits, not observable in fossils, were more central in causing reproductive isolation. Morphological change might then provide little information about rates of speciation.

Is it the case that the rapid transition between periods of morphological stasis is incompatible with the classical, gradual picture of evolution? Mathematical light has been shed on this question by recent studies of Newman, Cohen, and Kipnis (1985). These authors propose that the change in a population's mean phenotype is the result of the sum of a term due to selection, and of a random component that could reflect small population effects or environmental fluctuations. The trajectory of the mean phenotype in the population can exhibit transitions between stable equilibria indistinguishable from punctuations. The process here is qualitatively very similar to the "shifting balance" model of Wright (e.g., 1931, 1955). (Properties similar to those noted by Newman, Cohen, and Kipnis [1985] have been observed by Arthur, Ermolier, and Kaniovski [1987] in economic models with increasing returns.) It remains to resolve theoretically what characteristics of the fossil record can be used to distinguish dynamics described by Newman and his colleagues from that depicted by Stanley.

The theory of species selection and coevolution raises a class of mathematical questions that, to some degree, can be dealt with independently of the stasis issue. Slatkin (1981) began with a diffusion approach that included speciation and extinction factors as part of the assumptions used to derive the stochastic process. An alternative approach would regard species as "physical" objects subject to splits at random times or even at random places. The justification for this is that if the only empirical information is the number, type, and time of observation of the object, the biological processes that go on within the species can be subsumed within a single parameter set that describes its splitting properties. Alternatively, the probability that one type splits may depend on the strength of its interaction with entities of different types within the same stochastic process. All of this theory falls within the purview of multitype birth and death processes. The theory also allows for immigration. Now the nature of the interaction between the entities is *ecological;* these ecological interactions would determine the splitting rates and times of branching. Extinction characteristics, relative numbers of the different types, and other properties of the process that are completely analogous to those of paleobiological interest have been computed for these stochastic processes.

Mathematical modeling that addresses the fossil record in terms of ecological or even genetic processes, which are the essence of theoretical population biology, is an area of research opportunity. Perhaps we, as theoreticians, have become too complacent in our use of diffusion theory and our reference to adaptive topographies and shifting balances. I suspect that newer quantitative tools are likely to lead to a greater level of communication between population biologists and paleobiologists on the subject of the fossil record.

The focus of Stanley's paper is on the paucity of evolutionary change that he infers from the fossil record. Travis and Mueller focus on the forces that *promote* change and diversity within and among populations. They do this by surveying empirical and theoretical studies of ecological effects on fitness.

Classical population genetic theory is built around the frequencies of genotypes and the forces, mutation, random genetic drift, and fitness differences that cause these frequencies to change over time. In this classical theory, population size has no direct effect on selection, although it obviously does on the extent to which a population is inbred. But even with respect to inbreeding, classical theory does not assume that the probability that any one genotype inbreeds depends on the population size.

Ecological genetic theory introduces the numbers of individuals as a population variable, along with the frequencies of their types, and then allows the evolutionary forces to be functions of these numbers. Travis and Mueller focus on fitness, and on how ecological studies can illuminate the problem of phenotypic fitness determination. Most of their discussion introduces genotypic fitness determination only through heritability, the classical linear model that relates genotype and phenotype. I would like to stress the importance of Timothy Prout's early work (Prout 1965) in setting the agenda for a great deal of the subsequent work in estimation of the components of fitness.

The studies Travis and Mueller cite certainly illustrate the connections between ecological forces, such as competition and predation, and selection on phenotypes. They focus largely on mortality and fertility. Their review points to the difficulty of ecological studies of fertility differences, and also to the great importance of such studies. In population genetics (see, e.g., Lewontin 1974), the measurement of fertility has always been a stumbling block because of (1) its great variability, and (2) the need to consider lifetime production in comparisons between genotypes. Theoretically, fertility poses even greater difficulties; most of the pleasant mathematical properties of viability selection, which we have come to know and love, are violated. It is a great challenge for future evolutionists to determine how the statics and dynamics of populations can provide information on fertility differences.

Travis and Mueller do not address the ecology of sexual behavior beyond a single reference to O'Donald and Majerus. I believe this to be an area of opportunity both experimentally and theoretically. It may, in the long run, also provide one of the connecting points to the problem of speciation. The existence of mate choice based on melanic patterning that is relatively simple in its genetic determination has been shown by O'Donald and his colleagues to be an empirical and theoretical goldmine. We need further investigations of this evolutionarily central force, especially when it is not so simply determined.

I take my rapporteur's privilege one step further by suggesting an addendum to the empirical and theoretical studies of life histories. A key element in ecological genetics is dispersal—in its reduced term, gene flow. We have little demographic theory about age-structured migration. In fact, the microecology of migration is not often considered in evolutionary genetics. In human ecology, for example, the reproductive status of migrating individuals is of paramount importance for the subsequent dynamics of the age structure. I suspect that the same is true for other species.

The original models that built the interface between ecology and genetics, those of Roughgarden (1971) and Clarke (1972), did so by making viability density-dependent. Later studies by Matessi and Jayakar (1976) and Roughgarden (1976) extended these initial studies to more general intra- and interspecific competitive situations. But what about density-dependent fecundity and mate choice? Mueller's and Andy Clark's (Clark and Feldman 1981) studies suggest that the density at which females grow has a profound effect on their later fecundity. How general is this finding, and what are its theoretical ramifications? Based on our experience with density-independent fertilities, I would predict fascinating dynamical properties for models with density-dependent fecundity or mate choice. Prout's (1980) review points out the pitfalls that lurk behind estimation of the extent of density-dependent effects at various life-cycle stages.

The past fifteen years have seen a tremendous growth in behavioral genetic theory. This theory has, in effect, become the theoretical underpinning of behavioral ecology. But it really does not include ecology in the sense of Mueller and Travis. We have little theory on age-dependent donation and reception of altruism, nor on situations where both the population size and the frequencies of the separate behavioral phenotypes change over time.

Ecological evolutionary theory is very young. We have only recently seen epidemiology brought into ecological thinking, for example. The challenge remains to relate as much of this as possible to fundamental population genetic theory. If we consider such highly variable genetic systems as those involved in the immune response, it is not sufficient to wave our hands and say "frequency-dependent selection" or "resistance to parasites." We need to determine which rigorously specified and carefully analyzed models for the interactions between species and within a species can reasonably account for this enormous variability.

REFERENCES

Arthur, W. B., Y. M. Ermoliev, and Y. M. Kaniovski. 1987. Path-dependent processes and the emergence of macro-structure. *Eur. J. Oper. Res.* 30:294–303.

Clark, A. G., and Feldman, M. W. 1981. Density dependent fertility selection in experimental populations of *Drosophila melanogaster. Genetics* 98:849–69.

Clarke, B. 1972. Density-dependent selection. *Amer. Natur.* 106:1–13.

Lewontin, R. C. 1974. *The Genetic Basis of Evolutionary Change.* Columbia University Press, New York.

Matessi, C., and S. D. Jayakar. 1976. Models of density-frequency dependent selection for exploitation of resources. In S. Karlin and E. Nevo, eds., *Population Genetics and Ecology,* pp. 707–21. Academic Press, New York.

Newman, C. M., J. E. Cohen, and C. Kipnis. 1985. Neo-darwinian evolution implies punctuated equilibria. *Nature* 315:400–401.

Prout, T. 1965. The estimation of fitnesses from genotypic frequencies. *Evolution* 19:546–51.

Prout, T. 1980. Some relation between density independent selection and density dependent growth. *Evol. Biol.* 13:1–96.

Roughgarden, J. 1971. Density-dependent natural selection. *Ecology* 52:453–68.

Roughgarden, J. 1976. Resource partitioning among competing species—a coevolutionary approach. *Theor. Pop. Biol.* 9:388–424.

Slatkin, M. 1981. A diffusion model of species selection. *Paleobiology* 7:421–25.

Wright, S. 1932. Evolution in Mendelian populations. *Genetics* 16:97–159.

Wright, S. 1955. Classification of the factors of evolution. *Cold Spring Harbor Symp. Quant. Biol.* 20:16–24.

IV SCALE AND COUPLING IN ECOLOGICAL SYSTEMS

Chapter 10

Perspectives in Hierarchy and Scale

R. V. O'NEILL

The past two decades have seen a rapid change in the scale of ecological problems. In 1969, the National Environmental Policy Act required ecological assessment of impacts on surrounding ecosystems. From this local perspective we have moved to landscapes ecology, to the continental effects of acid precipitation, and to problems of nuclear winter, global carbon cycling, and climate change. The problems demand that we accelerate our ability to translate small-scale ecological principles to higher levels.

It is fortunate that the demand for answers at higher levels has been accompanied by the development of new tools for the investigation of scale. The combination of geographic information systems and remote-sensing capabilities permits us to work with large spatial data bases (e.g., Krummel et al. 1987). Aquatic ecologists have developed tools for rapid sampling of fine spatial scales (Auclair et al. 1982; Boyd 1973; Haury 1976) and remote sensing of ocean gyres. Statistical tools such as time-series analysis provide for analysis of pattern and scale (Fasham 1978; Mackas and Boyd 1979; Platt and Denman 1975). The statistical tools are augmented by models that attempt to explain pattern (Dubois 1975; Levin and Segel 1976; Scavia 1980). Finally, from the early explorations of Overton (1972) and Reichle, O'Neill, and Harris (1975), the theoretical basis for the hierarchy concept has been well established with texts by Allen and Starr (1982), Salthe (1985), Eldredge (1985), and O'Neill et al. (1986).

It is perhaps the juxtaposition of new problems and new techniques that has led to the remarkable convergence of the ecological community on the need to study problems of scale and hierarchy. Realization of this need is reflected in research on soil processes (Sollins, Spycher, and Topik 1983), vegetation analysis (Delcourt, Delcourt, and Webb 1983), and aquatic ecosystems (Bainbridge 1957; Steele 1978; Levasseur, Therriault, and Legendre 1984). In terrestrial systems, McIntosh (1985) points out that current trends reemphasize an interest in pattern and scale that dates back at least to Watt (1925, 1947) and the work of early biogeographers (O'Neill et al. 1986). In aquatic systems, the interest in scale (Powell et al. 1975; Cox, Haury, and Simpson 1982) is driven by a recognition of the physical constraints that shape aquatic hierarchies (Gower, Denman, and Holyer 1980; Legendre and Demers 1984) and by a desire to explain control mechanisms operating across temporal scales (Carpenter, in press). Such unparalleled convergence may represent a maturing of our science and unquestionably calls for the continued development and testing of hierarchy concepts.

PARALLEL LINES OF DEVELOPMENT

The current vitality of hierarchy considerations is indicated by four separate lines of development that are occurring simultaneously. By choosing only four, I do not mean to minimize other efforts but rather to tease out some trends in a rapidly developing field.

The first line of development, which I will characterize as the Empirical Approach, is actually a continuation of a longstanding interest among ecologists in detecting scales and pattern in ecological data (McIntosh 1985). Terrestrial studies that have focused on scale and pattern are exemplified by Bormann and Likens (1979). These authors show that, although a woodlot may never reach equilibrium, a landscape of sufficient scale can converge to an equilibrium mosaic with relative constancy in the fraction of the area occupied by various successional states. Increasingly, terrestrial studies are focusing on problems of scale (e.g., Addicott et al., in press; Maurer 1985; Wiens 1986; Wiens, Crawford, and Gosz 1985). The interest in scale is particularly evident in new fields of emphasis such as landscape ecology (Risser, Karr, and Forman 1984; Forman and Godron 1986). Pielou (1977), Ripley (1981), and Cliff et al. (1975) provide good points of entry into the theory and methodology that have developed for pattern and scale analysis.

A similar interest in scale is evident in geosciences and geomorphology. Schumm (1965, and in press) has pointed out that physical forces shaping the course of a river are disruptive at a fine time scale, continuously moving the system away from its current state. At larger scales, the forces act to equilibrate the overall course of the river, which remains constant for long periods of time. Stommel (1963) has made similar observations for oceanographic processes and demonstrates how the phenomena of interest change as the scale of observation changes.

The interest in understanding scale is widespread in aquatic ecology and often focuses on physical forces, such as hydrodynamics, that impose constraints on the ecosystem (Evans and Taylor 1980; Therriault and Platt 1981). The implications of pattern for aquatic ecosystem dynamics have been developed by a number of authors (Levasseur, Therriault, and Legendre 1984; Carpenter and Kitchell 1987) with particular emphasis on the need to look at larger-scale processes. Steele (1985), for example, has contrasted the scales of variability in terrestrial and marine ecosystems and drawn out the implications for long-term changes in community structure.

This Empirical Approach is so pervasive that it may appear to be all there is to the hierarchy concept. However, I will develop the premise that this is but one of a number of concurrent lines of development. Thus, for example, there is a second and completely independent line of research occurring in evolutionary biology.

In companion volumes, Salthe (1985) and Eldredge (1985) develop a hierarchical basis for a new and broader view of biological evolution. Their work, which I will characterize as the Evolutionary Approach, focuses on concrete biological entities, that is, organisms. They use this focus as a basis for explaining biological change at all levels of organization. They are particularly interested in biological change at higher levels, beyond the local population and beyond the explanatory power of population genetics. They point to large-scale phenomena such as stasis and mass extinction as part of the evolutionary history of the biosphere. One of their major points is that hierarchical principles provide an intellectual framework for dealing with change on all scales. This theme is also developed by Conrad (1983), who presents a mathematical theory for trade-offs in adaptability on different hierarchical levels. These authors see the hierarchical framework as essential for broadening the base of explanatory tools available to deal with evolutionary phenomena. The emphasis is on developing a philosophical base for evolution rather than on making specific testable predictions.

The mathematical development of Network Theory by Patten and his colleagues (1976; Hannon 1979; Patten 1982; Ulanowicz 1986) stands in marked contrast to the philosophical tone of Salthe (1985) and Eldredge (1985). This third branch of hierarchical thinking focuses on a general mathematical foundation for ecology. Fundamental to this approach is the insight that interactions between entities (e.g., organisms or populations) determine the dynamics of higher levels (Patten 1982). This emphasis leads to graph theoretic analyses of component interactions. The approach generates, for example, new indices of functional complexity (Ulanowicz 1983). In addition, the theory argues for the importance of indirect causality, that is, influences that go beyond one-on-one competition or predation interactions (Patten 1982). Indirect causality may require important changes in our approach to community structure and ecosystem processes by emphasizing interactions rather than component behavior. In general, Network Theory develops a set of concepts for organizing ecological understanding rather than generating testable hypotheses.

The fourth line of development emphasizes the juxtaposition of theory and testing. Formalized by Allen and Starr (1982), it shares with Evolutionary and Network approaches a direct dependence on General Systems Theory (e.g., Simon 1962) but differs in seeking empirical tests of predictions. It shares with the Empirical Approach a commitment to ecological data and analysis and the generation of testable hypotheses (Allen, O'Neill, and Hoekstra 1984) but differs in developing a comprehensive conceptual framework. Emphasis has been placed on beginning with a specific observation set (Allen, O'Neill, and Hoekstra 1984; O'Neill et al. 1986) and taking advantage of the hierarchical structure that emerges. This fourth approach has been particularly useful in applying hierarchical implications at larger scales such as the landscape (Urban, O'Neill, and Shugart 1987) and biosphere (O'Neill, in press).

SOME PRINCIPLES OF HIERARCHY

At the core of the four lines of development is a body of principles begun in the 1960s (Simon 1962; Whyte, Wilson, and Wilson 1969; Pattee 1973) as an approach to analyzing complex systems. The chief insight was that systems often contain an endogenous organization, a hierarchy of levels. The investigator can take advantage of this organization to isolate behavior that can be studied by classic approaches.

Hierarchical organization results from differences in process rates. Complex systems, such as ecosystems, operate over a wide spectrum of rates. Behaviors can be grouped into classes with similar rates that constitute levels in the hierarchy. Phenomena on a given level will be relatively isolated from lower and higher levels. Lower levels have relatively rapid rates and appear as background static or variability that is filtered or averaged out at the level of interest. Higher-level behavior is relatively slow and appears as constant from the level of interest, that is, as a set of constraints or boundary conditions on the phenomenon of interest. The hierarchical organization permits one to dissect phenomena out of the total complexity of the system. Studied on the appropriate time and space scale, the phenomenon can be viewed as the behavior of a relatively simple system and approached with traditional scientific methods.

There is a clear analogy between this conceptual framework and a controlled experiment, say, at the population level. Here we would "control" higher-level behaviors such as temperature and light by keeping them constant. We handle lower-level "noise" by averaging across many individuals. In this way we try to isolate a single level and a simple set of phenomena.

One consequence of this hierarchical structure is that behavior must be explained on three adjacent levels (O'Neill 1966; Koestler 1967). The focal level is the level of interest in a particular study. The focal level changes as the research problem changes and may be the population in one study and the landscape in another. Once the appropriate focal level is identified, dynamics are explained by looking to the next lower level. For example, if we are focusing on a population, then behavior is explicable in terms of the interactions of its components, that is, individual organisms at the next lower level. The significance of behavior, its functional or adaptive relevance, is explicable by reference to a higher-level system. Thus, if we are focusing on the individual level, sexual behavior finds its significance in the persistence of the population. To date, all of the hierarchical developments in ecology have found insight in this Triadic structure (Allen and Starr 1982; Patten 1982; Salthe 1985).

Dynamics at a particular level are structured by what happens at higher and lower levels. Higher levels impose constraints or boundary conditions, for example, temperature or precipitation. The level of interest is constrained to operate within the bounds set by the system of which it is a part. An intuitive example might be a consumer whose growth is constrained by the productivity of the ecosystem.

Lower levels impose additional limitations, termed "initiating conditions" by Salthe (1985). An intuitive example is the flight speed of a flock of birds that is limited by the speed of individual birds in the flock. Initiating conditions represent a type of constraint that is imposed on a higher level by its components. A level can only display behavior that is the feasible resultant of its components. An ecosystem cannot fix atmospheric nitrogen if there are no nitrogen fixers within that ecosystem.

Current theories vary in the extent to which they rely on the concept of constraint. But constraint appears to be a useful concept in explaining ecological dynamics. Developments in aquatic ecology clearly reflect a growing awareness of how hydrodynamics impose scaled constraints on the planktonic level of organization (Denman and Platt 1976; Harris 1980). Similarly, Carpenter (in press) shows that food-web dynamics are constrained by time scales imposed by top carnivores.

Current developments in hierarchy share this sparse set of philosophic concepts. The hierarchical concepts themselves are nothing new. They are Neoplatonic and can be found in Plotinus (ca. 250 A.D.) and Proclus (546 A.D.). They have been rediscovered periodically throughout the history of Western culture. It is interesting to note that, although the theories share these concepts, they differ on the reality of hierarchical structure. Salthe (1985) stands at one extreme with a stated commitment to concrete reality. He is talking about *the* hierarchy of life. Other authors (Webster 1979; MacMahon et al. 1978, 1981) seem to agree with the concreteness of hierarchical levels, at least implicitly. Network Theory seems neutral on the point, beginning its analysis with an existing network of interactions rather than considering the reality of the defined network. Allen (in Allen and Starr 1982) and O'Neill (in O'Neill et al. 1986) share the view that the ontological question is a moot point for the scientist. We begin with phenomena viewed in a particular observation set. If a hierarchical structure can be detected, then the system can be treated as hierarchical and the principles of the theory will be expected to apply.

EFFECTS OF THE NEW AWARENESS OF SCALE

The shared set of philosophic concepts has an intuitive appeal but is thin on hypotheses. It is a new way of looking at systems rather than a set of equations and theorems. In fact, for many ecologists, hierarchical considerations appear more as a conceptual framework than as a predictive theory. It is in the development of derivative mathematical theories yielding specific predictions, in the stimulation of new field and laboratory investigations, and in the generations of testable hypotheses that the current concern for scale will find its fruition. However, it is already becoming clear that the conceptual framework itself will prove valuable to ecology, and I would like to review some of the more notable applications.

The first and simplest application has been an increased consciousness of the extent to which our models and measurements are scale dependent. It is easy to view hierarchy as nothing more than insightful experimental design. But this becomes increasingly hard to maintain as more and more scale problems are identified even in the work of insightful experimentalists. The archetypic example involves aquatic ecologists who towed their nets in random transects trying to measure plankton that were lined up in turbulence cells like beads on a string (Abbott, Powell, and Richardson 1979). Only with the development of fine-scaled measurement and spectral analysis did the effect of scale on their measurements become apparent (Levasseur, Therriault, and Legendre 1984).

A similar story occurred in stream ecology. Streams were analyzed for nutrient efficiency, comparing inputs and outputs for a reach of stream. Only with the development of the scale-free measures of nutrient spiraling (Newbold et al. 1983) did it become apparent that the existing measures of efficiency were inadequate, being critically dependent on the dimensions of the reach being measured.

Yet another instance is recounted by Gingerich (1983). He reviews paleontological studies comparing rates of morphological change in darwins (change by a factor of *e* per million years, where *e* is the base of the natural logorithm). Recent mammal assemblages are abundant, allowing observations every few thousand years. Many minor changes are observed at a rate of about 4 darwins. Older invertebrate assemblages are rare, permitting observation only every million years. Most minor changes are not observed and the rate is given as 0.1 darwins. But in fact the mammals did not evolve more rapidly, and when the data are corrected for the scale bias, the invertebrates probably evolved more rapidly.

Walters (1986) provides an insightful analysis of why fisheries management sometimes fails. The resource is modeled on one scale as a single population over several generations. The fisherman operate at much longer time scales because they are concerned with discounted capitalization of ships and equipment. The result is that the models do not do a good job of predicting the resource, and the fishermen do not act as "logically" as the model says they should to optimize yield.

One of the best tests of a new conceptual framework is its ability to resolve controversies. An example of the utility of hierarchical concepts has emerged from recent controversies over the mechanisms that control aquatic food webs. There is evidence for "top-down" control showing that fish, at the top of the food web, constrain the food web. Other evidence indicates a "bottom-up" control showing that phytoplankton, at the bottom of the food web, control productivity throughout the food web. Hierarchical principles have been found to provide a simple explanation for the conflicting data (Carpenter, in press). Phytoplankton operate on short time scales and represent a lower hierarchical level. They constrain the food web over short time scales and explain much of the intraseasonal variability in the system. Fish operate on longer time scales and represent a higher hierarchical level. They constrain the food web over longer time scales and explain much of the interseasonal variability in the system. Thus, both "top-down" and "bottom-up" controls exist but operate on different time scales, as predicted from their hierarchical position.

The examples could be expanded almost indefinitely. In most fields of ecology, the new awareness of scale has led to new insights, new concepts, and proposed solutions to old and scale-naive dilemmas. The case can be made, even on existing literature, that the concepts of scale and hierarchy have proven their value. Investigators are beginning to take measurements across several scales (Perez et al., in press), to consider how simultaneous constraints from several hierarchical levels can influence community structure (Ricklefs 1987) and geographic distributions (Neilson and Wullstein 1983, 1985; Neilson 1986). However, given the active development of hierarchical concepts, I believe that we have only seen the tip of the iceberg. To convey the true promise of the hierarchical approach, I would like to touch on several current developments, emphasizing how the conceptual development is leading to new and testable hypotheses.

SOME CURRENT DEVELOPMENTS IN HIERARCHY

Spatial Hierarchies

A consideration of scale leads one to question whether an arbitrary circle drawn on the ground will capture the dynamics of all the organisms contained therein. The question suggests a radical Gleasonian approach with each population, or perhaps each life form, operating on a different spatial scale.

If a species, under a particular environmental regime, can win one-on-one competitive interactions with other species, a small-scaled strategy should be selected for. A plant, for example, would tend to become a dominant, shading out competitors and using vegetative reproduction to establish and maintain its spatial position. If an organism outcompetes its immediate neighbors then it gains a selective advantage at this small scale. But if a species loses ninety-nine out of every one hundred interactions, then selective pressures might lead it to operate on a scale hundreds of times larger than the dominant. If the species succeeds in operating at a sufficiently larger scale, it will operate on a different hierarchical level, and hierarchical principles predict that it will thereby isolate itself, no longer directly interacting with the dominant. The species would have to win only on rare occasions and still persist in the environment. This strategy would lead to shorter life cycles, smaller stature, and widely dispersed seeds rather than vegetative reproduction.

According to current ecological theory, plants operating at a large scale would be considered rare and classed as losers in the competitive battle. If persistence is the criterion, however, such plants are very successful. A simple consideration of scale might find a final resting place for St. Rosalia's goat bones. The speculation is interesting because it involves very simple and testable assumptions: the probability of success in competition experiments should be inversely related to the scale at which a species operates and to life history characteristics such as dispersal ability. While such an experiment might make little sense when comparing trees with ferns, it should be feasible in old field communities. Such experiments may permit us to reevaluate competitive exclusion as a scaled phenomenon of greatest interest in greenhouse pots and Ehrlenmeyer flasks. In the field, competition plus scaling strategies may be more relevant for explaining community structure. Consider also whether the thorny question of the distribution of the abundances of organisms is the simple consequence of the distribution of the probability of competitive success combined with scaling strategies to avoid exclusion.

Spatial Scales on the Landscape

A key prediction of the hierarchical concept is that process rates should be grouped into distinct levels. At a particular spatiotemporal scale, interacting components operate at similar rates and are relatively isolated from higher and lower levels that operate at much slower or faster rates. For hierarchical principles to be helpful in analyzing ecological systems, it is necessary to identify such levels in observations of the system. Hierarchical considerations predict that a complex system will not show rates uniformly distributed between the fastest and slowest rates. Rather, the rate processes will be grouped into distinct levels.

Application of time-series analysis to environmental data has already indicated that the required levels can be detected in ecological systems. Van Voris et al.

(1980) and Perez et al. (in press) have demonstrated a number of distinct levels in microcosm systems. Distinct levels have also been clearly demonstrated in aquatic ecosystems (e.g., Stommel 1963).

Krummel et al. (1987) have recently demonstrated hierarchical structuring in spatially distributed systems. They found that small shapes on the landscape had simple outlines, probably determined by human activities. Large shapes had complex outlines determined by a distinctively different set of constraints, such as topography. Processes determining shapes on the landscape fell into only two distinct levels rather than being uniformly distributed at all scales.

Similar-level structures can be detected in land-use data for landscapes. O'Neill et al. (in press) have examined the percentage of a map transect occupied by a particular land use. By replicating the transect it is possible to estimate the variance on the land-use measure. Adjacent sites along the transect are then pooled and the aggregates are assigned to the land-use type occupying the majority of the sites. The aggregation is repeated so that the same land-use data is examined at a set of increasing scales and the variance is calculated at each scale.

Greig-Smith (1957) points out that the sample variance in such pooled data should decrease as the size of the sample (i.e., aggregate size) increases. If land use on adjacent sites is uncorrelated, a plot of log variance against log scale will show a slope of −1.0. Correlations, implying some process operating on the landscape (Levin and Buttel 1986), will lead to slopes lying between −1.0 and 0.0.

O'Neill et al. (in press) found that the variance analysis revealed a hierarchical structure in landscape data. Over some scales, the variance decreased with a slope much less than −1.0 (−0.003 to −0.27 for different landscapes). At adjacent scales, the slopes changed abruptly and approached −1.0. For different landscapes, the analysis showed as many as three distinct levels (i.e., regions of shallow slope) separated by regions of slopes approaching −1.0. The implications of the analysis are that processes determining land-use patterns on the landscape are segregated into distinct spatiotemporal levels as predicted by hierarchical considerations.

Ecosystem Models and Stiff Systems

Aquatic ecosystem models typically assume that consumers are moving through a well-mixed soup of food organisms. Current scale studies invalidate this assumption. Piscivorous fish must operate at large scales relative to planktivorous fish that, in turn, must operate at larger scales than plankton. Hierarchical considerations indicate that dealing with all of these scales in a homogeneous volume of water may be the wrong approach. Fish may not interact with plankton, they may constrain them. The difference is that model formulation is radically changed and requires stiff system models that explicitly consider more than one time scale. Thus the growing concern for scale may result in radically new models of aquatic ecosystems.

Concern for modeling multiple scales leads us to some of the theoretical work being done in hierarchy. If a model explicitly considers two time scales, then a common scenario shows the slower time scale establishing a manifold, that is, a slowly changing trajectory, to which the faster dynamics are asymptotically stable. However, the trajectory may move the system toward unstable regions.

As the system moves toward a major instability, it is a reasonable conjecture that the rate of return to the manifold will decrease as the point of instability is approached. This possibility suggests a way of monitoring large-scale changes, even at the global level (O'Neill, in press). Large-scale changes are difficult to measure directly, and it may not be possible to determine whether a change is good or bad. However, it may be possible to detect whether slow changes are leading the system to a point of instability by monitoring short-term recovery experiments. If recovery time increases, it may indicate that the total system is moving toward instability and remedial action is called for. We are moving toward testing the feasibility of this approach in microcosms. We will look for a pattern in recovery from small perturbations as the system is moved along a temperature/light trajectory toward a known point of instability determined in preliminary experiments.

Nonequilibrium Thermodynamics

A related result is derived from our explorations of nonequilibrium thermodynamics. The theory states that, far from equilibrium, a system will tend toward a state of minimum entropy production. It will approach this state along a manifold described by a potential function. For present purposes, the most interesting aspect of the theory is what happens as a bifurcation point is approached. Far from a bifurcation point, the potential function increases rapidly with deviations from minimum entropy production. Thus, specific exemplars of the system, tending to minimize entropy production, will lie close to the minimum entropy state and close to each other. Near a bifurcation point, the potential function increases slowly and the manifold "flattens." Therefore, a random sample of systems would show greater deviations from the minimum entropy state.

These considerations lead to the conjecture that the variance measured among systems with similar perturbation regimes should be greater for systems near a bifurcation point. Further, the variance should increase through time for systems approaching a bifurcation point. For example, African grasslands close to the line of progressive desertification in the Sahel should show greater variability than similar grasslands with similar climatic variability further to the south. Aspects of this conjecture will be tested in aquatic microcosms where replication is possible, as is manipulation of environmental conditions that determine the state of the system relative to a bifurcation point.

PERSPECTIVES ON FUTURE DEVELOPMENTS

Given the current hectic pace of developments in hierarchy and scale, it is apparent that this will remain an active and fruitful field for some time to come. Because of the kaleidoscopic dimensions of the field, it is impossible to chart the precise course. However, it is interesting to speculate on the possibilities and the major challenges facing hierarchy theorists.

The most obvious challenge is the continued development of methods for measuring and analyzing spatial and temporal patterns. We can expect continued developments in quantifying scales with indices such as fractal dimension (Burrough 1981; Levin and Buttell 1986; Krummel et al. 1987). In terrestrial ecology we can expect continued emphasis on landscape and global scales. In aquatic ecology, we can expect new approaches to modeling and continued insight on the implications of hydrodynamic and food-web constraints. I am particularly anxious to see an explicit wedding of empirical insights with theoretical developments in hierarchy.

Understanding scaled ecological systems will require translation of information between scales (O'Neill 1979) and methods to relate findings at different scales (O'Neill, in press). Considerable effort has already gone into understanding scale translations (e.g., Allen, O'Neill, and Hoekstra 1984), but much work remains to be done. Important aspects will be the effects of aggregating dynamic behavior at finer scales (Gardner, Cale, and O'Neill 1982; Cale, O'Neill, and Gardner 1983) to higher levels, the use of scale functions to translate across scales (Milne et al., in press), and the explicit use of hierarchical principles for extrapolation (King 1986).

We can expect active development of the Network Approach to hierarchy. At present, I am aware of four books in preparation that provide a comprehensive and forceful presentation of this approach. We can look forward to this approach providing significant new insights into the way populations interact to cause ecosystem dynamics.

At present, hierarchical principles constitute a loosely integrated set of insights on scale phenomena. The conceptual framework is based on the inductive observation that complex systems often show a hierarchical structuring of rate processes. The concepts show their greatest utility in synthesizing available information and in reconciling apparent dilemmas in both theory and measurement. As illustrated by numerous examples in this paper, many workers find the principles valuable in investigating and explaining ecological systems. Others find the conceptual framework difficult to apply and call for development of mathematical theories that will allow precise and objective derivation of predictions.

The challenge is to progress from the present generalizations to theory in the sense of systematic principles and methods. This progress will be most useful if it is accompanied by mathematical theory that permits rigorous derivation of testable hypotheses. As a result, we can expect to see many aspects of the conceptual framework developed into specific theories. Such a development would greatly extend the influence of the hierarchical concept.

The most important challenge facing hierarchy theorists is the timely development of testable hypotheses. However, even before a systematic theory is achieved, we can expect the strong empirical interest in scale to result in a rapid development of field and laboratory testing of the implications of current concepts. We will see the development of large-scale measuring instruments such as FTIR (Fourier Transform InfraRed) being used to test hierarchical concepts in the field (Gosz, pers. comm.). In short, I think we are entering a period of great excitement and activity that will permanently alter the way we conceptualize and study ecological systems. We can remain optimistic that the current interest in questions of scale will stimulate efforts to develop a more precise mathematical theory that will facilitate even more rapid progress in the future.

ACKNOWLEDGMENTS

The author's research and preparation of this manuscript was supported in part by the Office of Health and Environmental Research, United States Department of Energy, under Contract No. DE-AC05-840R21400 with Martin Marietta Energy Systems, Inc., and in part by the Ecosystem Studies Program, National Science Foundation, under Interagency Agreement BSR-8021024 with the United States Department of Energy. I would particularly like to thank George Sugihara and Steve Carpenter, who helped clarify many points through their thoughtful review of the manuscript.

REFERENCES

Abbott, M. R., T. M. Powell, and P. J. Richardson. 1979. The effects of transect direction on observed spatial patterns of chlorophyll in Lake Tahoe. *Limnol. Oceanogr.* 25:534–37.

Addicott, J. F., J. M. Aho, M. F. Antolin, D. K. Padilla, J. S. Richardson, and D. A. Soluk. Ecological neighborhoods: Scaling environmental patterns. *Oikos.* In press.

Allen, T.F.H., and R. V. O'Neill. In press. Linking networks, hierarchies and landscapes. In T. Burns and K. Higashi, eds., *Hierarchy Theory.* Columbia University Press, New York.

Allen, T.F.H., and T. B. Starr. 1982. *Hierarchy: Perspectives for Ecological Complexity.* University of Chicago Press, Chicago.

Allen, T.F.H., R. V. O'Neill, and T. W. Hoekstra. 1984. Interlevel relations in ecological research and management: Some working principles from hierarchy theory. Gen. Tech. Rept. RM-110, Rocky Mountain Experiment Station, Fort Collins, Colorado.

Auclair, J. C., S. Demers, M. Frechette, L. Legendre, and C. L. Trump. 1982. High frequency endogenous periodicities of chlorophyll synthesis in estuarine phytoplankton. *Limnol. Oceanogr.* 27:348–52.

Bainbridge, R. 1957. The size, shape, and density of marine phytoplankton concentrations. *Biol. Rev.* 32:91–115.

Bormann, F. H., and G. E. Likens. 1979. *Pattern and Process in a Forested Ecosystem.* Springer-Verlag, New York.

Boyd, C. M. 1973. Small scale spatial patterns of marine zooplankton examined by an electronic in situ zooplankton detecting device. *Neth. J. Sea Res.* 7:103–11.

Burrough, P. A. 1981. Fractal dimensions of landscapes and other environmental data. *Nature* 294:241–43.

Cale, W. G., R. V. O'Neill, and R. H. Gardner. 1983. Aggregation error in nonlinear ecological models. *J. Theor. Biol.* 100:539–50.

Carpenter, S. R. In press. Transmission of variance through lake food webs. In S. R. Carpenter, ed., *Complex Interactions in Lake Communities.* Springer-Verlag, New York.

Carpenter, S. R., and J. F. Kitchell. 1987. The temporal scale of variance in limnetic primary production. *Amer. Natur.* 129:417–33.

Cliff, A. D., P. Haggett, J. K. Ord, K. Bassett, and R. B. Davies. 1975. *Elements of Spatial Structure.* Cambridge University Press, Cambridge, England.

Conrad, M. 1983. *Adaptability: The Significance of Variability from Molecule to Ecosystem.* Plenum Press, New York.

Cox, J. L., L. R. Haury, and J. J. Simpson. 1982. Spatial patterns of grazing-related parameters in California coastal surface waters, July 1979. *J. Mar. Res.* 40:1127–53.

Delcourt, H. R., P. A. Delcourt, and T. Webb. 1983. Dynamic plant ecology: The spectrum of vegetation change in space and time. *Quaternary Sci. Rev.* 1:153–75.

Denman, K. L., and T. Platt. 1976. The variance spectrum of phytoplankton in a turbulent ocean. *J. Mar. Res.* 34:593–601.

Dubois, D. M. 1975. A model of patchiness for predator-prey plankton populations. *Ecol. Model.* 1:67–80.

Eldredge, N. 1985. *Unfinished Synthesis: Biological Hierarchies and Modern Evolutionary Thought.* Oxford University Press, New York.

Evans, G. T., and F.J.R. Taylor. 1980. Phytoplankton accumulation in Langmuir cells. *Limnol. Oceanogr.* 25:840–45.

Fasham, M.J.R. 1978. The statistical and mathematical analysis of plankton patchiness. *Oceanogr. Mar. Biol. Ann. Rev.* 16:43–79.

Forman, R.T.T., and M. Godron. 1986. *Landscape Ecology.* Wiley, New York.

Gardner, R. H., W. G. Cale, and R. V. O'Neill. 1982. Robust analysis of aggregation problems. *Ecology* 63:1771–79.

Gingerich, P. D. 1983. Rates of evolution: Effects of time and temporal scaling. *Science* 222:159–61.

Gower, J.F.R., K. L. Denman, and R. J. Holyer. 1980. Phytoplankton patchiness indicates the fluctuation spectrum of mesoscale oceanic structure. *Nature* 288:157–59.

Greig-Smith, P. 1957. *Quantitative Plant Ecology.* Butterworth Scientific Publications, London.

Hannon, B. 1979. Total energy costs in ecosystems. *J. Theor. Biol.* 80:271–93.

Harris, G. P. 1980. Temporal and spatial scales in phytoplankton ecology. *Can. J. Fish. Aquat. Sci.* 37:877–900.

Haury, L. R. 1976. Small scale pattern of a California Current zooplankton assemblage. *Mar. Biol.* 37:137–57.

King, A. W. 1986. The seasonal exchange of carbon dioxide between the atmosphere and the terrestrial biosphere: Extrapolation from site-specific models to regional models. Ph.D. dissertation, University of Tennessee, Knoxville.

Koestler, A. 1967. *The Ghost in the Machine.* Macmillan, New York.

Krummel, J. R., R. H. Gardner, G. Sugihara, R. V. O'Neill, and P. R. Coleman. 1987. Landscape patterns in a disturbed environment. *Oikos* 48:321–34.

Legendre, L., and S. Demers. 1984. Towards dynamic biological oceanography and Limnology. *Can. J. Fish. Aquat. Sci.* 41:2–19.

Levasseur, M., J. C. Therriault, and L. Legendre. 1984. Hierarchical control of phytoplankton succession by physical factors. *Mar. Ecol. Prog. Ser.* 19:211–22.

Levin, S. A., and L. Buttel. 1986. Measures of patchiness in ecological systems. Ecosystems Research Publication ERC-130. Cornell University, Ithaca, N.Y.

Levin, S. A., and L. A. Segel. 1976. Hypothesis for the origin of planktonic patchiness. *Nature* 259:659.

McIntosh, R. P. 1985. *The Background of Ecology.* Cambridge University Press, Cambridge, England.

Mackas, D. L., and C. M. Boyd. 1979. Spectral analysis of zooplankton spatial heterogeneity. *Science* 204:62–64.

MacMahon, J. A., D. L. Phillips, J. V. Robinson, and D. J. Schimpf. 1978. Levels of biological organization: An organism centered approach. *Bioscience* 28:700–704.

MacMahon, J. A., D. J. Schimpf, D. C. Andessen, K. G. Smith, and R. L. Bayn. 1981. An organism-centered approach to some community and ecosystem concepts. *J. Theor. Biol.* 88:287–307.

Maurer, B. A. 1985. Avian community dynamics in desert grasslands: Observational scale and hierarchical structure. *Ecol. Monogr.* 55:295–312.

Milne, B. T., R. V. O'Neill, R. H. Gardner, M. G. Turner, and B. Jackson. The utility of scale-independent relationships. *Landscape Ecology.* In press.

Neilson, R. P. 1986. High resolution climatic analysis and southwest biogeography. *Science* 232:27–34.

Neilson, R. P., and L. H. Wullstein. 1983. Biogeography of two southwest American oaks in relation to atmospheric dynamics. *J. Biogeogr.* 10:275–97.

Neilson, R. P., and L. H. Wullstein. 1985. Comparative drought physiology and biogeography of *Quercus gambelii* and *Quercus turbinella*. *Amer. Natur.* 114:259–71.

Newbold, J. D., J. W. Elwood, R. V. O'Neill, and A. L. Sheldon. 1983. Phosphorus dynamics in a woodland stream ecosystem, a study of nutrient spiralling. *Ecology* 64:1249–65.

O'Neill, R. V. 1966. Possibility for advance in the organism concept. *Biologist* 48:62–67.

O'Neill, R. V. 1979. Transmutations across hierarchical levels. In G. S. Innis and R. V. O'Neill, eds., *Systems Analysis of Ecosystems,* pp. 58–78. International Cooperative Publishing House, Fairland, Maryland.

O'Neill, R. V. In press. Hierarchy theory and global change. In T. Rosswell, R. G. Woodmansee, and P. G. Risser, eds., *Spatial and Temporal Variability in Biospheric and Geospheric Processes.* SCOPE Report XX, Scientific Committee on Problems of the Environment, Paris.

O'Neill, R. V., D. L. DeAngelis, J. B. Waide, and T.F.H. Allen. 1986. *A Hierarchical Concept of Ecosystems.* Princeton University Press, Princeton, N.J.

O'Neill, R. V., R. H. Gardner, B. T. Milne, M. G. Turner, B. Jackson. In press. Heterogeneity on patterned landscapes: A test of hierarchy theory. *Landscape Ecology.*

Overton, W. S. 1972. Toward a general model structure for forest ecosystems. In J. F. Franklin, ed., *Proc. Symposium on Coniferous Forest Ecosystem.* Northwest Forest Range Station, Portland, Oregon.

Pattee, H. H., ed. 1973. *Hierarchy Theory.* Braziller, New York.

Patten, B. C. 1982. Environs: Relativistic elementary particles for ecology. *Amer. Natur.* 119:179–219.

Patten, B. C., R. W. Bosserman, J. T. Finn, and W. G. Cale. 1976. Propagation of cause in ecosystems. In B. C. Patten, ed., Systems Analysis and Simulation in Ecology, Vol. 4, pp. 457–79. Academic Press, New York.

Perez, K. T., G. E. Morrison, E. W. Davey, N. F. Lackie, A. E. Soper, R. J. Blasco, D. L. Winslow, R. L. Johnson, S. A. Marino, and J. F. Heltshe. Is bigger better? Physical simulation models of a marine system perturbed by a contaminant. *Ecology.* In press.

Pielou, E. C. 1977. *Mathematical Ecology.* Wiley, New York.

Platt, T., and K. L. Denman. 1975. Spectral analysis in ecology. *Ann. Rev. Ecol. Syst.* 6:189–210.

Powell, T. M., P. J. Richerson, T. M. Dillon, B. A. Agee, B. J. Dozier, D. A. Godden, and L. O. Myrup. 1975. Spatial scales of current speed and phytoplankton biomass fluctuations in Lake Tahoe. *Science* 189:1088–90.

Plotinus. *Enneads.* 1956. Trans. by S. MacKenna. Faber and Faber, New York.

Proclus. *Elements of Theology.* 1963. Trans. by E. R. Dodds. Oxford University Press, New York.

Reichle, D. E., R. V. O'Neill, and W. F. Harris. 1975. Principles of energy and material exchange in ecosystems. In W. H. von Dobben and R. H. Lowe-McConnell, eds., *Unifying Concepts in Ecology,* pp. 27–43. Dr. W. Junk, The Hague.

Ricklefs, R. E. 1987. Community diversity: Relative roles of local and regional processes. *Science* 235:167–71.

Ripley, B. D. 1981. *Spatial Statistics.* Wiley, New York.

Risser, P. G., J. R. Karr, and R.T.T. Forman. 1984. *Landscape Ecology.* Illinois Nat. Hist. Survey, Special Publication 2. Champaign, Illinois.

Salthe, S. N. 1985. *Evolving Hierarchical Systems: Their Structure and Representation.* Columbia University Press, New York.

Scavia, D. 1980. Conceptual model of phosphorus cycling. In D. Scavia and R. Moll, eds., *Nutrient Cycling in the Great Lakes,* pp. 119–40. Special Report 83, Great Lakes Research Division, University of Michigan, Ann Arbor.

Schumm, S. A. Variability of the fluvial system in space and time. In T. Rosswell, R. G. Woodmansee, and P. G. Risser, eds., *Spatial and Temporal Variability in Biospheric and Geospheric Processes.* SCOPE XX, Scientific Committee on Problems of the Environment, Paris. In press.

Schumm, S. A., and R. W. Lichty. 1965. Time, space, and causality in geomorphology. *Amer. J. Sci.* 263:110–19.

Simon, H. A. 1962. The architecture of complexity. *Proc. Amer. Phil. Soc.* 106:467–82.

Sollins, P., G. Spycher, and C. Topik. 1983. Processes of soil organic-matter accretion at a mudflow chronosequence, Mt. Shasta, California. *Ecology* 64:1273–82.

Steele, J. H., ed. 1978. *Spatial Pattern in Plankton Communities.* Plenum Press, New York.

Steele, J. H. 1985. A comparison of terrestrial and marine ecological systems. *Nature* 313:355–58.

Stommel, H. 1963. Varieties of oceanographic experience. *Science* 139:572–76.

Therriault, J. C., and T. Platt. 1981. Environmental control of phytoplankton patchiness. *Can. J. Fish. Aquat. Sci.* 38:638–41.

Ulanowicz, R. E. 1983. Identifying the structure of cycling in ecosystems. *Math. Biosci.* 65:219–37.

Ulanowicz, R. E. 1986. *Growth and Development.* Springer-Verlag, New York.

Urban, D. L., R. V. O'Neill, and H. H. Shugart, 1987. Landscape ecology. *Bioscience* 37:119–27.

Van Voris, P., R. V. O'Neill, W. R. Emanuel, and H. H. Shugart. 1980. Functional complexity and ecosystem stability. *Ecology* 61:1352–60.

Walters, C. 1986. *Adaptive Management of Renewable Resources.* Macmillan, New York.

Watt, A. S. 1925. On the ecology of British beechwoods with special reference to their regeneration. *J. Ecology* 13:27–73.

Watt, A. S. 1947. Pattern and process in the plant community. *J. Ecology* 35:1–22.

Webster, J. R. 1979. Hierarchical organization of ecosystems. In E. Halfon, ed., *Theoretical Systems Ecology,* pp. 119–31. Academic Press, New York.

Wiens, J. A. 1986. Spatial scale and temporal variation in studies of shrubsteppe birds. In J. Diamond and T. J. Case, eds., *Community Ecology,* pp. 154–72. Harper and Row, New York.

Wiens, J. A., C. S. Crawford, and J. R. Gosz. 1985. Boundary dynamics: A conceptual framework for studying landscape ecosystems. *Oikos* 45:421–27.

Whyte, L. L., A. G. Wilson, and D. Wilson, eds. 1969. *Hierarchical Structures.* Elsevier, New York.

Chapter 11

Physical and Biological Scales of Variability in Lakes, Estuaries, and the Coastal Ocean

THOMAS M. POWELL

What is meant by "scale"? Why is it an important concept in ecology? Why does scale play a particularly prominent role when one speaks of "coupling"? Can one make nontrivial, general statements, if phrased in terms of scale alone, that apply to several seemingly different systems? Can other generalizations that highlight "scale" and "coupling" help us understand why some ecological systems differ so greatly from others? I shall address aspects of these questions here, with particular emphasis on results from lakes, estuaries, and the coastal ocean. I begin with some elementary notions, then review how scale considerations enter the coupling between physical and biological systems in the size scale regime between 100 m and 100 km—a regime of great ecological interest and the scale of most lakes, estuaries, and the coastal ocean. Other papers in this volume focus on, for example, terrestrial systems and the general question of how large, complex systems are structured; I end, then, by advancing some speculations about these and other areas I have neglected. Finally, some publications on theory in biological oceanography (and related disciplines) have recently appeared: one is nearly a tutorial (Platt, Mann, and Ulanowicz 1981) and the other assays the use and prospects for ecosystem theory (Ulanowicz and Platt 1985). I therefore largely avoid the topics that these authors have addressed so well.

ELEMENTARY CONSIDERATIONS

Intuitively, one speaks of the spatial scale of a problem as the distance one must travel before some quantity of interest changes significantly. Two adjacent parcels of water separated by 1 mm are likely to have very similar concentrations of chlorophyll, for example, or similar concentrations of a single phytoplankton species (within the limits of sampling error, of course). Conversely, water parcels separated by 1 km might be expected to differ greatly in these quantities. An observer would understand that the relevant spatial scale was greater than 1 mm and less than 1 km. This example raises two points. First, when one speaks of scale one is considering variability. A spatial scale cannot be assigned to a process, or quantity, that is uniform in space. The terms "scale" and "scale of variability" are synonymous. Second, the spatial scale may depend upon the varying quantity, or process, under consideration. For example, the chlorophyll concentration in each of the two parcels of water separated by 1 km could be identical, while the phytoplankton species content could differ substantially. The "biological" scale (determined by the varying phytoplankton concentration field) can, in principle, differ from another "biological" scale, as determined by chlorophyll variation.

The intuitive discussion about spatial scales applies equally to time, time scales, and temporal variability. The time scale, then, is that period over which one waits to see a significant change in some quantity of interest.

The entire discussion can be formalized. Let $q(x)$ be any quantity that varies with spatial distance, x. Let $<q>_{\mathrm{RMS}} = <(q - <q>)^2>^{1/2}$ and $<dq/dx>_{\mathrm{RMS}} = <(dq/dx)^2>^{1/2}$. The brackets ($< >$) indicate a spatial average. Then one defines L_q as the spatial scale over which $q(x)$ varies significantly, or, simply, the spatial scale:

$$L_q = [<dq/dx>_{\mathrm{RMS}}]^{-1}<q>_{\mathrm{RMS}}. \qquad (11.1)$$

Similarly, defining $\bar{q}_{\mathrm{RMS}} = [\overline{(q - \bar{q})^2}]^{1/2}$ and $\overline{dq/dt}_{\mathrm{RMS}} = [\overline{(dq/dt)^2}]^{1/2}$, where the overbar indicates a time average, one defines T_q as the time scale over which $q(t)$ varies significantly, or, simply, the time scale:

$$T_q = [\overline{dq/dt}_{\mathrm{RMS}}]^{-1}\bar{q}_{\mathrm{RMS}}. \qquad (11.2)$$

Note the connection to variance (i.e., variability) and quantities like the variance spectrum. Other similar definitions are, of course, possible.

We often speak loosely of the spatial scale as the area (or length or volume) in space that we are considering; the time scale as the duration of our attention to a phenomenon. We can be found admonishing our colleagues to remember that ecological interactions may differ if we consider larger (or smaller) space or time scales (I translate this roughly to mean that if the observer had looked at a larger area, or for a longer time, different results might have been obtained). The mere size of the area under consideration is rarely the quantity of direct interest, however. We are usually much more concerned with knowing the largest size scale that one must observe to capture "most" of the variance. Though it is often true that at the very largest space and time scales more or new variance enters a system, we are often restricted to smaller intervals in space and time for unavoidable logistic and practical reasons. Accordingly, dominant processes may act over considerably smaller time or space scales than the full durations, or distances, over which we observe. A satisfactory formalism to quantify the variability—variance—as a function of spatial or temporal scale is necessary in order to determine what the relevant scales are in a problem. Time-series analysis (and its analogue in space), including spectral techniques, is one approach (Williamson 1975; Shugart 1978, Chatfield 1984). Other approaches, for example multivariate statistics such as principal components or empirical orthogonal functions (Williamson 1972; Denman and Mackas 1978; Chatfield 1980; Gauch 1982), are also useful and may not be so demanding (or so powerful) as the time-series techniques. We shall not be surprised, then, to find that model calculations of variance spectra are those that may tell us most about scale questions (and those most easily compared to actual data concerned with these questions).

Why are considerations of scale important in ecology, especially in theoretical ecology? Because they hold out the prospect of simplification—how simplification of our models might arise and, equally important, when and why one might not expect such simplification.

The small-scale physical environment in aquatic systems differs from that at large scales. Life "in the small" is dominated by viscous effects (Purcell 1977; Vogel 1981; Berg 1983); the Coriolis "force" is the prominent actor at large scales (Pedlosky 1979; Gill 1982). A formalism to utilize these facts—dimensional analysis—plays a major role in determining those forces which dominate physical phenomena at given space and time scales. Further, and most important, there is often a pronounced separation of scales—that is, scales where one effect dominates may not overlap the range of scales where another holds sway. One may disregard certain effects within certain time and space scales, thus achieving a simplification. Finally, though separation of different effects at differing scales is the "zeroth order" model, some coupling between scales is always present. For example, a fluctuation in velocity, that is, variance—mechanical energy in this case—enters most physical systems at large scale. Ultimately, its impact is dissipated to heat at small scale via viscosity (see fig. 11.1). Nonetheless, the coupling is sufficiently weak that one may neglect viscous effects when considering phenomena over ocean basin scales. In summary, the physical environment in the sea is divided into separate spatial and temporal regimes; though coupled to one another, the regimes maintain their integrity because the coupling is weak enough. Finally, the coupling between the separate regimes must be via nonlinear terms in the governing equations of motion.

A similar separation describes the ecological structure in the sea (and in lakes and estuaries). Figure 11.2a shows a measure of the spatial scales of variability for phytoplankton community composition, zooplankton community composition,

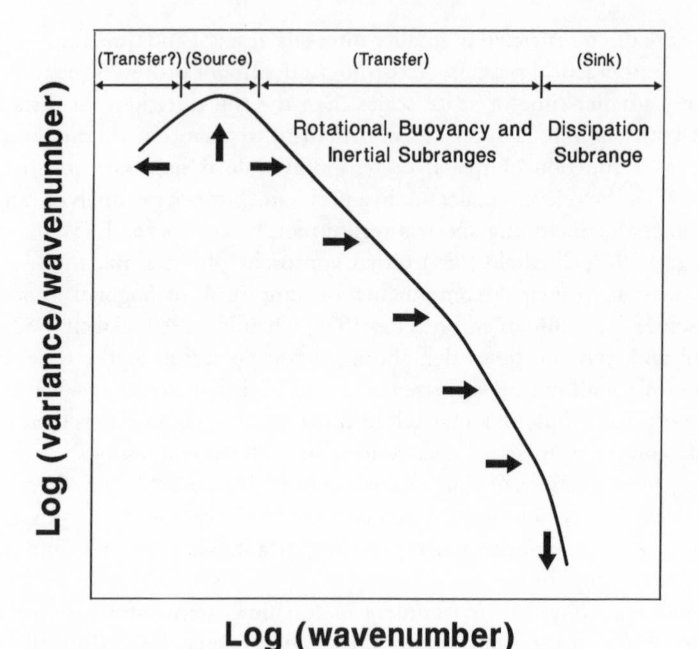

FIGURE 11.1. Spectrum of a scalar tracer vs. (size scale)$^{-1}$ = wavenumber = 2π/wavelength. "Energy," or variance, enters at large scales, cascades from large scale to small, and is lost through dissipation at small scales. Rotational, buoyancy, or inertial forces dominate subranges of the cascade/transfer region. (After Mackas et al. 1985.)

and zooplankton biomass. Note the separation between dominant scales for each quantity. Note also the differences between the alongshore and cross-shelf scales; this anisotropy is evidence for the control the physical environment exerts on the spatial distribution of these planktonic organisms. Figure 11.2b shows estimates of particle size and doubling times for phytoplankton, zooplankton, invertebrate carnivores or omnivores, and fish communities. With the provisos noted above, these are taken as surrogates for the scales of variability. Note again the separation of spatial and temporal regimes. We anticipate that these separations point to simplifications that ecological theorists can advantageously incorporate, following modelers of the physical environment.

PHYSICAL-BIOLOGICAL COUPLING

Figure 11.3 shows a prototype "box model" for the budget of any quantity, q (which might be phytoplankton biomass, the mass of a nutrient such as nitrate, or

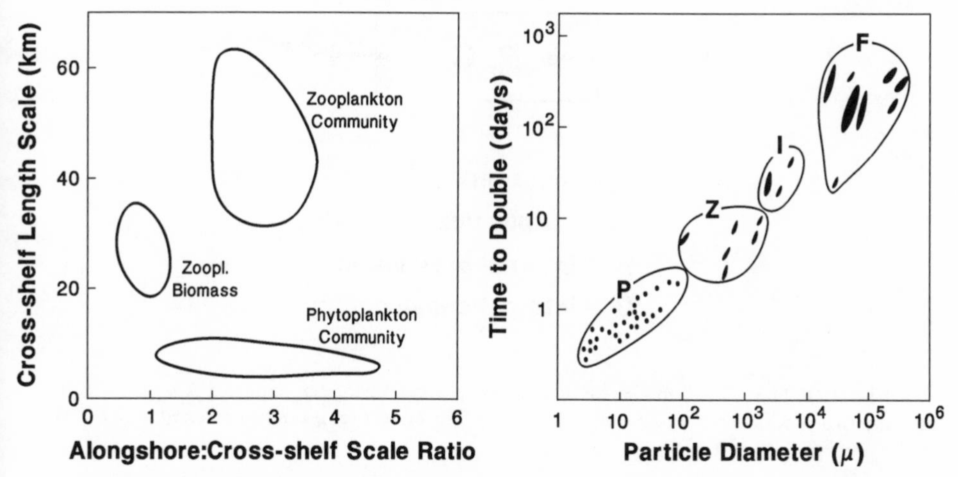

FIGURE 11.2. *Left:* 75% confidence regions for the scales of variability of zooplankton biomass, zooplankton community composition, and phytoplankton community composition along the coast of British Columbia. (After Mackas 1984.) *Right:* Doubling times vs. particle size distributions for phytoplankton (P), zooplankton (Z), invertebrate carnivores or omnivores (I), and fish (F). (After Sheldon et al. 1972, and Steele 1978.)

the density of a particular species of zooplankton). The governing equations in Steele's (1974) model are of this form, for example. The rate of change of q obeys a balance equation of the form

$$dq/dt = F_{\mathrm{I}} - F_{\mathrm{O}} + P - C. \tag{11.3}$$

F_I and F_O are input and output fluxes of q, respectively; P and C are the rates of internal production and consumption of q within the "box"—any volume of water. In aquatic systems dominated by plankton, the input and output fluxes, F_I and F_O, are usually the result of physical transport processes (though biological transport via swimming, for example, can enter); often, but not always, they can be an order of magnitude or more greater than the *in situ* rates, P − C. Accordingly, the first task is to express the coupling of the physical environment to the ecological structures in these aquatic systems. The general question of physical-biological coupling has been reviewed by Denman and Powell (1984) and Mackas, Denman, and Abbott (1985). Other references can be found in these two pieces. Several useful generalizations emerged. First, the dominant temporal scales seemed to be set by biological processes, but the characteristic spatial scales were largely determined by the physics. Second, a "linear, first-order" hypothesis is that if a concentration of "energy" (variance) were to be found at a certain spatial

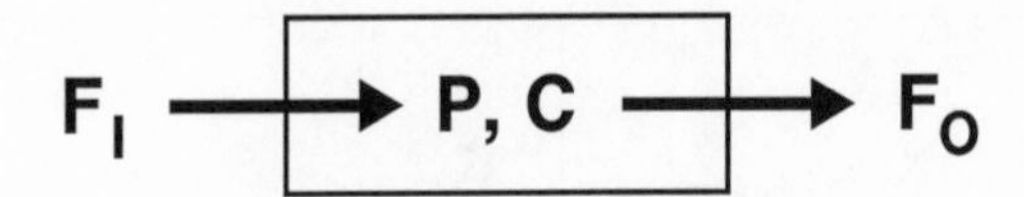

FIGURE 11.3. "Box" model for a quantity, q. q can change by inputs (F_I), outputs (F_O), internal production (P), and consumption (C). The budget for q can be expressed as $dq/dt = F_I - F_O + P - C$.

scale in a physical phenomenon—say, a front—then one might expect to find a concentration of variance at the same scale in biologically interesting quantities. (A biological oceanographer told me that he made it a point to be on board ship, taking data, when two of his physical oceanographer colleagues were on cruises. "If those two have found something interesting," he said, "there's bound to be something interesting for me, too.") This expectation seemed to be met over a very wide range of spatial scales. The second hypothesis says nothing about the coupling of one scale to another—only about the source regions for biological variance.

The coupling of ecological phenomena occurring at different scales is an area that is particularly ripe for theoretical investigation. For logistic reasons, experimenters can rarely cover a sufficiently large area in a short enough time to truly measure synoptically over several scale ranges. Have things changed because of the changes in observed scale, or have they "merely" changed in situ with time? We are stuck in our parochial little ponds—even on the ocean! Perhaps the only way to investigate these questions now is with theoretical calculations, until satellite remote sensing can provide a greater number of more useful biological indicators. Because some, and perhaps most, of the coupling between scales is accomplished by physical processes, we will consider these models in detail.

100 Meter–100 Kilometer Scales

Early patch models in the sea—plankton growing with constant growth rates and diffusing with constant diffusivity, called KISS models after an appealing grouping of the first letters of the names of early investigators (*Ki*erstead, *S*lobodkin, and *S*kellam)—are reviewed by Okubo (1980, 1984). Models for the power spectrum of spatial fluctuations in phytoplankton biomass that assume a constant growth rate in isotropic turbulence (Denman and Platt 1976; Denman, Okubo, and Platt 1977) lead to predictions that agree with a large number of measurements in lakes, estuaries, and the ocean (Platt 1978). Some years later Denman (1983) realized that so long as the growth rate was only a function of time, not space (including a constant function, of course), then the growth rate could be transformed out of equations for the spectral distribution. Unhappily, then, these models proclaimed that the magnitude of growth rates of phytoplankton, *if spatially homogeneous,* played no role in determining patchiness; phytoplankton acted only as passive tracers, like dye. With a Lagrangian particle statistics representation of the large scale, turbulent velocity field, Bennett and Denman (1985) modeled phytoplankton concentrations in a spatially varying random growth rate field, $r(x,t)$. Results were obtained for two cases: $r(x,t)$ held fixed in space, and $r(x,t)$ advected with the flow. In each case the initial spectral distribution was peaked around the characteristic spatial scale set by the initial phytoplankton concentration; i.e., it was "patchy." The spectrum "decayed," however, to a noisy spectrum proportional to k^{-1} (k is 2π/"wavelength," i.e., 2π/size scale). (See fig. 11.4.) Three results are noteworthy. First, the model provided a pathway for variance to show up at small scales once initiated at large scales—a variance cascade to smaller scales as in the physical turbulent transfer. Second, a growth-rate field that varied spatially, but in a random sense, was insufficient to maintain an initially patchy distribution of phytoplankton. Third, the few synoptic measurements at these large scales (e.g., Gower, Denman, and Holyer, 1980; Denman and Abbott 1988) show a spectral distribution closer to k^{-2}, not k^{-1}. The present intense exploration of satellite records may help resolve this discrepancy.

The Bennett/Denman calculation presents a much more realistic description of the essentially 2-D turbulence field at the larger scales of interest than do studies by previous workers (i.e., isotropic turbulence occurs only at small scales, < 10 m). Further, the dominant features of the model can be phrased largely in terms of only the temporal and spatial scales of the processes at work. It also has the potential to incorporate more complex biological dynamics than previous efforts, i.e., calculations that involve two coupled fields—a nutrient and a phytoplankton, for example, or a phytoplankton species and a zooplankton predator. The biological dynamics need not be restricted to linear functions. It is not clear that calculations incorporating such complications will lead to results similar to the Bennett/Denman calculations, i.e., "running down" to noise. For example, we expect that a predator-prey model of the form

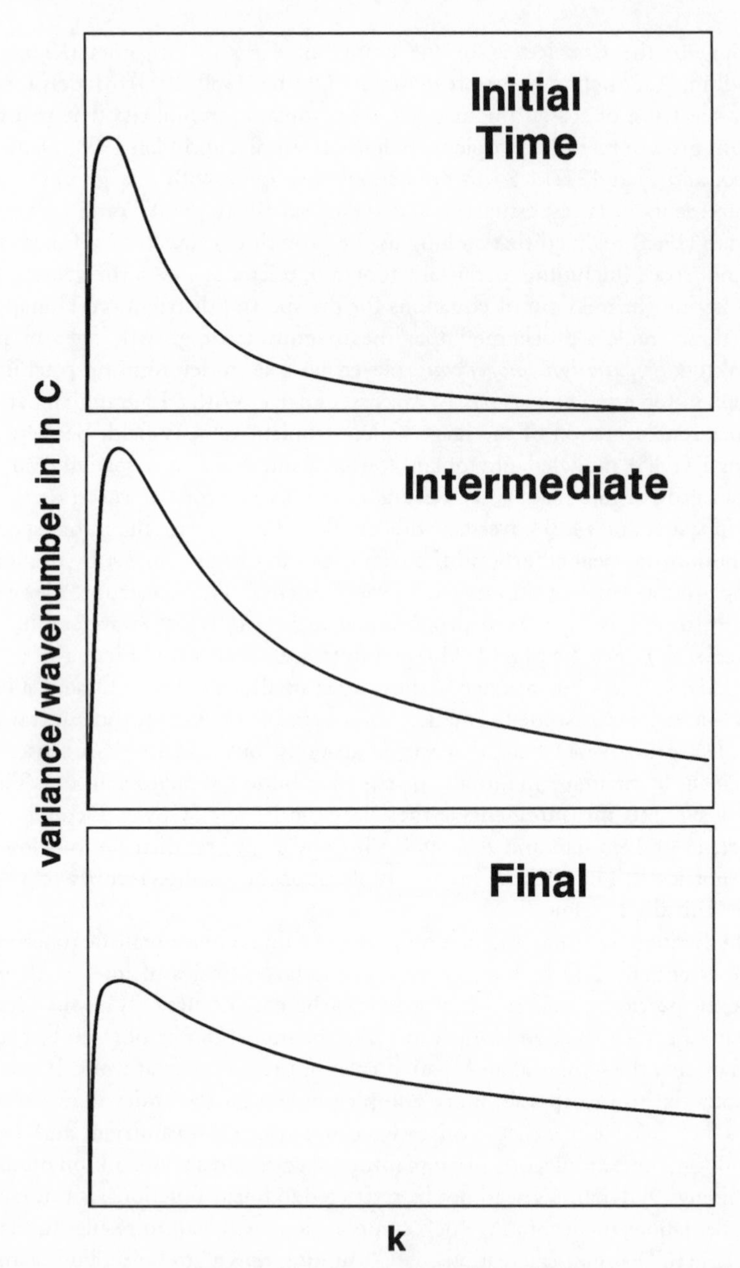

FIGURE 11.4. Spectrum of (transformed) concentration of phytoplankton versus k (which is (size scale)$^{-1}$ = 2π/wavelength) for initial, intermediate, and final times. Note the transfer of variance from large to small scales. (After Bennett and Denman 1985.)

$$\frac{dp}{dt} = rp\left(1 - \frac{p}{K}\right) - A\left(\frac{p^2}{\Gamma + p^2}\right)h = F(p,h) \tag{11.4}$$

$$\frac{dh}{dt} = B\left(\frac{p^2}{\Gamma + p^2}\right)h - dh = G(p,h), \tag{11.5}$$

with logistic growth of the prey, p, and Holling type III functional (and numerical) response to the predator, h, has three types of solutions: smooth approach to equilibrium; damped, oscillatory approach to equilibrium; and limit cycle activity (see, e.g., Roughgarden 1979 for an analysis with a similar, but not identical, Holling type II model). The *time scale* of the oscillatory behavior depends on the parameters of the model and could be shorter than, of the same order, or longer than the physical time scale of relaxation to equilibrium. Accordingly, if the oscillatory time scale is short, one might see large bursts of phytoplankton concentration early in the process of "physical decay," when large spatial scales dominate. Such activity arises from the internal biological dynamics of the system; it provides a mechanism of adding variance to the system at large scales, before the system has had a chance to "run down." Bennett and Denman (1988) have already shown how phytoplankton patchiness can arise from the addition of an annual cycle to a simple 2-D turbulence model. The calculated spectral distribution could be more like the observed k^{-2}, not k^{-1}. A simple, 1-D simulation model with Lotka-Volterra dynamics and random injections of variance (Steele and Henderson 1977) captured some of this flavor, with calculated spectral shapes between k^{-2} and k^{-3}.

Calculations of the Bennett/Denman type with more complex biological dynamics are possible in principle, but would be extremely tedious. The models would have to allow several or all of the parameters (r, K, A, B, Γ, and d) to behave stochastically. (It should be easier to let only r, A, B, and d vary stochastically; eqs. (11.4) and (11.5) are linear in those quantities.) Another heuristic approach would be to relax some of the physical realism and parametrize the "turbulence" with constant, but unequal, Fickian diffusivities, D_1 and D_2. 2-D reaction-diffusion equations like (11.6) and (11.7) would result in

$$\frac{\partial p}{\partial t} = D_1\nabla^2 p + F(p,h) \tag{11.6}$$

$$\frac{\partial h}{\partial t} = D_2\nabla^2 h + G(p,h) \tag{11.7}$$

The advantage of such a "mimic" of reality is tractability. It would be extremely interesting to see what Bennett/Denman-type results would be obtained for the parameter groups that lead to diffusive instability. Though the turbulence mechanism in the reaction diffusion calculations differs substantially from the Lagrangian statistics approach, perhaps enough of the "turbulence" acts in Fickian fashion that some smeared, non-uniform "steady" states might emerge (at least transiently). This could be another source of large-scale variance.

The reaction-diffusion calculations also promise to help researchers develop some intuition about the much more difficult statistical endeavors, like those of Bennett and Denman. We know a great deal about 1-D reaction diffusion systems, and investigation continues to be very intense. We know less about 2-D systems, but some signal studies are emerging (Mimura, Kan-on, and Nichiura 1988). If the pace of discovery continues in 2-D systems as rapidly as it has in 1-D systems, then we shall soon have a number of results to use for reference. Comparison studies between the reaction-diffusion formulations and statistical calculations (à la Bennett and Denman) could be extremely valuable.

Two observations are pertinent. First, eqs. (11.6) and (11.7) contain several parameters that must be estimated. Ecologists are often too ready to apologize for the lack of solid information about "constants" that appear in models. Compared to many other biological disciplines, ecologists can have more confidence about the magnitude of such numbers. This is not the case in developmental biology, for example, where coefficients enter models (quite useful, distinguished, and sophisticated models; see, e.g., Murray and Oster 1984) for which no information whatsoever exists. Estimates of all the constants in eqs. (11.6) and (11.7), certainly to within an order of magnitude, could be made for a phytoplankton-zooplankton predator-prey interaction. Even the earlier dynamical models in the sea (e.g., Steele 1974) tracked actual systems (i.e., the North Sea) to a surprising degree. Second, there is little field evidence that reaction-diffusion models explain the development of spatial patterns, particularly in plankton systems. There may be a reason for this. Virtually all of the field investigations consider only biomass, not the distribution of individual species (or functional groups of species, perhaps). The subtle distributional patterns of individual species relative to one another could be missed by the coarser measures of mere biomass (e.g., fluorescence in aquatic systems). There is some preliminary, unpublished support for this view. In a large lake—Lake Tahoe, bordering California and Nevada—Peter Richerson and I collected samples along 25 km transects and, with the aid of a small squad of phytoplankton counters, enumerated the samples down to the species. In one representative transect the numerically dominant diatom species, *Cyclotella stelligera* and *Fragilaria crotenensis,* showed little pattern beyond randomness—their spatial autocorrelations are well described as a Markov process with a very short fall-off distance. Conversely, several of the less numerous chlorophyte species show much more large-scale structure and have significant cross-correlations with other species at large-scale separations. (See fig. 11.5.) That the dominant species have little spatial pattern beyond that determined by the "physics" confirms our earlier analyses of chlorophyll in this large lake

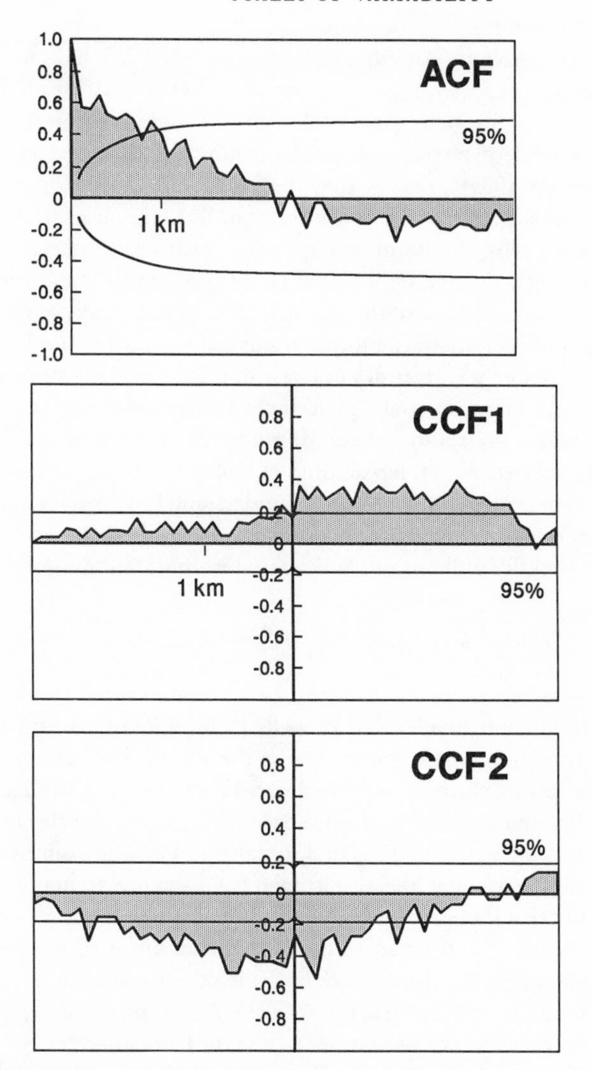

FIGURE 11.5

ACF. The autocorrelation function (ACF) of a small chlorophyte (phytoplankton) species vs. (lag) distance. Note the large scale distribution—the ACF falls to e^{-1} of its zero-lag value at approximately 1 km. Similar e^{-1} lengths for the dominant species are no more than 100 m. 95% confidence limits are shown.

CCF1. The cross correlation function (CCF) of the small chlorophyte with a less numerous diatom, *Fragilaria pinnata,* vs. distance. Note the significant positive correlation of the two species abundances when displaced by 1–2 km—large scale.

CCF2. The cross correlation function of the first chlorophyte with another relatively rare chlorophyte. Note the significant negative correlation between the two species abundances, again at large scale.

(Richerson et al. 1978; Abbott, Powell, and Richerson 1982). We have no ready explanation why the rarer species, significant contributors to lacustrine planktonic diversity, do seem to show some large-scale pattern, while the dominants do not. At least one model of phytoplankton competing for dissolved nutrients (Powell and Richerson 1985) predicts that the reaction-diffusion mechanism could lead to patterns at precisely these size scales in this large lake. It remains unexplained, however, why the dominant species, which are presumably also competing for nutrients, do not show these patterns as well. These remarks underscore an important conclusion: the spatial scale of community composition may differ from that of biomass variability, a result not yet entirely understood. (See also fig. 11.2a.) A solely empirical approach to this crucial question would involve a monumental effort to count planktonic species from net and bottle samples, an extremely tedious and time-consuming task because of our lack of automatic counting devices. An understanding of biology at the level of interacting species is critical to ecology, and we need to understand how species partition space (if indeed they do). Careful theoretical guidance is in order before we can go sailing off to count literally millions of plankters, consuming millions of dollars.

Smaller (< 100 m) and Larger (> 100 km) Scales

The nature of the patchy marine environment at very small scales (the scale of phytoplankton—100 m and smaller) has propelled a number of investigators to ask if phytoplankton can maintain themselves on the average level of, for example, nitrogen in the sea (McCarthy and Goldman 1979; Jackson 1980; Lehman and Scavia 1982). The question turns on the spatial and temporal scales of, first, physical variability in the ocean and, second, uptake and assimilation by algae. Everyone agrees that the average level of nitrogen is far too low to maintain the observed growth, and that there are bursts of high-nutrient water at sea. Are these bursts long-lived enough for the algae to take them up and utilize them for growth? The question remains unresolved, but at least one calculation (Jackson 1985) can account for all measurements, both in very oligotrophic waters, like the midocean, and in very eutrophic environments, like the laboratory vessels where some of these experiments were carried out. Theory contributed to this endeavor by calling into question the initial explanations of McCarthy and Goldman, forcing them and others to seek new explanations for the observed phenomena. Jackson's initial calculations were authoritative enough for researchers, especially J. Goldman, to seek other small-scale concentrating mechanisms, focusing on the role played by bacterial communities on small particles. The rapid and stimulating interplay between theoretical calculation and field measurement is particularly noteworthy, since the effect of small-scale variance and the nature of its presumed transfer to larger scales have extraordinary consequences for the entire large-scale marine ecosystem.

Similar remarks can be made about the small-scale details of zooplankton feeding that depend critically on the small-scale, low Reynolds number hydrodynamics—"the physics" (Koehl and Strickler 1981). The details of this feeding mechanism has not yet been incorporated into models for zooplankton foraging (e.g., Lehman 1976; Lam and Frost 1976); such a calculation would quantify how the small-scale behavior of individuals controls large-scale phenomena, since zooplankton grazers are believed to control phytoplankton standing crop in large parts of the sea and lakes. If this connection could be made, perhaps we would have a more solid basis on which to judge the alternative hypothesis that more control is exerted by previously unsuspected microzooplankton grazers, even bacteria, than by the traditional calanoid copepod zooplankton.

At very large scales—whole-ocean basin and global—one of the "linear, first-order" generalizations of Denman and Powell (1984) is seriously violated. In the coupled physical/biological systems, the dominant time scales are no longer controlled by biological processes. Interannual variability of weather patterns and even longer climatic variations are prominent. Figures 11.6a,b from Clark (1985)

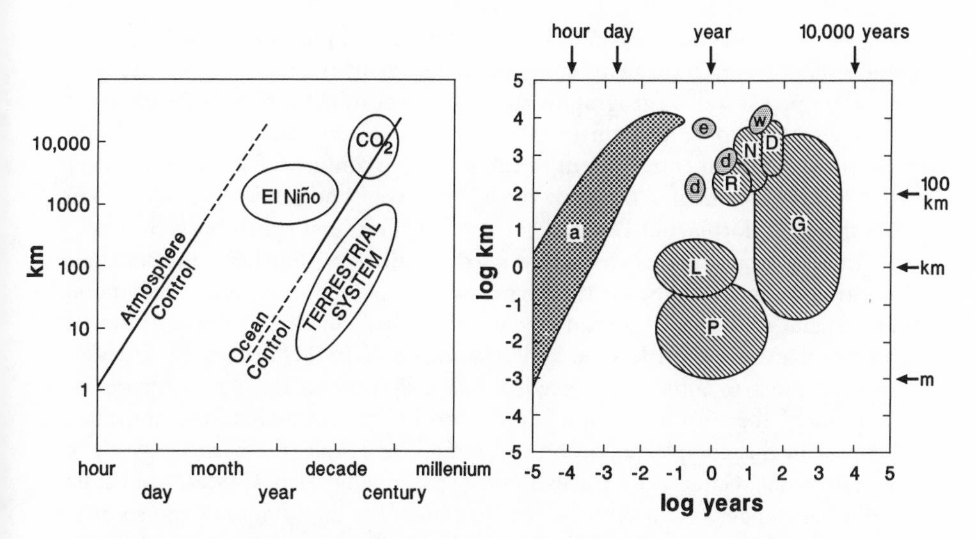

FIGURE 11.6. *Left:* Space-time regimes (see fig. 11.2, *right*). Events with large spatial scales generally have associated long time scales. The atmosphere exerts control over time scales of weeks or less. The ocean exerts control over years or more. (After Steele 1984.) *Right:* Space-time regimes, continued, including man's impact. (a) atmospheric phenomena; (e) El Niño; (d) drought; (w) global warming trend; (P) ecological processes at the population level; (G) ecological processes over larger geographical extents, i.e., at the landscape or regional level; (L) local farming activities; (R) regional farming activities; (N) national industrial activity; (D) changing global patterns of political and demographic activities. (After Clark 1985.)

and Steele (1984) present the space-time regimes of dominant processes on a diagram such as the one in figure 11.2b. We have excellent evidence that these large-scale events exert considerable influence on ecological processes. The impact of the 1983 El Niño on biological processes in the ocean has been reviewed by Barber and Chavez (1983, 1986). We have increasing evidence that these large-scale, long-term events are also coupled to shorter-term, small-scale variability as well. Strub, Powell, and Goldman (1985) documented the impact of the El Niño phenomenon over the previous twenty years on a small, subalpine lake in Northern California; Scavia et al. (1986) showed the impact of interannual variability on processes in Lake Michigan. The mechanisms generating these longer-term events are poorly known, if at all, and uncovering the details of their coupling to smaller-scale, shorter-duration processes promises to be an area of intense activity in the decades ahead. Clearly, theoretical activity will be focused on long-duration, large-scale regimes because relevant measurement programs have such very long time horizons. We simply cannot afford to wait for the data to pour in to determine what the future may bring for our planet.

Oceanic-Terrestrial Coupling

Figure 11.6 shows how oceanic events affect terrestrial processes at the large scale and over the long term. This coupling is important at smaller scales and over shorter times as well. A geographic zone where one might anticipate the effects of such coupling to be prominent is at the land-sea interface—at subtidal and intertidal environments. The empirical situation is confused. Dayton and Tegner (1984) have documented the effects of El Niño events on subtidal kelp forests in Southern California, but Paine (1986) sees few effects on the benthic community at Tatoosh Island, which he has studied for decades. Further, "intermediate disturbance" workers (Sousa 1984) have documented the importance of temporal and spatial scale in the intertidal zone. Additional coupled biological/physical models that expand on Jackson and Strathmann's (1981) effort would be a welcome adjunct to some much-needed data collection on the actual physics of "exposure," that is, the physical mechanism of energy, momentum, and mass transfer in this complex environment. On a larger scale, the barnacle work of Roughgarden, Gaines, and collaborators (e.g., Roughgarden, Gaines, and Pacala 1987) has signaled the need for a reevaluation of the importance of transport in recruitment processes. Space and time-scale considerations play a critical role in their studies and speculations; moreover, simple models coupled to detailed experimental studies are a hallmark of this group's influential work. Their concerns are mirrored in the work of Botsford and collaborators on the Dungeness crab (*Cancer magister*). Again, critical timing of transport of larval stages is thought to be controlled by wind-driven transport in the upper mixed layer along the entire west coast of North America. These authors argue that cyclic variations in wind stress lead to the cyclical nature of this process (Johnson et al. 1986). An interesting aspect of Botsford's work is the application of a theoretical model to dismiss a competing hypothesis, namely, that the cycles can be explained solely by the density-dependence in the predator-prey dynamics within larval crab populations. These studies point to a coupling of events at different scales in addition to aquatic-terrestrial coupling.

TERRESTRIAL SYSTEMS AND TROPHIC LEVELS: CONCLUSION

It is now useful to note some aspects of terrestrial and marine systems as separate entities. For example, there are Steele's (1985) model studies, which have pointed out that the two systems should differ at the longest scales because of the very different climates each encounters. And we can compare the two different systems at shorter time and smaller space scales. Kareiva (e.g., 1982, 1984, 1986) has made effective use of models that apply equally to plankton systems in his insect studies. A more formal comparison between models of dispersing, interacting populations of insects and dispersing, interacting planktonic populations also bears consideration. One important difference between the models concerns dispersal: insect dispersal depends on the population level (Okubo 1980), and plankton are at the mercy of water movements, resulting in less behavioral control over their "dispersal."

Interesting comparisons could also be made between some of the plankton patchiness work and the recent studies of Fahrig and collaborators (e.g., Fahrig 1988; Fahrig and Paloheimo 1987). In these, the arrangement of patches in the insect communities might be akin to the "arrangement" that the physical environment "performs" on plankton. Since dispersal distances scale both the insect and plankton problems, some comparison could be made at the same (and/or different) *dynamical* scales for both systems.

Perhaps even more intriguing are the recent supercomputer analyses of Levin and Buttel (1988). Following a patch model of competing intertidal organisms (Levin and Paine 1974), Levin and Buttel showed that the population variance *decreased* as size scale increased. This behavior is profoundly different from that seen in aquatic communities, where there is much more variance at larger scale. (See figs. 11.1 and 11.4.) One might expect that in intertidal communities, as one begins to encompass increasingly different habitats, the population variance must increase at some size scale. Is there a scale where a mimimum occurs? The details of the transfer of variance between scales in the Levin/Buttel model may be instructive. Lake/pond habitats may also provide interesting comparisons. They are subject to climatic fluctuations (i.e., atmospheric time scales) to some degree, and yet they exhibit many of the aspects of variability seen in marine systems (Richerson et al. 1978).

I have avoided discussing coupling between trophic levels in deference to other authors in this volume who have, no doubt, forgotten more about this topic than I may ever know. One subject, however, has attracted attention from freshwater ecologists. Following the ideas of Paine (1966), model investigations by Carpenter and Kitchell (1984) and later field investigations (Carpenter et al. 1987) have demonstrated that higher vertebrate predators can exert considerable control over primary productivity at the lowest autotrophic level in small lake systems—top-down control. Other model studies (Carpenter 1987) suggest that through the "top-down" mechanism significant variance will appear in primary productivity at the time scale associated with the generation time of the dominant predator—a reasonable, if unsupported, assertion. The top-down suggestion does jostle the conventional wisdom on this issue—the large body of knowledge that has demonstrated the control of primary production by environmental fluctuations (e.g., Harris 1986; Harris and Trimbee 1986). Note, however, that the time scales of environmental fluctuation are surely linked to the time scales of predator abundance—perhaps through large-scale die-offs or migrations and introductions. Both "top-down" and environmental control must be at work in all lakes, though the degree to which each dominates will surely differ from lake to lake. The time scales of interest are long in this case, and further theoretical analyses are certain to sharpen and restrict the applicability of the top-down hypothesis before enough data will be available to decide the issue.

I conclude by reminding the reader of another kind of coupling—the productive coupling between theory and field measurement and experimentation in all of the areas I have surveyed. The importance of theoretical suggestion (and refutation) cannot be overestimated, especially as we increasingly focus on large-scale, long-term processes for which measurement programs may be so vast and time consuming that we can investigate some questions only theoretically. It is difficult to conceive of further progress on the questions I have raised without a substantial infusion of theoretical guidance.

ACKNOWLEDGMENTS

Andrew Bennett, Ken Denman, John Steele, and especially Dave Mackas gave me extensive, thoughtful, and good-humored comments. My reliance on their work best expresses my gratitude. Bennett and Denman pointed out an error in the Gower et al. (1980) study: accordingly, one should compare spectral densities to a k^{-2} form, not k^{-3} as in the original article. John Steele made two figures available; he also composed a masterful and helpful summary (chapter 12) from a wide-ranging discussion in which the strong views of individual conference participants continually threatened to overwhelm the search for consensus. Lenore Fahrig made her work available in advance of publication, as did Bennett and Denman. Conversations with Peter Kareiva, Bob Paine, and Wayne Sousa were especially useful. Over the fifteen years I have investigated these and closely related questions, the work of Si Levin, Akira Okubo, Trevor Platt, and John Steele has inspired and challenged me; it is a pleasure to acknowledge them here.

This work has been supported by the National Science Foundation through grant OCE-8613749 to Project SUPER.

REFERENCES

Abbott, M. R., T. M. Powell, and P. J. Richerson. 1982. The relationship of environmental variability to the spatial patterns of phytoplankton in Lake Tahoe. *J. Plankton Res.* 4:927–41.

Barber, R. T., and F. P. Chavez. 1983. Biological consequences of El Niño. *Science* 222:1203–10.

Barber, R. T., and F. P. Chavez. 1986. Ocean variability in relation to living resources during the 1982–1983 El Niño. *Nature* 319:279–85.

Bennett, A. F., and K. L. Denman. 1985. Phytoplankton patchiness: Inferences from particle statistics. *J. Mar. Res.* 43:307–35.

Bennett, A. F., and K. L. Denman. 1988. Large-scale patchiness due to an annual cycle. *Trans. Amer. Geophys. Union* (EOS). 68:1696. Abstract only, and unpublished manuscript.

Berg, H. 1983. *Random Motion in Biology.* Princeton University Press, Princeton, N.J.

Carpenter, S. R. 1987. Transmission of variance through lake food webs. Unpublished manuscript for workshop on complex interactions in lake communities. Dept. of Biol. Sci., University of Notre Dame. March 1987.

Carpenter, S. R., and J. F. Kitchell. 1984. Plankton community structure and limnetic primary production. *Amer. Natur.* 124:159–72.

Carpenter, S. R., J. F. Kitchell, J. R. Hodgson, P. A. Cochran, J. J. Elser, D. M. Lodge, D. Kretchmer, and X. He. 1987. Regulation of lake primary productivity by food-web structure. *Ecology* 68:1863–76.

Chatfield, C. 1980. *Introduction to Multivariate Analysis.* Chapman and Hall, New York.

Chatfield, C. 1984. *The Analysis of Time Series,* 3d ed. Chapman and Hall, New York.

Clark, W. C. 1985. Scales of climate impacts. *Climatic Change* 7:5–27.

Dayton, P. K., and M. J. Tegner. 1984. Catastrophic storms, El Niño, and patch stability in a Southern California kelp forest. *Science* 224:283–85.

Denman, K. L. 1983. Predictability of the marine planktonic ecosystem. In G. Holloway and B. J. West, eds., pp. 601–602. *The Predictability of Fluid Motions.* American Institute of Physics, Conference Proceedings No. 106, New York.

Denman, K. L., and M. R. Abbott. 1988. Time evolution of surface chlorophyll patterns from cross-spectrum analysis of satellite color images. *J. Mar. Res.* In press.

Denman, K. L., and D. L. Mackas. 1978. Collection and analysis of underway data. In

J. H. Steele, ed., *Spatial Pattern in Plankton Communities,* pp. 85–109. Plenum, New York.

Denman, K. L., and T. Platt. 1976. The variance spectrum of phytoplankton in a turbulent ocean. *J. Mar. Res.* 34:593–601.

Denman, K. L., and T. M. Powell. 1984. Effects of physical processes on planktonic ecosystems in the coastal ocean. *Oceanogr. Mar. Biol. Ann. Rev.* 22:125–68.

Denman, K. L., A. Okubo, and T. Platt. 1977. The chlorophyll fluctuation spectrum in the sea. *Limnol. Oceanogr.* 22:1033–38.

Fahrig, L. 1988. A general model of populations in patchy habitats. *J. Appl. Math. and Comp.* In press.

Fahrig, L., and J. E. Paloheimo. 1987. Determinants of local population abundance in patchy habitats. Unpublished manuscript.

Gauch, H. G. 1982. *Multivariate Analysis in Community Ecology.* Cambridge University Press, Cambridge, England.

Gill, A. E. 1982. *Atmosphere-Ocean Dynamics.* Academic Press, New York.

Gower, J.F.R., K. L. Denman, and R. J. Holyer. 1980. Phytoplankton patchiness indicates the fluctuation spectrum of mesoscale oceanic structure. *Nature* 288:157–59.

Harris, G. P. 1986. *Phytoplankton ecology: Structure, function, and fluctuation.* Chapman and Hall, London.

Harris, G. P., and A. M. Trimbee. 1986. Phytoplankton population dynamics of a small reservoir: Physical/biological coupling and the time scales of community change. *J. Plankton Res.* 8:1011–25.

Jackson, G. A. 1980. Phytoplankton growth and zooplankton grazing in oligotrophic oceans. *Nature* 284:439–41.

Jackson, G. A. 1985. Pulses, aggregates, and diffusion: Life in the small. Forty-eighth annual meeting Amer. Soc. Limnol. Oceanogr., Minneapolis. Abstract only.

Jackson, G. A., and R. R. Strathmann. 1981. Larval mortality from offshore mixing as a link between precompetent and competent periods of development. *Amer. Natur.* 118:16–26.

Johnson, D. F., L. W. Botsford, R. D. Methot, Jr., and T. C. Wainwright. 1986. Wind stress and cycles in Dungeness crab (*Cancer magister*) catch off California, Oregon, and Washington. *Can. J. Fish. Aquat. Sci.* 43:838–45.

Kareiva, P. 1982. Experimental and mathematical analyses of herbivore movement: Quantifying the influence of plant spacing and quality on foraging discrimination. *Ecol. Monogr.* 52:261–82.

Kareiva, P. 1984. Predator-prey dynamics in spatially structured populations: Manipulating dispersal in a coccinellid-aphid interaction. In S. A. Levin and T. G. Hallam, eds., *Mathematical Ecology.* Lecture Notes in Biomathematics, vol. 54. Springer-Verlag, Berlin.

Kareiva, P. 1986. Patchiness, dispersal and species interactions: Consequences for communites of herbivorous insects. In J. Diamond and T. J. Case, eds., *Community Ecology,* pp. 192–206. Harper and Row, New York.

Koehl, M.A.R., and J. R. Strickler. 1981. Copepod feeding currents: Food capture at low Reynolds number. *Limnol. Oceanogr.* 26:1062–73.

Lam, R. K., and B. W. Frost. 1976. Model of copepod filtering response to changes in size and concentration of food. *Limnol. Oceanogr.* 21:490–500.

Lehman, J. T. 1976. The filter feeder as an optimal forager, and the predicted shapes of feeding curves. *Limnol. Oceanogr.* 21:501–16.

Lehman, J. T., and D. Scavia. 1982. Microscale patchiness of nutrients in plankton communities. *Science* 216:729–30.

Levin, S. A., and L. Buttel. 1988. Measures of patchiness in ecological system. *Ecology.* In press.

Levin, S. A., and R. T. Paine. 1974. Disturbance, patch formation, and community structure. *Proc. Natl. Acad. Sci. USA* 71:2744–47.

McCarthy, J. J., and J. C. Goldman. 1979. Nitrogenous nutrition of marine phytoplankton in nutrient-depleted waters. *Science* 203:670–72.

Mackas, D. L. 1984. Spatial autocorrelation of plankton community composition in a continental shelf ecosystem. *Limnol. Oceanogr.* 29:451–71.

Mackas, D. L., K. L. Denman, and M. R. Abbott. 1985. Plankton patchiness: Biology in the physical vernacular. *Bull. Mar. Sci.* 37:652–74.

Mimura, M., Y. Kan-on, and Y. Nichiura. 1988. Oscillations in segregation of competing populations. In T. G. Hallam, L. J. Gross, and S. A. Levin, eds., *1986 Proceedings of the Trieste Research Conference on Mathematical Ecology,* pp. 711–17. World Scientific Publishing, Singapore. In press.

Murray, J. D., and G. F. Oster. 1984. Generation of biological pattern and form. Institute of Mathematics and Its Applications. *J. Math. Appl. Med. and Biol.* 1:51–75.

Okubo, A. 1980. *Diffusion and Ecological Models.* Springer-Verlag, New York.

Okubo, A. 1984. Critical patch size for plankton and patchiness. In S. A. Levin and T. G. Hallam, eds., *Mathematical Ecology,* pp. 456–77. Lecture Notes in Biomathematics, vol. 54. Springer-Verlag, Berlin.

Paine, R. T. 1966. Food web complexity and species diversity. *Amer. Natur.* 100:65–75.

Paine, R. T. 1986. Benthic community—water column coupling during the 1982–83 El Niño: Are community changes at high latitudes attributable to cause or coincidence? *Limnol. Oceanogr.* 31:351–60.

Pedlosky, J. 1979. *Geophysical Fluid Dynamics.* Springer-Verlag, New York.

Platt, T. 1978. Spectral analysis of spatial structure in phytoplankton populations. In J. H. Steele, ed., *Spatial Pattern in Plankton Communities,* pp. 73–84. Plenum Press, New York.

Platt, T., K. H. Mann, and R. E. Ulanowicz, eds. 1981. *Mathematical Models in Biological Oceanography. UNESCO* Press, France.

Powell, T., and P. J. Richerson. 1985. Temporal variation, spatial heterogeneity, and competition for resources in plankton systems: A theoretical model. *Amer. Natur.* 125:431–64.

Purcell, E. M. 1977. Life at low Reynolds numbers. *Amer. J. Phys.* 45:3–11.

Richerson, P. J., T. M. Powell, M. R. Leigh-Abbott, and J. A. Coil. 1978. Spatial heterogeneity in closed basins. In J. H. Steele, ed., *Spatial Pattern in Plankton Communities,* pp. 239–76. Plenum Press, New York.

Roughgarden, J. 1979. Theory of population genetics and evolutionary ecology: An introduction. Macmillan, New York.

Roughgarden, J., S. Gaines, and S. Pacala. 1987. Supply side ecology: The role of physical transport processes. In P. Giller and J. Gee, eds., *Organization of Communities: Past and Present.* Proc. Brit. Ecol. Soc. Symp., Aberystwyth, Wales, April 1986. Blackwell Scientific, London.

Scavia, D., G. Fahnenstiel, M. Evans, D. Jude, and J. Lehman. 1986. Influence of salmonid predation and weather on long-term water quality in Lake Michigan. *Can. J. Fish. Aquat. Sci.* 43:435–43.

Sheldon, R., A Prakash, and W. Sutcliffe. 1972. The size distribution of particles in the ocean. *Limnol. Oceanogr.* 17:327–40.

Shugart, H. H., ed. 1978. *Time Series and Ecological Processes. SIAM,* Philadelphia.

Sousa, W. P. 1984. The role of disturbance in natural communities. *Ann. Rev. Ecol. Syst.* 15:353–91.

Steele, J. H. 1974. *The Structure of Marine Ecosystems.* Harvard University Press, Cambridge, Mass.

Steele, J. H. 1978. Some comments on plankton patches. In J. H. Steele, ed., *Spatial Patterns in Plankton Communities,* pp. 1–20. Plenum Press, New York.

Steele, J. H. 1984. Long-range plans. Woods Hole Oceanographic Institution, 1984 Annual Meeting. Woods Hole, Mass.

Steele, J. H. 1985. A comparison of terrestrial and marine ecological systems. *Nature* 313:355–58.

Steele, J. H., and E. W. Henderson. 1977. Plankton patches in the northern North Sea. In J. H. Steele, ed., *Fisheries Mathematics,* pp. 1–19. Academic Press, New York.

Strub, P. T., T. M. Powell, and C. R. Goldman. 1985. Climatic forcing: Effects of El Niño on a small, temperate lake. *Science* 227:55–57.

Ulanowicz, R. E., and T. Platt, eds. 1985. Ecosystem theory for biological oceanography. *Can. Bull. Fish. Aquat. Sci.* 213:260.

Vogel, S. 1981. *Life in Moving Fluids.* Princeton University Press, Princeton, N.J.

Williamson, M. 1972. The relation of principal component analysis to the analysis of variance. *Int. J. Math. Educ. Sci. Technol.* 3:35–42.

Williamson, M. 1975. The biological interpretation of time series analysis. *Bull. Inst. Math. Appl.* 11:67–69.

Chapter 12

Discussion: Scale and Coupling in Ecological Systems

JOHN H. STEELE

The first discussion on hierarchies at the conference focused on two questions: (1) What kinds of systems are hierarchies? and (2) Are they the basis for theories or rather for the ordering of concepts?

The traditional hierarchy is the organism, patch, population, and community. This interacts with a second structure—a population or community, its food supply, and the predators within it. Several participants pointed out that in understanding these interactions, we cannot operate only at the level of the average population density; we must consider how the individual organism behaves when feeding and when avoiding predators. Especially we need to consider the effects of patchiness, which not only affect the statistical variability but have a profound effect on the nature of the dynamic interactions.

A major point of emphasis during the discussion was the reaction of both of these categories to spatial and temporal scales. Many of the participants stressed that the definition of "scales" is a major necessity in any ecological study. It is apparent that inherent scales are associated with the elements in each of these hierarchies, but there can be several different scales for each interaction between these elements. For example, each predator will first have a behavioral interaction with each prey organism, then a functional response with a patch of prey, and finally a numerical response to the total population. Each of these responses involves different space and time scales. The importance of nonequilibrium patch dynamics was pointed out. At certain scales the system can appear to be far from equilibrium. This aspect does not preclude equilibrium on larger space and time scales, but it can determine the nature of that equilibrium state. These concerns led to a more general discussion of coupling between scales. Any interaction between groups (i.e., prey to predator) implies a relation between different space and time scales. Further, such relations are nonlinear so that the consequences of the interactions can be seen at yet other scales. Thus the physical cascade of turbulence in aquatic systems leads to redistribution of variance to smaller scales, but the ecological system can, potentially, reverse that direction and transfer biological variance to larger scales. The problems of constructing adequate theories were discussed. They arise because of the differences between physical systems, especially fluid ones in which distribution of variance between scales is relatively clear, and biological ones where the process of transfer between scales is still not well understood.

The consideration of scale transfer of variance led to a discussion of how "smooth" these processes were. The importance of episodic events as a dominant factor changing community structure was noted. It was also pointed out that spatial distributions are not random but display "mosaics" (i.e., irregular patterns), whether on land or sea. Thus scale effects cannot be considered as purely random sequences determined by power spectra or autocorrelations. A plea was made to develop methods to incorporate pattern into the definition of scale processes and to develop a theory of landscape mosaics. Patch models were noted as a first step in this direction, but other promising approaches were mentioned, e.g., percolation theory and fractal measures.

The question of whether "hierarchy theory" is really a theory was raised repeatedly. The discussion emphasized the need to develop it from a conceptual framework to a more powerful source of hypotheses, including the development of a more mathematical basis. A number of uses were cited—microcosm experiments, field studies at different scales, development of "neutral" models.

The discussion to this point showed agreement on the following: (1) the usefulness of having a "hierarchal" framework as a basis for discussing ecological problems; (2) the necessity of introducing space and time scales as essential parameters; and (3) the range and complexity of the consequent interactions.

To illustrate how ecological theories can be depicted, we considered three "dimensions" (fig. 12.1): the species units (x), the trophic interactions (y), and the scales (z). Although ecological interaction is normally portrayed in a space/time domain (e.g., chapter 11, fig. 11.1), we lumped them here into one axis, partly to limit the presentation to three "dimensions," and partly because space and time scales tend to increase together, making the second scale largely redundant in this simple presentation.

This representation is not only a useful way to portray the hierarchy scale relation, but, more important, the volume within these three axes appears to be well filled by all the possible relationships between organisms. In this sense, the papers by O'Neill (chapter 10) and Powell (chapter 11), and the discussions, capture the major aspects of ecological systems and problems within these three dimensions. However, the results do not reduce to a line or even to a surface within this volume. One can argue—and we did—whether this approach is a description or a theory. This is a question that cannot yet be answered. The concensus appeared to be, however, that a representation of ecological questions in this format, as shown in figure 12.1, is a useful way to structure the overall system and then determine how the system could be directed into conceptually (theoretically) and logistically (experimentally) manageable components. (As in geophysical fluid dynamics, we assume that even if we knew the basic equations, the total system would be insoluble in a numerical sense, and, even if it were possible, a computer "solution" could be "opaque" conceptually.)

For both the mathematical and practical aspects, the critical question is the same: Are there surfaces within this volume across which there is relatively little transfer? In other words, is the natural world divisible into separable, bounded regions within this space? This question affects our design or selection of experimental results. (A later discussion centered around whether a small number of "canonical" experiments is sufficient.)

Within this framework, theories can be regarded as surfaces, usually bounded, within this volume. For example, food webs, optimal foraging models, or spectral theories could be represented as surfaces intersecting the overall system.

Similarly, data sets are a sampling of this space. In our discussion, special attention was given to the need for samples covering a large range of scales (z). The fact that the world is spatially mosaic and episodic at all scales (i.e., fractal or

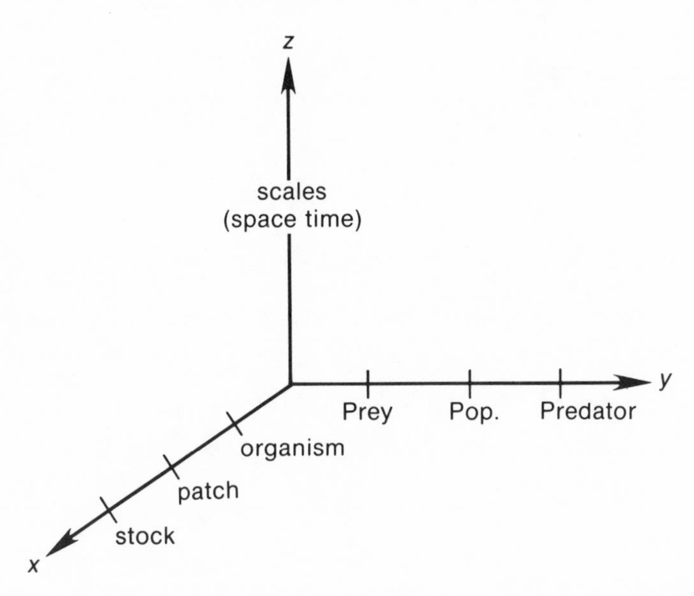

FIGURE 12.1. A representation of the three "dimensions" in which ecological theories can be depicted.

rough) requires that we sample along the z-axis as fully as possible. Such a sampling protocol is very different from that of geographic-grid sampling. At present, only remote-sensed (e.g., satellite) data can provide this coverage of terrestrial, biological systems at the data densities needed by some theories. Such data sets can be regarded as lines through the space in figure 12.1 which can be projected onto the theoretical surfaces. Thus the critical problem is to make such "lines" longer.

The utility of verbal concepts such as hierarchy was another topic in this discussion. The point was made that such concepts are a useful and possibly a necessary link between experimental studies and mathematical theories. Initial syntheses about ecology may be expressed verbally (as with Darwinism and plate tectonics). Verbally formulated theory is then improved in scope and testability when mathematics is added.

In summary, the consensus was that issues of scale and hierarchy are important in ecology. Although important empirical problems in collecting data on a wide range of scales remain (only remote sensing methods are capable of collecting data on some of the important large scales, and such data are extremely limited in what they can measure), there are even more pressing theoretical problems. The development of theoretical models which incorporate multiple scales, and which will guide the collection and interpretation of data, is a major and exciting theoretical task.

STRUCTURE AND ASSEMBLY OF COMMUNITIES V

Chapter 13

Food Webs and Community Structure

JOEL E. COHEN

A central problem of biology is to devise helpful concepts (e.g., genes) and tested quantitative models (e.g., Mendel's laws) to describe, explain, and predict biological variation. The problem of characterizing variation arises in different guises in population genetics (genetic variation), demography (variation by age, sex, location, etc.), epidemiology (variation by risk factors and disease status), and ecology (variation in species composition and interactions in communities). In each field, there is variation over time, in space, and among units of observation (individuals, populations, or comparable habitats).

This paper reviews some recent efforts to describe, explain, and predict variation in the food webs of ecological communities. There are many notions of an ecological community and many approaches to describing and understanding community ecology. Panoramic reviews of community ecology are available (e.g., Diamond and Case 1986; Kikkawa and Anderson 1986; National Research Council 1986). For present purposes, a community is whatever lives in a habitat (lake, forest, sea floor) that some ecologist wants to study.

Once the physical boundaries of a habitat are defined, it is natural to study flows of matter and energy across and within the boundaries. A partial description of these flows is provided by food webs, which used to be called food cycles (Elton 1927).

A food web describes which kinds of organisms in a community eat which other kinds, if any. A community food web (hereafter simply "web") describes the feeding habits of a set of organisms chosen on the basis of taxonomy, location, or other criteria without prior regard to the feeding habits among the organisms. Webs were invented in the natural-historical approach to community ecology as a descriptive summary of which species were observed to eat which others.

If an ecological community is like a city, a web is like a street map of the city: it shows where road traffic can and does go. A street map usually omits many important details, for example, the flow of pedestrian and bicycle traffic, the amount of traffic that flows along the available streets, the kind of vehicular traffic it is, the reasons for the traffic, the laws governing traffic flow, rush hours, and the place of origin of the vehicles. By analogy, a web often omits small flows of food or predation on minor species, the quantities of food or energy consumed, the chemical composition of food flows, the behavioral and physical constraints on predation, temporal variations (periodic or stochastic) in eating, and the population dynamics of the species involved. Thus a web gives at best very sketchy information about the functioning of a community. But just as a map provides a helpful framework for organizing more detailed information, a web helps picture how a community works.

Many approaches to studying webs are available. I will not attempt here a comprehensive review of food webs, since such reviews are available (see Pimm 1982 and in press; DeAngelis, Post, and Sugihara 1983). A difference of temperament, training, and language seems to divide those who prefer to study webs in physical and chemical terms (e.g., Lotka 1925; Lindemann 1942; Wiegert 1976; Budyko 1980; Margalef 1984; Remmert 1984) from those who prefer to study webs in terms of the natural history of species of living organisms (e.g., many authors in the collection by Hazen 1964). Here "natural history" comprises morphological, genetic, physiological, behavioral, and demographic characteristics of species. Recent natural-historical approaches have focused on combinatorial aspects of web structure (Cohen 1978; Sugihara 1982, 1983, 1984), on the theory of interactions between web structure and the stability of dynamic models (May 1973; Pimm and Lawton 1977; Pimm 1982, 1984; Sugihara 1982), and on empirical generalizations (Paine 1980; Briand 1983a,b; Beaver 1983, 1985).

Fortunately, nature is serenely indifferent to the prejudices ecologists bring her. It will eventually be necessary to integrate the physico-chemical and natural-historical approaches to community ecology. I hope that the food-web models reviewed here will help bring about that integration.

This paper reviews some recent discoveries about webs, suggests opportunities for further empirical and theoretical study, and sketches some uses for actual and potential knowledge about webs. I attempt to describe in a simple way some recent discoveries that, in their original presentations, may appear forbiddingly technical, and to place these discoveries in a larger scientific and practical setting.

So far as I know, webs were first described in scientific detail at the beginning of this century. Simplifications that they were, the webs appeared extremely complex relative to the concepts available for understanding them. The webs differed strikingly from one habitat to another. Now enough webs have been patiently observed and recorded to demonstrate that ensembles or collections of webs display simple general properties that are not evident from any single web.

Building on a collection of webs that I initiated (Cohen 1978), F. Briand assembled and edited 113 community webs from 89 distinct published studies. Thus many field ecologists contributed to the discoveries reviewed here. Most of the world's biomes are represented among these webs. There are 55 continental (23 terrestrial and 32 aquatic), 45 coastal, and 13 oceanic webs, ranging from arctic to antarctic regions. The sources and major characteristics of these webs are listed by Briand and Cohen (1987). Forty have been fully documented (Briand 1983a) and the remainder soon will be (Cohen, Briand, and Newman, in press).

In what follows, I will illustrate what a web is and how a web is described. I will present some recent quantitative empirical generalizations about webs. Then I will present a simple model, called the "cascade model," that unifies the quantitative generalizations. Though this model does not purport to represent everything field ecologists know is happening in webs, no other model at present connects and explains quantitatively what is observed. The cascade model also makes novel predictions that can be tested. Then I describe problems from other parts of ecology that can be analyzed using the cascade model and the facts on which it is based. Finally, I sketch some potential uses of facts and theories about webs.

TERMS

Let me introduce some terms and illustrate them with an example. A trophic species is a collection of organisms that has the same diets and the same predators. This definition combines Sugihara's definitions (1982, p. 19) that resources are trophically equivalent if they have identical consumers and that consumers are trophically equivalent if they have identical resources. A trophic species will sometimes, but not always, be a biological species in the usual sense of biological species: a collection of organisms with shared genetics. A trophic species may be a biological species of plant or animal, or several species, or a stage in the life cycle of one biological species. Hereafter the word "species" without further specification means "trophic species."

Independently of Sugihara (1982, p. 19), Briand and I (1984) introduced the concept of trophic species to find out if there was merit in a criticism that Pimm (1982, p. 168) made of my earlier finding (Cohen 1977), that webs generally had about four (biological) species of predators for every three (biological) species of prey. Pimm suggested that ecologists distinguish among species with fur or feathers, which are likely to be consumers, more often than among species with more difficult taxonomy, such as many plants, microorganisms, and insects, which are likely to be consumed. The excess of predators, he suggested, could be an artifact of the interests and knowledge of ecologists.

To test that possibility, Briand and I devised an automated lumping procedure that puts together those biological species or other biological units of a web that eat the same kinds of prey and have the same kinds of predators. We call each equivalence class that results from such lumping a trophic species. Our intent was to apply a uniform rule to distinguishing among the units of a web to see if this uniform rule altered the ratio of predators to prey. Indeed it did! A slight excess of predators remains, but the ratio of predators to prey counting lumped or trophic species (Briand and Cohen 1984) is much nearer 1 : 1 than the ratio based on the original data. Pimm's criticism had merit. We believe that using trophic species, as we shall do henceforth in this paper, corrects a bias of ecologists and gives a more realistic picture of the trophic structure of communities.

A web is a collection of trophic species, together with their feeding relations. Each arrow in a web goes from food to eater, or from prey to predator. I call each arrow a "link," short for "trophic link."

Figure 13.1 pictures the unlumped web on an island in the Pacific Ocean. Some species are top, meaning that no other species in the web eats them, for example, reef heron, starlings. Some species are intermediate, meaning that at least one species eats them, and they eat at least one species, for example, insects, skinks, fish. Some species are basal, meaning that they eat no other species, for example, algae, phytoplankton. The web omits decomposers. A crude way to

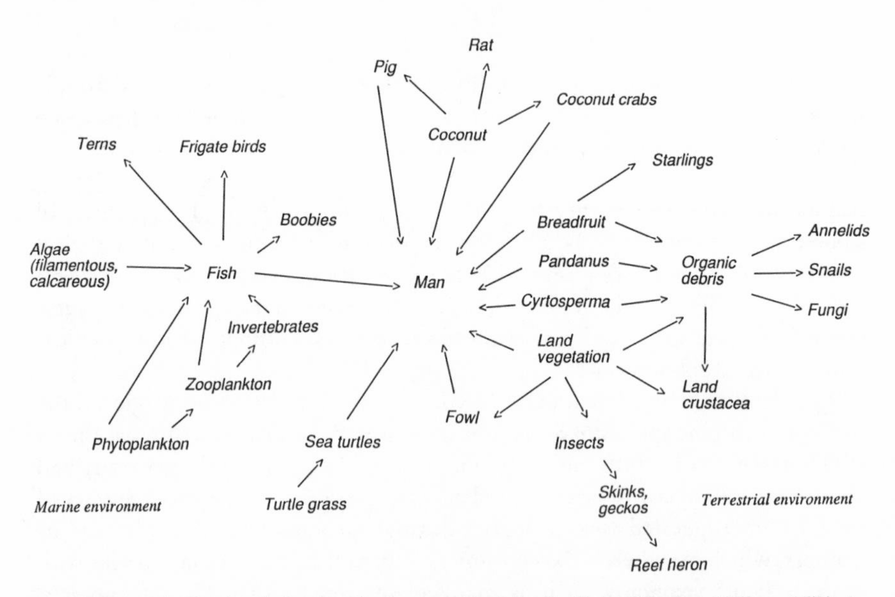

FIGURE 13.1. Food web in the Kapingamarangi Atoll. (From Niering 1963, p. 157.) As reported by Niering, the biological units in this figure range taxonomically from individual biological species (man, pig) to very large aggregates of species (phytoplankton, land vegetation), and do not correspond to trophic species.

quantify the structure of webs is to count the numbers of species that are top, intermediate, and basal.

These three kinds of species specify four kinds of links: basal-intermediate links, for example, phytoplankton to zooplankton; basal-top links, for example, coconut to man; intermediate-intermediate links, for example, zooplankton to fish; and intermediate-top links, for example, fish to frigate birds. Additional information about structure is given by the numbers of links of each of these four kinds.

A chain is a path of links from a basal species to a top species, e.g., phytoplankton to fish to terns. The length of a chain is the number of links in it. In figure 13.1 the longest chain has only four links, and there is only one chain of length 4.

A cycle is a directed sequence of one or more links starting from, and ending at, the same species. A cycle of length 1 describes cannibalism, in which a species eats itself. Cannibalism is common in nature. But ecologists report cannibalism so unreliably that we have suppressed it from all the data even where it is reported. A cycle of length 2 means that A eats B and B eats A. In this example, as in most webs, there are no cycles of length 2 or more.

In summary, the terms just defined are trophic species, including top, intermediate and basal; links, including basal-intermediate, basal-top, intermediate-intermediate, and intermediate-top; and chains, length (the number of links), and cycles.

In what follows, the terms "observed web" or "real web" mean a web edited to eliminate obvious errors, inconsistencies, and oversights, in which the original ecologist's biological units are replaced by trophic species. Are such webs really "real"?

Clearly the processed data are more constrained by reality than, for example, webs constructed a priori as model ecosystems. As a relative term, "real" means not that the data are perfect, but that they are not invented.

I think it may eventually be possible to claim much more for edited webs based on trophic species. By analogy, chemists have learned that it is more useful and economical to describe chemical "reality" in terms of chemical elements, which were once considered hypothetical, than in terms of gross phenomenology like color, taste, and density. Geneticists have learned that it is more useful and economical to describe the factors affecting inheritance in terms of genes, which were once considered hypothetical, than in terms of the gross phenomenology of certain macroscopic characters. I suggest that a web in which the units are trophic species may prove to be a more useful and economical description of the trophic organization of ecological communities than a description in terms of taxonomic phenomenology. Whether trophic species are closer to reality than the full glory of a naturalist's notebook will have to be determined by the eventual usefulness of the empirical and theoretical generalizations that develop using trophic species.

LAWS

Here are five laws or empirical generalizations about webs.

First, excluding cannibalism, cycles are rare. This generalization, without detailed supporting data, has been known for a long time (e.g., Gallopin 1972). Of 113 webs, 3 webs each contain a single cycle of length 2, and there are no other cycles (Cohen and Newman 1985, p. 426; Cohen, Briand, and Newman 1986, p. 333).

The rarity of cycles is not an artifact of using trophic species instead of the original units of observation, such as biological species, size classes, or aggregates of species. The reason is that the lumping procedure does not alter the connectivity of the web: the trophic species containing unit A is trophically linked to the trophic species containing unit B if and only if A was originally trophically linked to B. It follows that any cycle present in the original web must be represented by a cycle of the same length in the lumped web. Therefore, excluding cannibalism, if 110 of 113 lumped webs have no cycles, then 110 of the original webs had no cycles. The remaining three of the original webs had no cycles longer than length 2. There is no evidence that cycles occur in more webs if biological species are used instead of trophic species.

Second, chains are short (Hutchinson 1959). If one finds the maximum chain length within each web, then the median of this maximum in the 113 webs studied by Cohen, Briand, and Newman (1986) is four links and the upper quartile of the maximum chain length is five links. The longest chains in all 113 webs had ten links, and only one web had chains that long.

The last three laws deal with scale invariance (Cohen 1977; Briand and Cohen 1984; Cohen and Briand 1984). Scale invariance means that webs of different size have constant shape, in some sense.

Our third law is scale invariance in the proportions of all species that are top species, intermediate species, and basal species (fig. 13.2). There is evidently no increasing or decreasing trend in these proportions as the number of species increases (Briand and Cohen 1984). Here scale invariance describes the observation that as the number of species in 62 webs varies from 0 to 33, the proportions of top, intermediate, and basal species apparently remain invariant. This scale invariance explains my earlier observation (Cohen 1977) that the ratio of number of predators to number of prey has no systematic increasing or decreasing trend when webs with different numbers of species are compared. The number of predators is the sum of the numbers of top plus intermediate species, while the number of prey is the sum of the numbers of intermediate plus basal species. Mithen and Lawton (1986) and Tilman (1986) have developed other explanations for the same finding.

Our fourth law is scale invariance in the proportions of the different kinds of

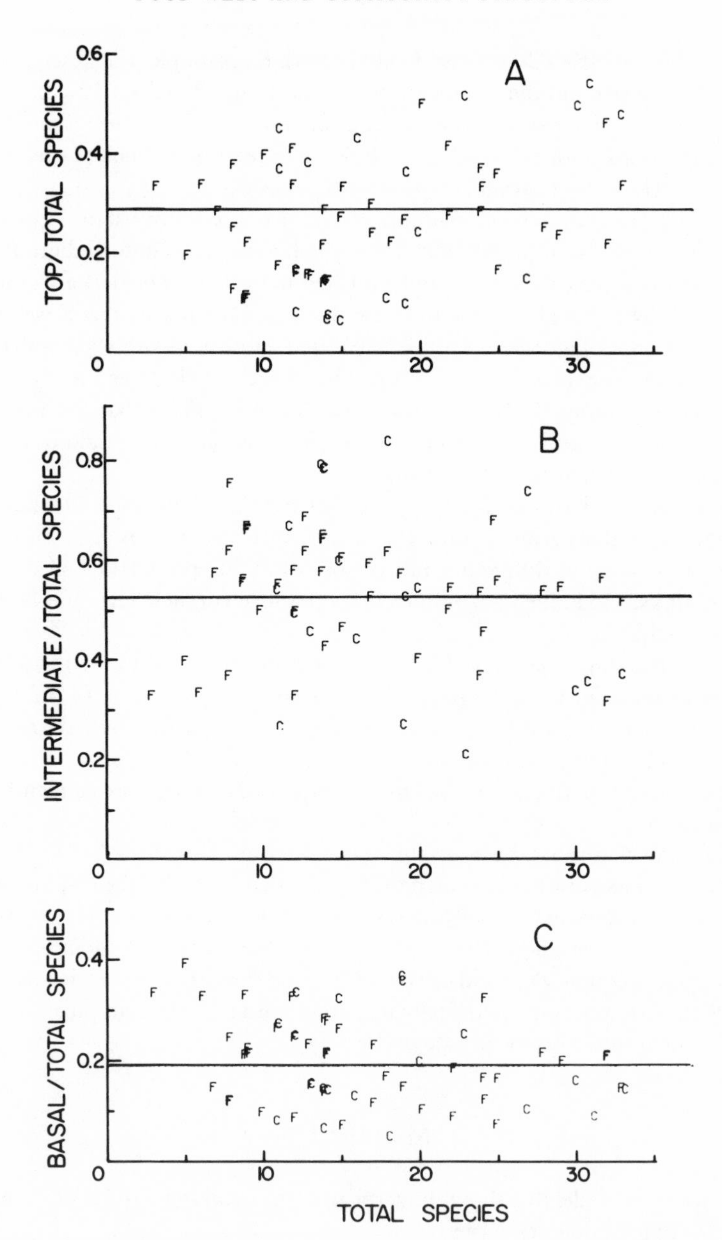

FIGURE 13.2. Three ratios, plotted as a function of the number of species, show scale-invariance in the proportions of species. The fitted lines are constrained to be horizontal. (a) Top species/total species; the height of the line is 0.2853. (b) Intermediate species/total species; the height of the line is 0.5251. (c) Basal species/total species; the height of the line is 0.1896. (From Briand and Cohen 1984, p. 265.)

links. In figure 13.3a (Cohen and Briand 1984), for example, the abscissa is the number of species and the ordinate is the proportion of basal-intermediate links among all links. There is no clear evidence of an increasing or decreasing trend. The proportions of different kinds of links, like the proportions of species, are approximately scale-invariant.

The fifth law is that the ratio of links to species is scale-invariant. Figure 13.4 plots the observed number of links in each web against the observed number of species, for 113 webs (Cohen, Briand, and Newman 1986). The data are approximated well by a straight line with slope about 2. That means that a web of 25 species has on average about 50 links. We first came across this generalization with 62 webs (Cohen and Briand 1984). Then Briand collected an additional 51 webs, and we found (Cohen, Briand, and Newman 1986) that the new data superimpose beautifully on the old. So far, this scale-invariant ratio of links to species is a consistent feature of nature.

In summary, I have reviewed evidence for five "laws" of webs. Qualitatively, these laws state that cycles are rare, chains are short, and there is scale-invariance in the proportions of different kinds of species, in the proportions of different kinds of links, and in the ratio of links to species. Each of these laws may be stated quantitatively.

By constructing hypothetical examples, it is not too hard to see that each of these laws may fail to hold while the remaining laws continue to hold. This means that the laws are logically independent. That all five laws characterize observed webs suggest that the laws are not empirically independent, and that it might be possible to find fewer than five assumptions that could explain and unify the five laws.

I make no claim that these are the only important empirical "laws" of webs. For example, I have omitted my own finding (Cohen 1978) that the trophic niches of predators in webs may usually be represented by intervals of a line, as well as Sugihara's related findings (1982, 1983, 1984) on the rarity of homological holes and the high frequency of rigid circuits. I selected the five "laws" reviewed above because they are phenomenologically important and because a simple model can connect them qualitatively and quantitatively.

MODELS

I turn now to a model that shows how the five empirical regularities described in the preceding section are related.

Let S denote the number of trophic species and L the number of links. List all the species along both the rows and columns of a "predation matrix," a square table of numbers with S rows and S columns. Name the matrix A. Put a 1 in the

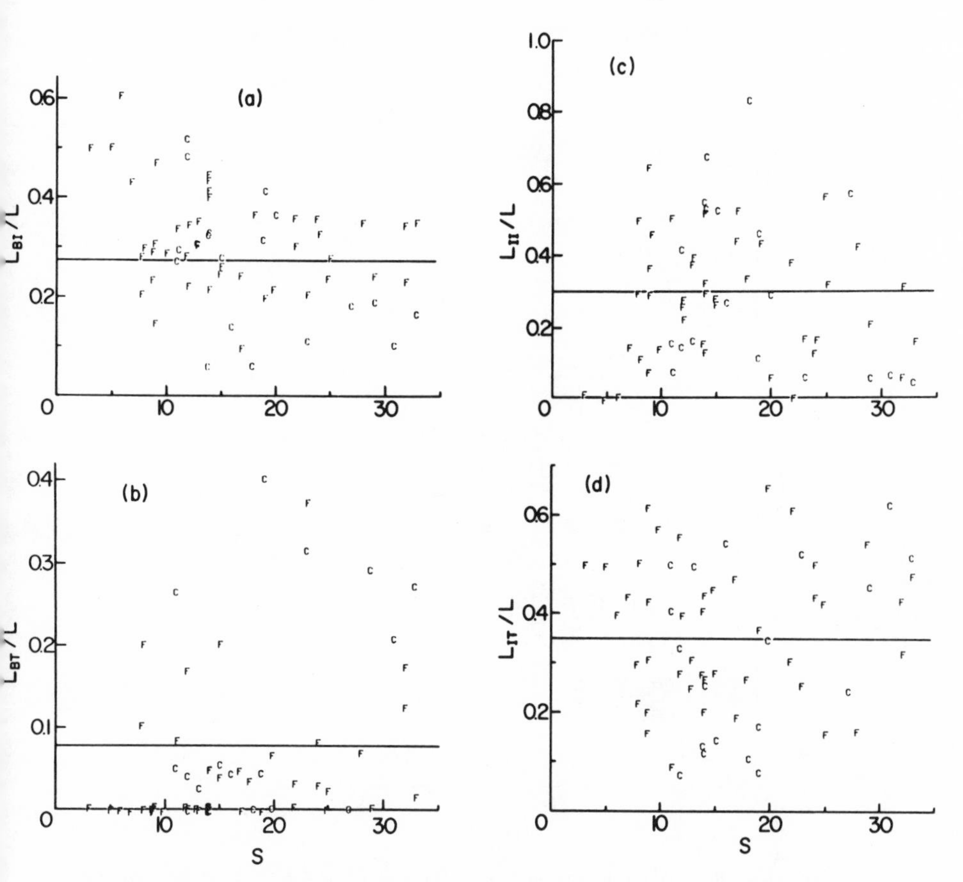

FIGURE 13.3. Four ratios, plotted as a function of the number of species, show scale invariance in the proportions of links. The fitted lines are constrained to be horizontal. (a) Basal-intermediate links/total links; the height of the line is 0.274. (b) Basal-top links/total links; the height of the line is 0.077. (c) Intermediate-intermediate links/total links; the height of the line is 0.301. (d) Intermediate-top links/total links; the height of the line is 0.348. The points in the upper left corner of (a) are based on very few links. (From Cohen and Briand 1984, p. 4107.)

intersection of row i and column j (element a_{ij} of the matrix A) if the species labeled j eats the species labeled i, and a 0 if species j does not eat species i. Since cannibalism is excluded from the data, all the diagonal elements (where $i = j$) are set equal to 0. In terms of this predation matrix, the total number of links is the sum of the elements of A. The sum picks up a 1 if there is a link from prey i to predator j and a 0 if there is no link.

The predation matrix also tells whether a species is top. If a species is top, then

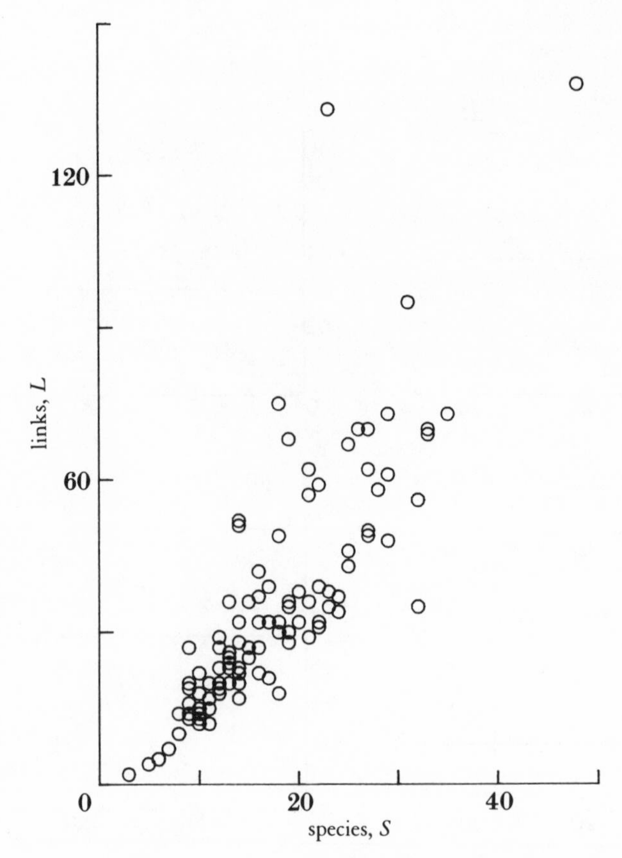

FIGURE 13.4. Observed number L of links as a function of the observed number S of species in 113 webs. (From Cohen, Briand, and Newman 1986, p. 335.)

nobody eats it. That means that the row of that species should be all 0's. So a 0-row corresponds to a top species. Similarly, a 0-column corresponds to a basal species because the species eats nothing. A species that has neither a 0-row nor a 0-column is intermediate.

I now describe the cascade model, but not the calculations required to squeeze results out of it. Some limnologists (e.g., Carpenter, Kitchell, and Hodgson 1985) use the term "cascade" with a different meaning, to describe the dynamics of limiting nutrients in webs. When the term "cascade" appears, it seems advisable to look for a definition. In this article, "cascade" refers only to the model in the next paragraph.

First, suppose nature numbers the S species in the community from 1 to S (without showing us the numbering), and suppose that the numbering specifies a pecking order for feeding, as follows. Any species j in this hierarchy or cascade can feed on any species i with a lower number $i < j$ (which does not mean that j *does* feed on i, only that j *can* feed on i). However, species j cannot feed on any species with a number k at least as large, $k \geq j$. Second, the cascade model assumes that each species actually eats any species below it according to this numbering with probability d/S, independently of whatever else is going on in the web. Thus the probability that species j does not eat species $i < j$ is $1 - d/S$. These two assumptions, of an ordering and of a probability of feeding proportional to $1/S$, are all there is to the cascade model.

In the predation matrix A, a_{ij} is always 0 if $i \geq j$. The predation matrix in the cascade model is strictly upper triangular, that is, every element on or below the main diagonal is 0. An element above the diagonal ($i < j$) is 1 with probability d/S and 0 with probability $1 - d/S$, and all elements are independent.

As is conventional, I use E to denote the average or expected number. I now show how to compute $E(L)$, the expected number of links, according to the cascade model. The expected number of links is the expectation of the sum of the predation matrix elements. There are S^2 elements in the predation matrix A, and the probability is d/S that an element a_{ij} ($i < j$) above the main diagonal equals 1. All other elements of A are 0 by construction. Since there are $S(S-1)/2$ elements above the main diagonal, the expected sum of the elements of A is $S(S-1)/2 \times d/S = d(S-1)/2 = E(L)$. Thus $E(L)$ is a linear function of S with slope $d/2$.

Since at present I have no theory to predict the slope, I have to estimate the slope from the data in figure 13.4. The slope of the line there is approximately 2, so I take $d = 4$ approximately. That is the only curve-fitting in this model. Everything else is derived. Thus $E(L) = 2(S-1) = 2S - 2$. Among webs with 26 species, the average number of links is predicted to be fifty. Since the number of species ranges from 3 to 48 in our data, the constant term -2 in this equation is negligible compared to the term $2S$ proportional to S. Qualitatively, the cascade model reflects the observation that the expected number of links is nearly proportional to the number of species. Quantitatively, the links-species scaling law fits because I made it fit by taking $d = 4$.

Roughly speaking, $d/2$ [more exactly, $d(S-1)/(2S)$] is the average number of predators per species and $d/2$ is the average number of prey per species. Here the average is taken over all webs with a given number of species and, more importantly, over all species within a web. Obviously, a species at the top of the cascade has no predators, while a species at the bottom of the cascade has no prey. However, averaged over all positions in the cascade, an average species has about two predators and about two prey.

As the number of species becomes large, the cascade model predicts 26% top species, 48% intermediate species, and 26% basal species. Thus the model predicts a 1 : 1 ratio of predators to prey. We observe 29% top species, 53% intermediate, and 19% basal (see fig. 13.2), giving roughly a 1.1 : 1 ratio of predators to prey. The model predicts the following percentages of basal-intermediate, basal-top, intermediate-intermediate, and intermediate-top links: 27, 13, 33, and 27. We observe, correspondingly, 27, 8, 30, and 35 (see fig. 13.3).

It is nice that the cascade model reproduces all the laws of scale invariance qualitatively, but it is far more striking that the cascade model gives a remarkable quantitative agreement between observed and predicted proportions. We put one number *d* into the cascade model and get out five independent numbers (because the three species proportions have to add up to 1 and the four link proportions have to add up to 1). I emphasize that these predictions use only the observed ratio of links to species.

For a finite number of species, we calculated from the cascade model the expected fraction of top species and the predicted variance. Figure 13.5 shows that the cascade model predicts not only the means but also the variability in the proportion of top species. We don't know whether the cascade model can predict the variability in proportions of links because we don't know how to calculate analytically what variability the cascade model predicts and have yet to do appropriate numerical simulations.

The cascade model was built to—and does—explain qualitatively and quantitatively the mean proportions of different kinds of species and links. Can the cascade model describe the number of chains of each length, counting all the possible routes from any basal species to any top species?

Let me illustrate with an artificial example (fig. 13.6) how to get a frequency histogram of chain length from a web. The link from 1 to 2 is a chain of length 1. The path 1, 3, 4 is a chain of length 2, and the path 1, 3, 5 is another chain of length 2. A numerical summary of the chain length distribution of the web in figure 13.6 is that it has one chain of length 1, two chains of length 2, and no longer chains.

Figure 13.7 shows the expected number of chains of each length, according to the cascade model, using parameters of a typical web, namely seventeen species and *d* close to 4. This figure also shows the results of one hundred computer simulations of the model using the same parameters. The sample mean numbers of chains of each length agree well with the theoretically expected number calculated from the model. That agreement is evidence that both the calculations and the simulations are right.

How well does the cascade model predict the observed distribution of chain length of a real web? To find out, we generated random webs according to the cascade model with the parameters of the observed web. We measured how often the chain-length distribution of a random web was further from the chain-length distribution predicted by the cascade model than the real observed chain-length distribution was from the predicted distribution. We used two measures of goodness of fit: the sum of squares of differences and a measure like Pearson's chi-squared. If the discrepancy between the observed and the expected frequency distributions was not larger than most of the discrepancies between webs ran-

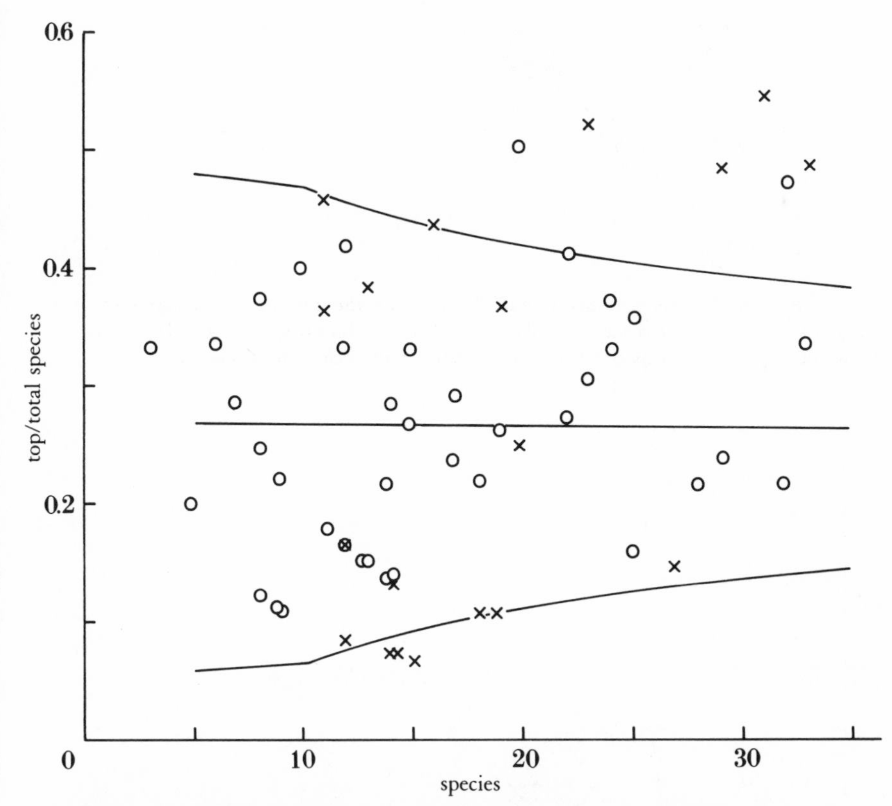

FIGURE 13.5. The predicted mean proportion of top species (middle line) and a confidence interval of ± 2 standard deviations (upper and lower lines) as a function of total species S, according to the cascade model. x is constant environment, o is fluctuating environment. The symbols x and o have been perturbed from their exact locations by a small random amount to indicate when several food webs have exactly the same coordinate. The data are replotted from Briand and Cohen 1984. (From Cohen and Newman 1985, p. 436.)

domly generated according to the cascade model and the mean frequency distribution expected from the model, we said the fit was good. If the discrepancy between observed and predicted chain length distributions was bigger than most simulated discrepancies, we said the fit was bad. Have no illusions about what a good fit means. In figure 13.8, food web 18 illustrates a good fit while food web 37 illustrates a poor fit.

Of 62 webs in Briand's original collection, the chain length distributions of 11 or 12 (depending on the measure of goodness of fit used) were badly described by the cascade model. The model's success with the chain-length distributions of 50 or 51 of these webs made us afraid that we had overfitted the model to the data. Perhaps by constructing the cascade model to explain the mean proportions of

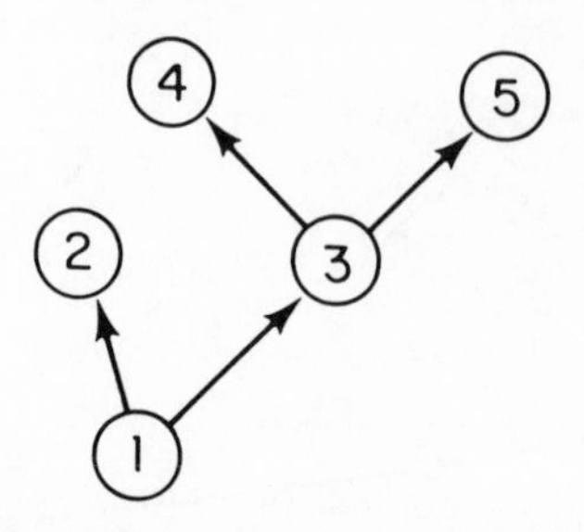

FIGURE 13.6. Hypothetical food web to illustrate how the frequency distribution of chain lengths is counted. There is one chain of length 1 (from species 1 to species 2), and there are two chains of length 2 (from species 1 to species 4 and from species 1 to species 5).

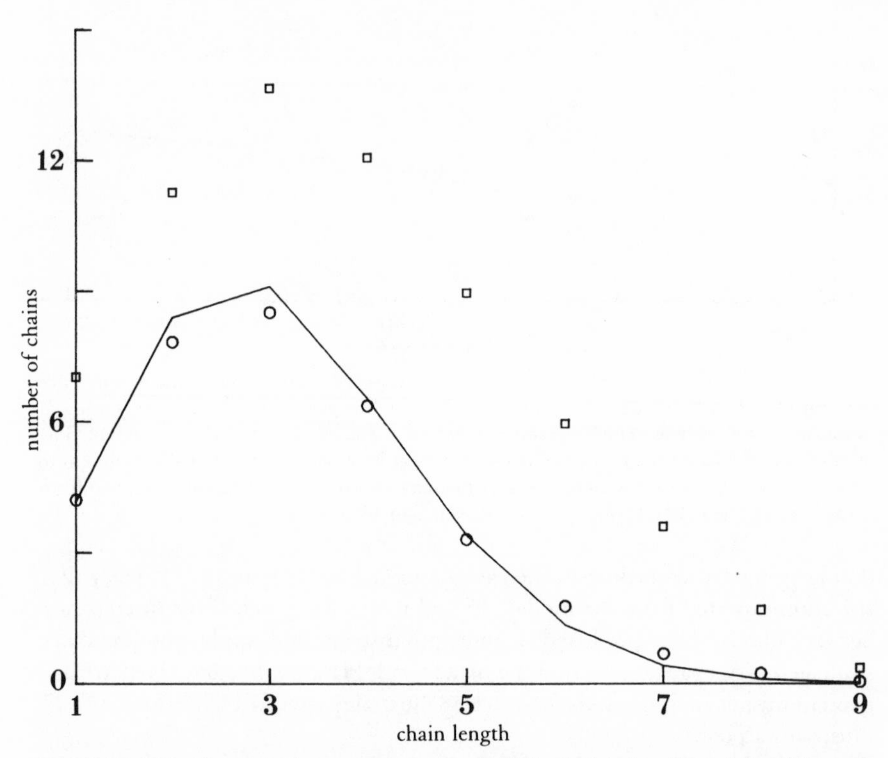

FIGURE 13.7. Theoretically expected number (solid line) of chains of length 1 to 9 in a web of $S = 17$ species, according to the cascade model with $d = 3.75$, sample mean number (o) of chains of each length in 100 simulations of the cascade model, and sample mean plus one sample standard deviation (□) in the number of chains of each length. No chains with more than nine links occurred in the simulations; the expected total number of such chains per simulation is 0.003. (From Cohen, Briand, and Newman 1986, p. 324.)

FIG. 13.8
Food web fits.

	Food web 18 "Acceptable" fit		*Food web 37 "Poor" fit*	
Chain length	*Predicted frequency*	*Observed frequency*	*Predicted frequency*	*Observed frequency*
1	6.5	13	6.0	0
2	9.9	10	10.3	21
3	8.6	5	10.1	23
4	5.2	4	6.8	8
≥5	3.6	0	5.3	0

FIGURE 13.8. Examples of "acceptable" and "poor" fits between the predicted (mean) numbers of chains of each length according to the cascade model and the observed numbers of chains of each length. In the serial numbering of Briand (1983), which is used here, number 18 is the Kapingamarangi Atoll food web (see fig. 13.1 above) of Niering (1963) and number 37 is the California sublittoral (sand bottom) food web of Clarke, Flechsig, and Grigg (1967). These webs correspond (see Briand 1983a), respectively, to food webs numbered 11 and 2 by Cohen (1978), who gives the predation matrices in full. For food web 18, four chains of length 4 are shown while figure 13.1 has one chain of four links. The reason for this discrepancy is that Cohen (1978) added, to the predation matrix for this web, links that Niering (1963) described in his text but omitted from his figure.

top, intermediate, and basal species and the proportions of different kinds of links, we had used so much information from the data that there was no possibility for the fits to the chain-length distribution to be bad, even though they were not used to build the model. This worried us. So Briand found and edited 51 additional webs which we had never analyzed before. The ratio of links to species was roughly the same for these new webs as for the old webs, as I mentioned already. With these fresh data, we found only 5 webs with poor fits to the cascade model's predicted frequency distribution of chain length. The proportion of poor fits, 5 of 51 webs, was smaller among the new webs than it had been among the original webs.

The cascade model uses no information about chain length to predict the frequency distributions of chain length. The predictions derive solely from the number of species and the number of links. No other parameters are free.

The cascade model needs to be tested further, tested until it fails, as it surely will. How well can the cascade model predict the moments of chain length (as Stuart Pimm has asked), or patterns of omnivory and intervality? Apparently, the niche overlap graph of most webs is an interval graph, that is, the overlaps of trophic niches revealed by most webs are consistent with the trophic niches being one-dimensional (Cohen 1978). Under what conditions, if any, does the cascade model predict that intervality should be common? Can the cascade model relate to the combinatorial web models of Sugihara (1982, 1983, 1984)? Much testing remains to be done.

The cascade model makes new predictions. In large webs ($S > 17$), the cascade model implies a rule of thumb that I have never seen stated in the ecological literature: The mean length of a chain should equal the mean number of prey species plus the mean number of predators of an average species (Newman and Cohen 1986). Both should equal a number near 4. This purported rule is open to empirical test.

The cascade model explains qualitatively why the longest chain in webs are typically short. Newman and I (1986) derived the relative expected frequency of various chain lengths as the number of species goes to infinity, according to the cascade model, and found that, with a realistic value of d, practically no chains have length 8, 9, or 10. The cascade model predicts that, in very large webs, the length of the longest chain grows like $(\log S)/(\log \log S)$. That is very slow growth. In a web with 10^8 species, which is probably an upper bound for the world, the cascade model predicts that the longest chain will almost never have more than twenty links.

CONNECTIONS

The cascade model connects with quantitative questions and theories elsewhere in ecology. I will sketch the connection of the cascade model with three topics: the species-area curve, the relative importance of predation and competition in communities, and allometric equations for the effects of body size.

First, one of the best-known quantitative empirical generalizations of ecology is the species-area curve (e.g., MacArthur and Wilson 1967; Schoener 1976, 1986; Diamond and May 1981). In its simplest form, the species-area curve asserts that the number of biological species on an island is proportional to the area of the island raised to some power near ¼. (When examined in detail [Schoener 1986], species-area curves are vastly more complicated.) The cascade model predicts, among other things, how the mean or maximal length of chains depends on the number of trophic species in a community. If the number of trophic species can be assumed or demonstrated (by a future empirical study of actual webs) to be proportional to the number of biological species, then a combination of the species-area curve and the cascade model predicts how chain lengths should vary on islands of different areas.

It is not necessary to go into the details of the formulas to see that if the number of species on an island increases very slowly with area, and if the maximal or mean chain length in a web increases very slowly with the number of species in a community, then the maximal or mean chain length should increase extremely slowly, or be practically constant, with increasing island area. The combination of the species-area curve and the cascade model explains, qualitatively at least, why there is *not* a known relation between the area a community occupies and the mean or maximal chain length of its web.

An alternative explanation, suggested by Robert T. Paine, is that there is no known relation between the area of a community and the mean or maximal chain length of its web because nobody has looked for such a relation. If the cascade model provokes an ecologist to examine the relation empirically, the model will have served a useful purpose, whether or not its predictions are confirmed. It remains to derive the formulas explicitly and to look for empirical tests of the assumptions and predictions.

Second, the cascade model relates to the roles of competition and predation in ecological communities. Hairston, Smith, and Slobodkin (1960), as described succinctly by Schoener (1982, p. 590), "argued that competition should prevail among top predators, whereas predation should prevail among organisms of intermediate trophic status, mainly herbivores. Because the herbivores are held down by competing top carnivores, competition should prevail again among the herbivore's [*sic*] food species, green plants." Menge and Sutherland (1976, p. 353) proposed, by contrast, that as trophic position goes from high to low within a community, the relative importance of predation should increase monotonically while the relative importance of competition should decline monotonically. Connell (1983) and Schoener (1983) reviewed at length field experiments on interspecific competition which bear on these generalizations, and Schoener (1985) analyzed the points of agreement and disagreement in the two reviews.

Predation and competition can be interpreted in terms of quantities computable from the cascade model. It is then possible to examine whether these quantities behave according to the generalizations of Hairston, Smith, and Slobodkin (1960) or Menge and Sutherland (1976). For example, a natural measure of the amount of predation on trophic species i in the cascade model is the expected (i.e., average) number of predators on trophic species i, which is easily seen to be $d(S-i)/S$. There are $S-i$ species above species i in the trophic pecking order, and the probability that any one of them will feed on species i is d/S, so the expected number of predators on species i is the product $d(S-i)/S$. Since $i = 1$ is the lowest trophic position in the cascade model and $i = S$ is the highest, the cascade model implies that this measure of predation should increase linearly as trophic position goes from high to low within a community, exactly as proposed by Menge and Sutherland (1976).

However, the generalizations of Hairston, Smith, and Slobodkin and Menge and Sutherland pertain to the *relative* importance of competition and predation. So the behavior of a measure of predation needs to be related to the behavior of a measure of competition, such as, for example, one used by Briand (1983a).

Third, physical interpretations of the ordering of trophic species assumed in the cascade model may make it possible to connect the study of webs with the study of allometry and physiological ecology. The combination might be called "ecological allometry." For example, extending to entire webs a qualitative suggestion of Elton (1927, pp. 68–70) for individual chains, suppose that each trophic species consists of individuals more or less homogeneous with respect to size or mass, and that the larger the species' label $i = 1, 2, \ldots, S$ in the cascade model (i.e., the higher the trophic position), the larger the mass of each individual in that species. (Food chains of parasites generally follow the opposite rule: parasites are much smaller than their hosts [Elton 1927, chap. 6].) The assumption that body mass increases with a species' label i in the cascade model can be tested empirically, since it implies that no (nonparasitic) trophic species can eat a species larger than itself. When trophic species in real webs are ordered by body mass, is the matrix that describes the trophic relations of the community generally upper triangular, as assumed by the cascade model?

If so, this simple assumption will permit the cascade model to connect facts about food webs with quantitative empirical generalizations that physiological ecologists have discovered about body size (e.g., Peters 1983; Calder 1984; Peterson, Page, and Dodge 1984; Peters and Raelson 1984; Vézina 1985; May and Rubinstein 1985). From preliminary calculations, it appears that several empirical ecological generalizations, which have previously lacked a physical explanation, may be derived from a combination of the cascade model with assumptions or facts about body size.

APPLICATIONS

This work may eventually contribute to human well-being in four ways.

First, environmental toxins cumulate along food chains. "Eating 0.5 kg of Lake Erie fish can cause as much PCB [polychlorinated biphenyl] intake as drinking 1.5×10^6 L of Lake Erie water" (National Research Council 1986). An understanding of the distribution of the length of food chains is necessary, though not sufficient, for understanding how toxins are concentrated by living organisms.

Second, people have not been very successful at anticipating all the consequences of introducing or eliminating species. Such perturbations of natural ecosystems are being practiced with increasing frequency in programs of biological control. An understanding of the invariant properties of webs is essential for anticipating the consequences of species' removals and introductions. For example, a perturbation that eliminated most of the top trophic species, or most of the basal trophic species, could be expected to be followed by major changes in the structure of the web if the community adjusts to reestablish invariant proportions of top, intermediate, and basal species. The cascade model or its successors may eventually make it possible to derive more quantitative predictions.

Third, an understanding of webs will help in the design of nature reserves and of those future ecosystems that will be required for long-term manned space flight and extraterrestrial colonies. A nature reserve with all top species would be expected to have trouble, according to the cascade model. For humans to survive and to be fed in space, we need to know more about the care and feeding of webs.

Fourth, and finally, since some webs include man, an understanding of webs may give us a better understanding of man's place in nature, here on earth. We have not detected any consistent differences between webs that contain man and webs that do not. Of course, we have not looked yet at webs of agricultural ecosystems strongly influenced by man. When we look at new classes of webs, we may expect to see new patterns.

ACKNOWLEDGMENTS

This work was supported in part by National Science Foundation grants BSR 84-07461 and BSR 87-05047 and by the hospitality of Mr. and Mrs. William T. Golden. Robert T. Paine and George Sugihara offered very helpful comments on a previous draft. Parts of this chapter overlap parts of my paper "Untangling 'an entangled bank': recent facts and theories about community food webs," prepared for a book on community ecology to be edited by Alan Hastings.

REFERENCES

Beaver, R. A. 1983. The communities living in Nepenthes pitcher plants: Fauna and food webs. In J. H. Frank, and L. P. Lounibos, eds., *Phytotelmata: Terrestrial Plants as Hosts for Aquatic Insect Communities,* pp. 129–59. Plexus Publishing, Medford, N.J.

Beaver, R. A. 1985. Geographical variation in food web structure in Nepenthes pitcher plants. *Ecol. Entomol.* (Lond.) 10:241–48.

Briand, F. 1983a. Environmental control of food web structure. *Ecology* 64:253–63.

Briand, F. 1983b. Biogeographic patterns in food web organization. In DeAngelis, Post, and Sugihara (1983), pp. 37–39.

Briand, F., and J. E. Cohen. 1984. Community food webs have scale-invariant structure. *Nature* 307:264–66.

Briand, F., and J. E. Cohen. 1987. Environmental correlates of food chain length. *Science* 238:956–60.

Budyko, M. I. 1980. *Global Ecology.* Progress Publishers, Moscow.

Calder, W. A. 1984. *Size, Function, and Life History.* Harvard University Press, Cambridge, Mass.

Carpenter, S. R. J. F. Kitchell, and J. F. Hodgson. 1985. Cascading trophic interactions and lake productivity. *BioScience* 35(10):634–39.

Clarke, T. A., A. O. Flechsig, and R. W. Grigg. 1967. Ecological studies during Project Sea Lab II. *Science* 157:1381–89.

Cohen, J. E. 1977. Ratio of prey to predators in community food webs. *Nature* 270:165–67.

Cohen, J. E. 1978. *Food Webs and Niche Space.* Princeton University Press, Princeton, N.J.

Cohen, J. E., and F. Briand. 1984. Trophic links of community food webs. *Proc. Nat. Acad. Sci. USA* 81:4105–4109.

Cohen, J. E., F. Briand, and C. M. Newman. 1986. A stochastic theory of community food webs. III: Predicted and observed lengths of food chains. *Proc. Roy. Soc. Lond.* (B) 228:317–53.

Cohen, J. E., F. Briand, and C. M. Newman. In press. *Community Food Webs: Data and Theory.* Springer-Verlag, New York.

Cohen, J. E., and C. M. Newman. 1985. A stochastic theory of community food webs. I: Models and aggregated data. *Proc. Roy. Soc. Lond.* (B) 224:421–48.

Cohen, J. E., C. M. Newman, and F. Briand. 1985. A stochastic theory of community food webs. II: Individual webs. *Proc. Roy. Soc. Lond.* (B) 224:449–61.

Connell, J. H. 1983. On the prevalence and relative importance of interspecific competition: Evidence from field experiments. *Amer. Natur.* 122:661–96.

Cooley, J. H., and F. B. Golley, eds. 1984. *Trends in Ecological Research for the 1980s.* Plenum Press, New York and London.

DeAngelis, D. L., W. M. Post, and G. Sugihara, eds. 1983. *Current Trends in Food Web Theory.* ORNL-5983. Oak Ridge National Laboratory, Oak Ridge, Tenn.

Diamond, J. M., and T. J. Case, eds. 1986. *Community Ecology.* Harper and Row, New York.

Diamond, J. M., and R. M. May. 1981. Island biogeography and the design of natural reserves. In May (1981), pp. 228–52.

Elton, C. 1927. *Animal Ecology.* (New impression with additional notes, 1935.) Macmillan, New York.

Gallopín, G. C. 1972. Structural properties of food webs. In B. C. Patten, ed., *Systems Analysis and Simulation in Ecology,* vol. 2, pp. 241–82. Academic Press, New York.

Hairston, N. G., F. E. Smith, and L. B. Slobodkin. 1960. Community structure, population control, and competition. *Amer. Natur.* 94:421–25.

Hazen, W. E., ed. 1964. *Readings in Population and Community Ecology.* Saunders, Philadelphia.

Hutchinson, G. E. 1959. Homage to Santa Rosalia, or why are there so many kinds of animals? *Amer. Natur.* 93:145–59.

Kikkawa, J., and D. J. Anderson, eds. 1986. *Community Ecology: Pattern and Process.* Blackwell Scientific, Melbourne and Oxford.

Lindemann, R. L. 1942. The trophic-dynamic aspect of ecology. *Ecology* 23:399–418.

Lotka, A. J. 1925. *Elements of Physical Biology.* Baltimore: Williams and Wilkins, Baltimore. (Reprinted 1956, *Elements of Mathematical Biology.* Dover, New York.)

MacArthur, R. H., and E. O. Wilson. 1967. *The Theory of Island Biogeography.* Princeton University Press, Princeton, N.J.

Margalef, R. 1984. Simple facts about life and environment not to forget in preparing schoolbooks for our grandchildren. In Cooley and Golley (1984), pp. 299–320.

May, R. M. 1973. *Stability and Complexity in Model Ecosystems.* Princeton University Press, Princeton, N.J.

May, R. M., ed. 1981. *Theoretical Ecology: Principles and Applications,* 2d ed. Sinauer, Sunderland, Mass.

May, R. M., and D. I. Rubenstein. 1985. Reproductive strategies. In C. R. Austin, and R. V. Short, eds., *Reproduction in Mammals,* Book 4: *Reproductive Fitness,* 2d ed. Cambridge University Press, Cambridge, England.

Menge, B. A., and J. P. Sutherland. 1976. Species diversity gradients: Synthesis of the roles of predation, competition, and temporal heterogeneity. *Amer. Natur.* 110:351–69.

Mithen, S. J., and J. H. Lawton. 1986. Food-web models that generate constant predator-prey ratios. *Oecologia* 69:542–50.

National Research Council, Commission of Life Sciences, Committee on the Applications of Ecological Theory to Environmental Problems. 1986. *Ecological Knowledge and Environmental Problem-Solving.* National Academy Press, Washington, D.C.

Newman, C. M., and J. E. Cohen. 1986. A stochastic theory of community food webs. IV: Theory of food chain lengths in large webs. *Proc. Roy. Soc. Lond.* (B) 228:355–77.

Niering, W. A. 1963. Terrestrial ecology of Kapingamarangi Atoll, Caroline Islands. *Ecol. Monogr.* 33:131–60.

Paine, R. T. 1980. Food webs: Linkage, interaction strength and community infrastructure. *J. Anim. Ecol.* 49:667–85.

Peters, R. H. 1983. *The Ecological Implications of Body Size.* Cambridge University Press, Cambridge, England.

Peters, R. H., and J. V. Raelson. 1984. Relations between individual size and mammalian population density. *Amer. Natur.* 124:498–517.

Peterson, R. O., R. E. Page, and K. M. Dodge. 1984. Wolves, moose, and the allometry of population cycles. *Science* 224:1350–52.

Pimm, S. L. 1982. *Food Webs.* Chapman and Hall, London.

Pimm, S. L. 1984. The complexity and stability of ecosystems. *Nature* 307:321–26.

Pimm, S. L. In press. The geometry of niches. In A. Hastings, ed., *Community Ecology: The Mathematical Theory.* Lecture Notes in Biomathematics. Springer-Verlag, New York.

Pimm, S. L., and R. L. Kitching. 1986. The determinants of food chain lengths. *Oikos,* special issue on trophic exploitation.

Pimm, S. L., and L. H. Lawton. 1977. The number of trophic levels in ecological communities. *Nature* 268:329–31.

Remmert, H. 1984. And now? Ecosystem research! In Cooley and Golley (1984), pp. 179–91.

Schoener, T. W. 1976. The species-area relation within archipelagos: Models and evidence from island land birds. *Proceedings of the 16th International Ornithological Congress* (Canberra), pp. 629–42.

Schoener, T. W. 1982. The controversy over interspecific competition. *Amer. Sci.* 70:586–95.

Schoener, T. W. 1983. Field experiments on interspecific competition. *Amer. Natur.* 122:240–85.

Schoener, T. W. 1985. Some comments on Connell's and my reviews of field experiments on interspecific competition. *Amer. Natur.* 125:730–40.

Schoener, T. W. 1986. Patterns in terrestrial vertebrate versus arthropod communities: Do systematic differences in regularity exist? In Diamond and Case (1986), pp. 556–86.

Sugihara, G. 1982. Niche hierarchy: Structure, organization, and assembly in natural communities. Ph.D. dissertation, Princeton University, Princeton, N.J.

Sugihara, G. 1983. Holes in niche space: A derived assembly rule and its relation to intervality. In DeAngelis, Post, and Sugihara (1983), pp. 25–35.

Sugihara, G. 1984. Graph theory, homology, and food webs. *Proc. Symp. Appl. Math.* 30:83–101. American Mathematical Society, Providence.

Tilman, D. 1986. A consumer-resource approach to community structure. *Amer. Zool.* 26:5–22.

Vézina, A. F. 1985. Empirical relationships between predator and prey size among terrestrial vertebrate predators. *Oecologia* 67:555–65.

Wiegert, R. G., ed. 1976. *Ecological Energetics.* Benchmark Papers in Ecology, 4. Dowden, Hutchinson and Ross, Stroudsburg, Penn.

Chapter 14

The Structure and Assembly of Communities

JONATHAN ROUGHGARDEN

The central question of community ecology was posed decades ago: Do the populations at a site consist of all those that happened to arrive there, or of only a special subset—those with properties allowing their coexistence? Elton (1933) wrote: "In any fairly limited area only a fraction of the forms that could theoretically do so actually form a community at any one time. . . . The animal community really is an organized community in that it apparently has 'limited membership.' " Alternatively, Gleason (1926) wrote that "the vegetation of an area is merely the resultant of two factors, the fluctuating and fortuitous immigration of plants and an equally fluctuating and variable environment. As a result, there is no . . . reason for adhering to our old ideas of the definiteness and distinctness of plant associations [communities]" (from Gleason 1953). In the decades since these views were enunciated, they have been explored theoretically and empirically. The journey has been long. It has involved the analysis of whether population interactions really occur, and if they do, what kind of structure they might produce. To this end, models have been formulated, solved, and discarded. The result has been a deeper and more synthetic understanding of what really goes on in ecological communities, and a substitution of new and contemporary questions for those of its infancy. This essay briefly recapitulates the development of theory in community ecology, especially models that address the dichotomy originally posed by Elton and Gleason. My emphasis is on animal communities, a bias that reflects my own field experience.

THE RECENT PAST

Competition and Limited Membership

The first generation of models traces to Hutchinson's (1959) conceptualization of the idea of limited membership. After reviewing data on birds and mammals, Hutchinson suggested that the ratio of the size (in units of length) of the smaller species to the larger species in a coexisting pair was 1 : 1.3, and this ratio was taken "as an indication of the kind of difference necessary to permit two species to co-occur at the same level of a food web." This difference in the average body size of two species was assumed to correspond to a difference in the average size of the prey consumed by the species, and thus to indicate some partitioning of the resources. This suggestion was quickly extended to the idea that the sizes of three or more co-occurring species should be regularly spaced along a size axis, with the ratio between consecutive sizes also being about 1 : 1.3. This extended idea was supported by studies in the rich avian community of New Guinea (Diamond 1973, 1975), and other bird communities (Cody 1974).

This context motivated MacArthur and Levins (1967) to use the Lotka-Volterra competition equations to search for a theoretical basis to the value of 1.3. The strength of competition between two species was assumed to increase with the closeness of their body sizes. The theoretical problem was to see if there exists a "limiting similarity" between coexisting species, that is, a minimum difference in body size consistent with coexistence. (The minimum ratio in body sizes becomes a minimum difference in body sizes when the body sizes are expressed in logarithmic units.) The setup of the model assumed three species equally spaced from each other, and the model was solved for the minimum separation that would allow all three to coexist. May and MacArthur (1972) later extended the analysis to N species equally spaced around a circle. Few appreciate that this early theory never "predicted" species to be evenly spaced; the species were simply arranged that way at the start. These formulations usually do predict a non-zero limiting similarity, although no "universal constant" emerged which could correspond to the value of 1.3.

Also at about this time, models were offered for the evolutionary consequence of competition. One branch of models concerned intraspecific competition and the question of whether a population should become polymorphic with many specialists or relatively monomorphic with a generalist phenotype. It was concluded that random mating and the constraints of inheritance in a sexual population prevent its spreading out effectively to use all resource types. Ecological "space" is therefore left over, allowing the community to consist of more than one species (Roughgarden 1972; Fenchel and Christiansen 1977). Another branch of models concerned interspecific competition and began to investigate whether morphological differences between species, like the ratio of 1 : 1.3, could represent the result of two species having evolved away from each other's resource use (Roughgarden 1976; Case 1979). This idea originates with Lack's (1947) discussion of the evolutionary effects of competition in the finches of the Galapagos Islands, and with comparisons of beak and bill sizes for bird species from places where their ranges do and do not overlap (Brown and Wilson 1956). The models were used to compute coevolutionarily stable equilibria for two or more competing species, but the trajectories leading to such equilibria were not considered.

Another highlight of this period was a reevaluation of the classic intuition relating species diversity to stability. It has been assumed since Elton (1933) that diversity promoted a community's stability to perturbation because of the compensatory interactions thought to be present in a diverse community. Yet May (1973) argued that diverse communities were typically less stable to perturbations than simple ones because, in effect, perturbations from all parts of the community are communicated to a species having many interactions compared with a species having few interactions. Indeed, it is also more difficult to devise a complex community than a simple one to begin with, because the strength of each interaction has to be "just right" to get all the members of a complex community to coexist, even if the environment is practically constant. But even if a complex community is devised, May's point is that it usually is not very stable either. The proposition that complexity is incompatible with diversity is not without exception, because counterexamples have been discovered (Cohen and Newman 1985). Nonetheless, the proposition remains valuable as an almost universal rule of thumb.

In a similar vein, the hypothesis that species interactions change during the course of coevolution to lead to greater community stability was investigated theoretically (Roughgarden 1977). The result was that a community after coevolution is, in theory, often less stable than it was initially, provided all the species remain in the community. However, species extinction frequently happens during coevolution. When it does, the community is usually more stable after coevolution than before, but only because it has become less diverse.

The 1970s witnessed the exploration of four more theoretical approaches involving competition that are topical today. First, MacArthur (1968) introduced an alternative to the Lotka-Volterra competition equations that incorporated explicit dynamics for the resources. The Lotka-Volterra equations subsume the dynamics of the resources being exploited in common by competing species into fixed parameters (the carrying capacity and competition coefficients). MacArthur derived expressions for these parameters as a function of the way the resource renews while it is being consumed. What is particularly interesting about this early analysis is the assumption that the resource dynamics are fast relative to the dynamics of the consuming species, so that the state of the resource tracks the relatively slow change in those dynamics. This is the first model in community ecology whose analysis relies on an inherent difference in scale for community components having different trophic positions (i.e., having different vertical positions in a food web). This approach to competition in which the dynamics of the resources are explicitly included in the model was extended by Lawlor and Maynard Smith (1976) and is the basis of the current studies by Abrams (1980, 1986) and Tilman (1982, 1986). These later studies do not, however, focus on scale differences between consumer and resource (or between predator and prey) and instead emphasize results relying on a close coupling in space and time between the dynamics of consumers and their resources.

Second, Horn and MacArthur (1972) and Slatkin (1974) introduced hierarchical population models where the species is divided into subpopulations occurring at sites throughout the species range. The subpopulation at each site may become extinct but the site can be recolonized by a nearby subpopulation. They obtained conditions for which two species populations can coexist on a regional scale even though the species cannot permanently coexist at any local site within the region.

Third, the role of spatial and temporal fluctuations in the environment, together with competition, as causes of spatial and temporal pattern in a population's abundance was explored by Roughgarden (1975, 1978) using spectral analysis on linearized population models. Today these results are relevant to characterizing the "landscape" aspect of a population's distribution in terms of the superposition of multiple scales, as discussed further in the papers by Powell and Levin in this volume.

Fourth, toward the end of the 1970s the presumed deleterious impact of environmental fluctuation was questioned. The classic view of environmental fluctuation, tracing to Elton (1933), was that it induced species extinctions and a corresponding loss of diversity. The view was that a community's stability had to be sufficient to withstand environmental perturbations, and if so, the community's biotic diversity would be retained. Sale (1977), in questioning the appropriateness of competition theory for coral-reef fish communities, suggested that somehow an unpredictable stochastic environment was a circumstance allowing more species to coexist than was possible in a constant environment. Sale's suggestion stimulated attempts to view stochasticity in a more positive light than before. Chesson and Warner (1981) introduced a model that showed how a kind of environmental stochasticity can promote coexistence. The new intuition to emerge was that stochastic fluctuation in birth rates (or in recruitment to vacant space) in a habitat with a fixed total number of individuals brings about a relative advantage to the rarer species because the common species in a saturated habitat cannot benefit from its lucky breaks, while the rare species can. Hence, the rare species is protected from extinction, and diversity tends to be conserved. This result is not true for stochastic fluctuation in death rates, however, and the qualitative impact of environmental fluctuation depends on quantitative aspects of the species' life history, as reviewed recently in Chesson (1986).

By the late 1970s, however, basic research in the mathematics of nonlinear dynamical systems gave pause to further exploration of models involving competing species. Hirsch and Smale (1974) pointed out that while the two-species Lotka-Volterra model is representative of other, more general two-species competition models, the Lotka-Volterra model for three or more species is not. Furthermore, Smale (1976) discovered that a general N-species competition model can lead to virtually any pattern of trajectories for the competing species. These results, based on differential equation models defined in continuous time for interacting populations, followed upon similar results by May (1974) that were based on difference equation models defined in discrete time for a single population. Hence, determining that N species compete does not, by itself, imply very much about their population dynamics. Indeed, results tracing to MacArthur and Levins' (1967) formulation rely on a special functional form to relate the strength of competition to the closeness of body size (cf. May 1974; Roughgarden 1979). Thus, in principle, competing species can show global equilibrial, multiple equilibrial, periodic, almost periodic, or chaotic dynamics, depending on quantitative details of how they compete with one another, and a purely a priori exploration of competition theory is not likely to have much meaning.

Predation and Limited Membership

Meanwhile, an independent conceptualization was developing for communities inhabiting the marine rocky intertidal zone. Here one visualizes a surface of finite area potentially covered by organisms of various species. Field experiments established that the species occupying a surface in this zone could be classified in a linear dominance hierarchy with respect to competitive success, say, $M > B > C > A$ (Connell, 1961a,b; Dayton 1971). In the absence of mortality, the system tends to culminate in a monoculture of M. But if mortality is aimed at M, then the surface cannot culminate in this monoculture, and instead retains a diversity of species in some mix at which the mortality rate matches the rate at which the hierarchical competition is progressing to a monoculture. Paine (1966) established that the agent of mortality against M (a mussel) along the coast of Washington State was a particular predator (a large starfish). This predator was termed a "keystone" predator, because its removal led to a decline in the species diversity of its prey. Other agents of mortality include the impact of debris, and scraping from ice. In all, these agents are referred to generically as "disturbance."

This conceptualization did not receive much theoretical research, and it is still important to understand better how the kinetics implicit in it operate. A start was made, however, by Levin and Paine (1974), who developed a model for the size distribution of the gaps left on the substrate after disturbance. This model, like the spectral-analysis models mentioned previously, address the landscape aspect to a community's appearance.

Unlimited Membership

The view that a community is open to all who arrive was developed most extensively in the theory of island biogeography proposed by MacArthur and Wilson (1963). The community is identified with the fauna on a small island. The number of species in the fauna is predicted from the rate of arrival of new species to the island and the rate of extinction of species already on the island. For any species, both the probability of its successful immigration to the island, and of its extinction once established, are taken as independent of the composition of the island's fauna. Even though the total species diversity approaches a stable steady state, species are continually coming and going. Although this model became known as the "equilibrium theory" of island biogeography, it is the least equilibrial of the models of that time. It embodies the Gleasonian concept of a community in that the island's fauna is largely a reflection of its colonization and extinction history. The MacArthur-Wilson model was further developed as a stochastic process, as summarized in Goel and Richter-Dyn (1974). Its applicability to the insect fauna of small mangrove islands was demonstrated experimentally by Simberloff and Wilson (1969, 1970). This type of model has been hypothesized to explain the species diversity in tropical rain forests (Hubbell and Foster 1986).

MacArthur and Wilson (1967), in an influential book, presented their earlier theory in an extended discussion that included many of the considerations now known to be important, and that are omitted in their mathematical model for species turnover. Therefore, when the "turnover" model is referred to later, reference is being made to the pure immigration-extinction model, without meaning to imply that the authors were unaware, even at the time, of the considerations that have now come to dominate discussions of island biogeography.

So, by the mid 1970s, theoretical ecology contained two formulations that viewed a community's membership as limited by the action of interspecific competition, and both views sought ways to ameliorate the action of competition. One view focused on resource partitioning, and the other on disturbance. And both agreed that the composition of a community is a special subset of those species that have crossed the community's boundaries, because not all can coexist. Yet theoretical ecology also contained one formulation viewing the community as open to all who might come.

THE PRESENT

The last ten years seem, in retrospect, to be a period whose focus was primarily empirical; there was some theoretical advance, but the action was mostly in the field. Because the period began with three major pictures of what goes on in a community, it was marked primarily by an analysis of whether these pictures were accurate. The conclusion seems to be that all three pictures are only part of the truth, even for the systems where they originated, and today all three seem somewhat obsolete. But at the same time, what really is going on in these systems is increasingly well understood.

Competition and Limited Membership Reconsidered

So first, the good news. In many systems, the idea of competition causing limited membership, and of resource partitioning as allowing coexistence, has been confirmed. Communities in which this picture is appropriate include North American desert rodents (Brown 1975), bumblebees in Colorado (Inouye 1977), lizards in the Greater Antilles (Williams 1983), hawks of the genus *Accipiter* (Schoener 1984), freshwater fish (Werner 1986), desert cactus and Mediterranean-climate scrub (Cody 1986), among others. The rules of membership, however, usually involve suites of characters (dubbed "ecomorphs" by Williams 1983) and not simply body size. Also, Wiens (1982) showed that data on size ratios among coexisting species do not support an idea of niche spacing regular enough to be analogized to the packing of atoms in a crystal, as had perhaps been hoped. Some of the data on limited membership were scathingly criticized on statistical grounds by Simberloff and Boecklen (1981), leading to debate on the legitimacy of the criticism and to the formulation of alternative statistical tests for membership rules (Grant and Abbott 1980; Bowers and Brown 1982; Schoener 1984; and others.) In this context then, it is worth emphasizing that membership rules in some communities are not abstract statistical patterns; one can, in the field, directly observe sharp species borders between ecologically equivalent species that are too similar to coexist (e.g., Terborgh 1985).

The reality of interspecific competition in some communities has now been demonstrated by field experiments independently of the membership rules that have been observed. Indeed, abundant experimental evidence now exists not only for the reality of competition, but increasingly also for how the strength of competition between two species depends on the difference in their body sizes (Pacala and Roughgarden 1985; Moulton and Pimm 1986). All in all, a huge research effort during the last decade to determine experimentally if interspecific competition occurs has been successfully carried out, with competition being detected in roughly one-third of the experiments, according to over 150 such studies surveyed by Schoener (1983) and Connell (1983), and the total is quite possibly over 200 today.

The theory for communities of competing species also progressed somewhat. The earlier analyses of limiting similarity simply did not speak to the degree of difference expected among the species that do coexist, except to say that they must differ by some minimum amount. The differences actually exhibited by coexisting species must somehow represent both the differences they had upon entry into the community and differences evolved while in the community. Therefore, models of faunal assembly were proposed and analyzed (Roughgarden, Heckel, and Fuentes 1983; Colwell and Winkler 1984; Rummel and Roughgarden 1985; Stenseth 1985). As an initial condition, the habitat is assumed empty, and then species are added sequentially until a large community is assembled.

A conclusion from this new theory is that the final community generally does not exhibit evenly spaced niches; only one theoretical case is even close to this picture—where the competition and carrying capacity distributions are symmetrical and where time is allowed for coevolutionary equilibrium to be attained between successive invasions. Another conclusion is that the divergent coevolution of two competitors away from each other's niche is a theoretically unlikely scenario because the initial condition for this requires the initial coexistence of two very similar competitors. Instead, the typical scenarios involve an island that starts with one species and is colonized by a second species quite different from the first, followed by parallel evolution of both species in the same direction—the first away from the second, and the second into space being vacated by the first. Depending on the symmetry in the situation, and the variety of resources available, either both species will eventually coexist with some niche separation, or one may converge upon the other, leading to its extinction. In the latter case, the system returns to a single species. Similar phenomena are also predicted in more complex communities. Thus on a historical biogeographic time scale, the community may be in flux, lending support to the ideas of the taxon cycle (Wilson 1961) and the Red Queen hypothesis (Van Valen 1973).

Empirical information about the assembly and coevolution of competing species is encouragingly similar to these theoretical predictions. The best and only example of character displacement as a result of divergent coevolution between two competing species involves snails of the Limfjord in Denmark (Fenchel 1975). The ranges of these species of marine snails do not overlap along the coast, and their sizes are the same. In 1825 storm damage produced a marine entrance to what had been a freshwater lake. The Limfjord was then colonized simultaneously by the two species of snails, and divergent coevolution in size subsequently occurred there.

In another system, the cyclic scenario mentioned above accords with many facts and seems correct in outline. On Caribbean islands, insectivorous lizards from the genus *Anolis* replace the ground-feeding insectivorous birds such as robins and jays that are so common in North America. And in the eastern Caribbean the small islands harbor one or two native species of anoles. For the islands between Guadeloupe and the Anegada Passage, analysis of Pleistocene fossil lizards and historical records point to the scenario diagrammed in figure 14.1 (Roughgarden et al. 1987). An island with a single species is invaded by a

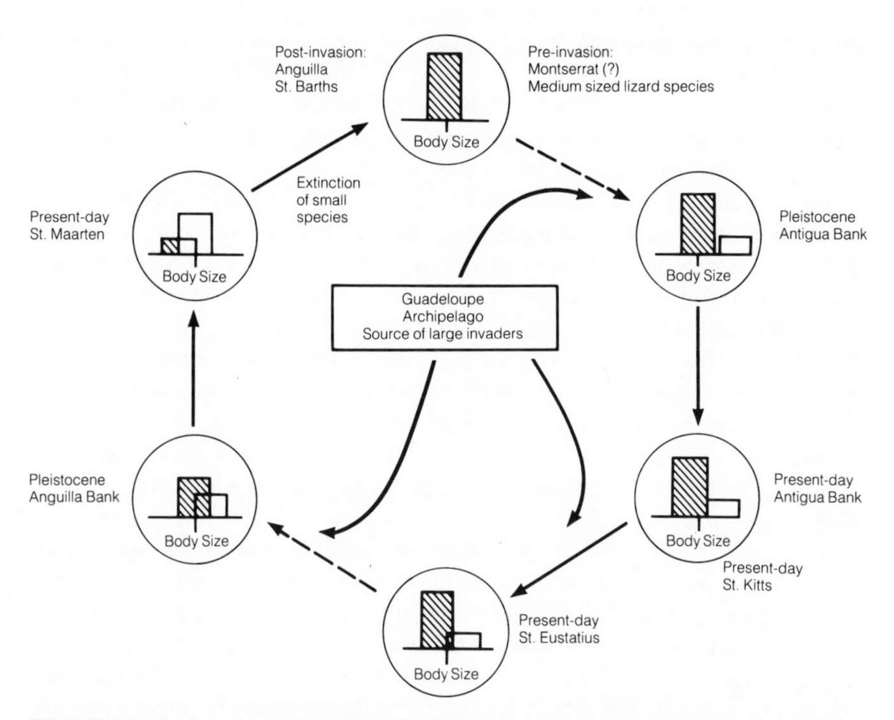

FIGURE 14.1. A taxon cycle involving *Anolis* lizard species in the northern Lesser Antilles. An island with one species evolves a characteristic "optimum" body size. It is invaded by a second species larger than the first. The second species evolves toward the size of the original resident, and the original resident itself becomes smaller. Then the species range of the original resident contracts, and it eventually becomes extinct. This hypothesis is presently supported by studies of fossil lizards, the ecology of colonization, present-day biogeography, and the phylogenetic relationships of anoles in the northern Lesser Antilles.

larger species. (Asymmetrical competition favoring large over small lizards implies that a successful invader is likely to be larger rather than smaller than the established resident.) Both then evolve a smaller size, and the larger species eventually converges upon the smaller one enough to contribute to its extinction. The classic hypothesis of character displacement from divergent coevolution is ruled out independently by the expermental studies of the ecology of colonization, the phylogenetic relationships of the lizards, and the fossil record.

A major theoretical highlight of community ecology during this period was the introduction by Cohen (1978) of theory pertaining to the entire food web of a community. This theory sought empirical generalizations from the food webs that had been published at that time, and relied on a graph-theoretic representation of the web. Causal explanations for these generalizations were suggested by

Pimm (1982) based on differential-equation models corresponding to specified food-web geometries. These topics are discussed in more detail by Cohen in this volume. Food-web theory marks an increasing interest in extending community theory beyond its early focus on the competing species at one trophic level, and this subject is discussed again later in this essay.

But now to the bad news. The extensive studies of competition during the last decade also revealed that the underlying mechanisms of competition are quite different among the systems in which limited-membership communities are documented. For example, in desert rodents, body size is not related to prey size, but to microhabitat use in a complex and as yet poorly understood way (Lemen 1978; Price 1978). Also, many communities were examined and found not to have clear membership rules at all, and those that do tend to be vertebrates with a closed population structure (cf. Roughgarden 1986; Schoener 1986). It became obvious that more than three qualitatively different pictures of community organization are needed, especially so in plant communities, where the issues of plasticity, microsite variation, and neighborhood interactions—for the uptake of nutrients, interception of light, and attraction of pollinators—need to be investigated. This is also true in animal communities, like those in ponds and streams, whose members live in different habitats at different phases of their life cycle, and in soft-bottom communities, where facilitating interactions are frequent.

More generally, the Lotka-Volterra competition equations, and their close relatives for other species interactions, are simply not concrete enough to use for more than the most abstract of metaphors about community organization, and when it comes to the nitty-gritty of how communities are really put together they fail to offer guidance. So, a major need now is to develop population models with more concrete reference, as discussed in the papers by Pacala and Kareiva in this volume, and to redo community theory on the basis of these. Undoubtedly some of the older mathematical results will be rederived in a new and more meaningful context, while new results will emerge as well, and the older treatments will come to be valued as practice exercises.

And finally, we are now seeing, I believe, the most fundamental challenge to community theory as we have known it: it leaves out what is often the rate-limiting step controlling community structure. This step is the mechanism transporting propagules to the community's boundary. Although for decades biogeographers have observed that certain taxa are prone to dispersal—for this is one of the classical traits of a good colonist—there is more to the matter. Most communities have many poor colonists, and they had to get there somehow, and the communities may be still undersaturated with respect to the niche space these taxa occupy. When communities are undersaturated, they become structured as much or more by the dynamics of the transport processes governing the introduction of propagules as by interactions among the residents. This is the missing Gleasonian dimension.

In the eastern Caribbean, for example, it is increasingly clear that even the Lesser Antilles, accepted by most biologists and geologists today as "obvious" exemplars of recently formed habitat that "must" have been colonized by overwater dispersal from large source faunas, is in fact a heterogeneous grouping of islands of different ages. Some are as old as parts of the Greater Antilles themselves, as illustrated in figure 14.2 (Roughgarden, Gaines, and Pacala 1987). The Guadeloupe region is the center of radiation for a group of lizards that is sister to a lineage spread throughout the Puerto Rico bank. Material in the Guadeloupe region broke away from Puerto Rico in the late Cretaceous or early Tertiary, carrying on it a fragment of the larger Puerto Rican fauna. Its placement in the center of the arc later led to its inoculating the relatively new habitat that formed nearby, between Guadeloupe and the Anegada Passage. Faunal buildup occurred on this newer habitat relatively recently. But the main point is that many Lesser Antillean islands are not early stages in the buildup of a complex fauna as had been believed. Instead, they are extractions from an already existing complex fauna that has developed in isolation for a long time.

The reluctance of scientists to consider the Gleasonian dimension sympathetically traces, I think, to its emphasis on fortuitousness, an emphasis that deflects attention away from transport itself. There is no reason, a priori, why the dynamics of dispersal and species transport is fortuitous or somehow intractable, and we whistle in the dark if we suppose that information taken within the boundaries of a community are sufficient to account for that community's composition and structure.

Predation and Limited Membership Reconsidered

A surprisingly similar evolution of thought is taking place in marine community ecology. The late 1970s provided several confirmations of the role of predation and disturbance in maintaining species diversity among the organisms attached to rocky substrates (Lubchenco and Menge 1978; Lubchenco 1978; Sousa 1979; Menge and Lubchenco 1981). Other studies, especially those in the subtidal zone, failed to confirm a dominance hierarchy in many cases (Sutherland and Karlson 1977; Osman 1977; Jackson 1979; Buss and Jackson 1979), but the role of predation and disturbance in ameliorating competitive exclusion and in promoting species diversity was largely unchallenged. As in the terrestrial competition studies, the phenomenology became richer—all wasn't as simple as originally hoped. Nonetheless, future research continued to be seen in terms of carrying out still more within-site experiments to determine the relative contributions of all the possible interactions among the residents at a site in the context of the predation and other mechanisms of disturbance impinging upon the site. This is the research program envisaged by Menge and Sutherland (1976) in a synthesis of the data available at that time.

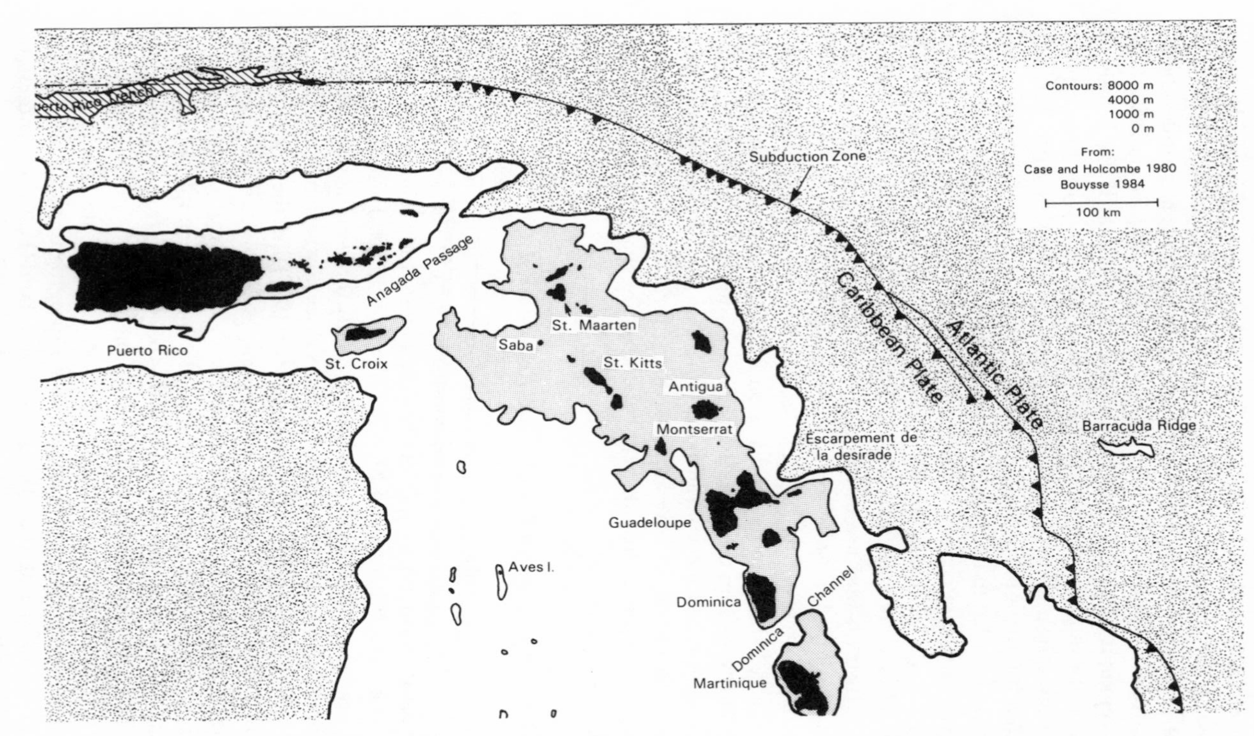

FIGURE 14.2. Tectonic map of the northern Lesser Antilles. The Guadeloupe area has deposits as old as the oldest in the Greater Antilles. It is also a center of endemism for the Bimaculatus group of *Anolis* lizards, a sister lineage of the Cristatellus group of the Puerto Rico bank. A part of the Guadeloupe archipelago apparently broke away from Puerto Rico, carrying a reptile fauna on it which then colonized the Northern Lesser Antilles as it formed, leading to the taxon cycle depicted in figure 14.1.

The party was spoiled by Underwood, Denley, and Moran (1983) who reported a negligible effect of predation on species diversity in the rocky intertidal zone of New South Wales in Australia, and that the actual composition of the communities was determined primarily by what happens to settle out of the water column onto the substrate. Disturbance, and interactions of any kind, did not offer much explanatory power. The exposition by Underwood and his colleagues expressed a strident condemnation of earlier studies, and sounded the dirge of hopeless and intractable stochasticity as underlying the phenomenon of larval settlement and its dominant role in determining community structure. And as with Gleason decades ago, a sympathetic scientific reaction to this finding has been slow on the uptake.

But today the importance of larval settlement in marine community ecology is receiving growing appreciation. Our laboratory at Hopkins Marine Station on Monterey Bay (Gaines and Roughgarden 1985) was perhaps the first American group to report similar findings to those of Underwood, Denley, and Moran (1983). We became sensitized to the importance of larval settlement by models tailor-made for marine invertebrates, such as barnacles, that have a sessile adult phase and a pelagic larval phase (Roughgarden, Iwasa, and Baxter 1985; Roughgarden and Iwasa 1986; Iwasa and Roughgarden 1986). From these models it quickly emerged that, within a local piece of intertidal habitat, the quantitative dynamics were strongly affected by the settlement rate, and that the qualitative properties, including its landscape appearance, were controlled by the settlement rate. Moreover, the dynamics of the species population, based on the cumulative production from many subpopulations of adults in habitats along the coast, may be governed by events and processes in the offshore water column as well. The idea is obvious in figure 14.3, where the kinetics of the life cycle are coupled at the larval-adult transition point with the dynamics of the offshore currents.

Indeed, the geographical distribution of the physical mechanisms that effect the return of offshore larvae to the parental stock seems to underlie the different results obtained by marine ecologists during the last decade. The West Coast of North America has a latitudinal gradient in the amount of coastal upwelling; most occurs in central and southern California, and becomes progressively less toward Oregon and Washington (Parrish, Nelson, and Bakum 1981). During upwelling, bottom water comes to the surface while surface waters move away from shore. Our studies in central California have shown that larval settlement in the barnacle, *Balanus glandula,* is high in years of limited upwelling, and low in years of high upwelling (Roughgarden, Gaines, and Pacala 1987). Moreover, in high-upwelling years the cause of low settlement is that larvae have been transported too far out to sea (up to 85 miles) for them ever to return (Roughgarden, Gaines, and Possingham 1988). In contrast, off the coast of Oregon and Washington, where some of the most influential studies of the 1970s in rocky intertidal ecology were carried out, the settlement rate is high, and settlement is

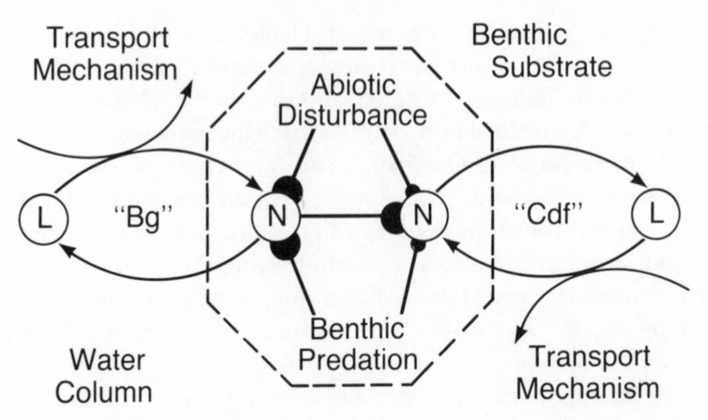

FIGURE 14.3. Schematic of interactions in a community of the rocky intertidal zone. Physical contact between sessile adult animals on the rocky substrate leads to competition representable as a hierarchy. As shown with a line terminating in a dot, an individual of species *Bg* overgrows or crushes an individual of species *Cdf* when they are in contact. Also, mortality from abiotic mechanisms, and from predation, affects *Bg* more than *Cdf.* Both species release a larval form to the water column that eventually must return and settle on vacent space to complete the life cycle. This return is mediated by physical transport mechanisms, as illustrated by the coupling of each life cycle to an arrow representing transport mechanisms. The importance of processes occurring on the substrate, such as competition between adults and the predation on them, relative to processes in the water column, such as predation on the larvae, their growth, or loss to offshore-moving currents, depends on how many larvae return to the adult habitat.

less often a rate-limiting process there. The importance of recruitment as a potential control of the community at a local site in the intertidal zone has recently been synthesized with previously identified factors by Menge and Sutherland (1987), and their review offers an excellent account of what is now known of the phenemonology of marine intertidal zone communities.

This new emphasis on the physical mechanisms of larval transport in marine ecology was christened "supply-side ecology" by Lewin (1986). It is the logical counterpart of the need to take account of the mechanisms of propagule introduction and community fragmentation in terrestrial community ecology. It is the Gleasonian dimension again, and in marine ecology is involved even at the population-dynamic level, not only the community level.

Unlimited Membership Reconsidered

Meanwhile, the theoretical picture of a community as a totally open system whose composition is determined by a balance of the immigration and extinction of species also came in for some rough sledding. Gilpin and Diamond (1976) established for the avifauna on the Solomon Islands that the immigration and extinction curves are strongly convex. They interpreted this as indicating that turnover is not uniformly distributed across the fauna; that turnover mostly involves the species prone to high immigration and extinction rates, while the species with inherently low immigration and extinction rates are relatively permanent residents on the island. An alternative explanation is that the community accumulates fauna until it is ecologically "full," and that any turnover represents only the failure of propagules to take once the community has filled up. Williamson (1983) presented census data for birds on the island of Skokholm west of Wales for the years 1928–1979 that showed no turnover in the established bird populations; he termed species turnover as "ecologically trivial." Lack (1976) had offered a similar view of island bird communities, especially in the West Indies. It is difficult, however, to decide on the basis of present evidence which of these alternatives is correct.

In amphibians and reptiles the species lists on islands form a nearly perfect pattern of nested subsets with respect to island area. (Schoener and Schoener 1983; Lazell 1983). Data from the Virgin Islands are illustrated in figure 14.4. This pattern is not consistent with random turnover because the islands are not random samples of the pooled fauna in the region encompassed by the dispersal capabilities of these organisms. In both the Virgin Islands and the Bahamas the pattern appears to be the result of extinction alone. The cays of the Virgin Islands were continuous with Puerto Rico, and the cays of the Bahamas were united in a few large banks, as recently as the last ice age, when the sea level was lower. As the sea level rose, small hill tops became the cays of today, retaining a special subset of its previous fauna—those capable of surviving in the face of habitat contraction.

While the nested-subset pattern is not consistent with turnover, it is consistent with, but does not necessarily indicate, species interactions. A noninteractional model for this pattern is simply that each species has a characteristic territory requirement per individual, and a threshold abundance to persist. The potential abundance of a species on an island is then the island's area divided by an individual's territory requirement. And if this is greater than the threshold, then it will be found on the island. Therefore, each species will have a characteristic island size at which it enters the fauna. The nested subset pattern then emerges when the whole fauna is considered as a function of island area. Still, any turnover observed would still represent the "trivial" failure of propagules to survive on an island that was too small for them.

As a whole, these contemporary studies of island biogeography show that a community is not simply a collection of all those who somehow arrived at the habitat and are competent to withstand the physical conditions in it. At the least, membership in the community is a matter of population dynamics, and this sometimes includes interactions with other members of the community.

So, after half a century of research, the need for a middle ground between the

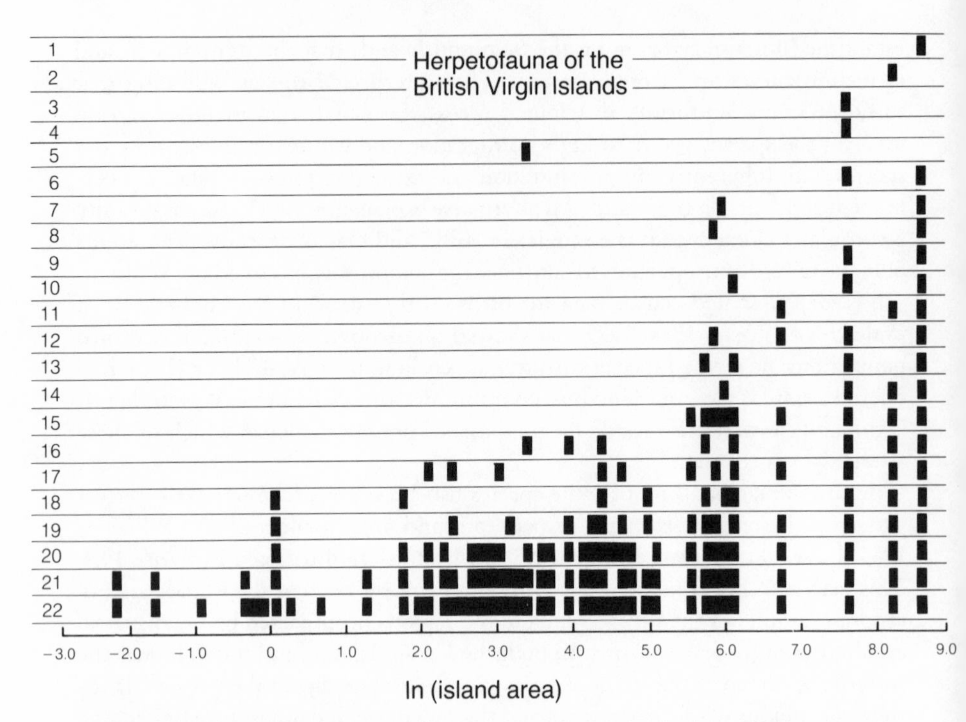

FIGURE 14.4. Nested subset pattern for the relation between island area and species occurrence for amphibians and reptiles in the British Virgin Islands. The vertical axis refers to different species (as noted below) and the natural logarithm of each island's area is plotted on the horizontal axis. Each dark rectangle indicates the presence of a species on a particular island. Species 22 is *Anolis cristatellus;* it survives on the smallest cays and has the best "bottlenecking" ability. Presumably for this reason the Cristatellus lineage is parental to the Bimaculatus lineage that is centered in the Northern Lesser Antilles, as noted in figure 14.2. The species corresponding to each of the numbers in the figure are 1: *Eleutherodactylus cochranae;* 2: *Iguana pinguis;* 3: *Sphaerodactylus parthenopion;* 4: *Bufo lemur;* 5: *Thecadactylus rapicaudus;* 6: *Eleutherodactylus antillensis;* 7: *Geochelone carbonaria;* 8: *Epicrates monensis;* 9: Eleutherodactylus schwartzi; 10: *Arrhyton exiguus;* 11: *Leptodactylus albilabris;* 12: *Amphisbaena fenestrata;* 13: *Iguana iguana;* 14: *Typhlops richardi;* 15: *Anolis pulchellus;* 16: *Alsophis portoricensis;* 17: *Mabuya sloanei;* 18: *Hemidactylus mabouia;* 19: *Anolis stratulus;* 20: *Ameiva exsul;* 21: *Sphaerodactylus macrolepis;* 22: *Anolis cristatellus.*

classically stated polarities was never clearer. A community's membership *is* structured by the transport processes that bring species to it, and it *is* structured by the population dynamics, including the species interactions of its members; and furthermore, it is rarely possible to focus on only one of these sides. That is to say, a community reflects both its applicant pool and its admission policies. And how these two sides contribute to community structure and composition has become well understood in an increasing number of cases in both marine and terrestrial habitats.

THE FUTURE

Of the many subjects needing theoretical research in the future, three seem worth emphasizing here.

1. We need to formulate a new generation of population and community models with the transport processes explicitly built in. In the marine context, one can imagine models in which the coupling among the local places within a species population is governed by equations from physical oceanography for fluid transport. Perhaps similar formulations could be developed for aerial transport in terrestrial contexts. Similarly, in community models, community disassembly should receive equal attention as community assembly, although a focus on assembly will continue to be valuable for understanding biological invasions and the fate of genetically engineered organisms released into the environment. Furthermore, the introduction of blocks of propagules should be considered, as when habitat is formed through the suturing of two previously isolated geological entities. One can then ask when two communities will "introgress," or alternatively, when an ecotone will remain between them coinciding with the geological suture zone. A related problem is to devise statistical methods for separating aspects of community structure determined by the identity and properties of the immigrants from those that result from interactions within the community.

2. Another major issue is to continue passing beyond one- and two-trophic level theory to theory for the entire community, as considered in the paper by Cohen in this volume. Perhaps a good place to begin is with an extension of the food web as a community's description. As a canonical representation of a community, a food web misrepresents what it does refer to and fails to include essential features of community structure and function; it is overdistilled. On the island of Anguilla, for example, the only predator on anoles is the kestrel. One pair occupies about 125 ha, an area containing about one million anoles. We estimate the chance of an anole being eaten by a kestrel each day is 1×10^{-5}. This surely compares with the probability that an anole dies from an arbitrary random hazard (flooding, tree falls, and so forth). Yet in any food web for the terrestrial community on Anguilla, the kestrel is at the top. All arrows lead there, and invite the hypothesis that kestrels somehow have a controlling role in the community. But they probably don't. The food web is misleading in this regard because the grossly disparate spatial scales for populations at different positions in the food web are omitted. Other omissions are sessile producers, mobile predators, and prey refuges. In short, organisms interact with others for reasons both of function and location, and these may be independent and not reducible to one or the other.

People, for example, interact with others professionally and also with their next-door neighbors. The food web represents only functional interactions, not positional interactions. The problem is not merely to look at more food webs at different places, but to enlarge the description itself to include space and position.

Another need is to allow for hierarchy in a food web, to allow nodes that are records, or webs themselves nested within the node and whose content is hidden from other nodes at a higher level. O'Neill's paper in this volume has raised the issue of hierarchy in ecosystem theory, and community theory needs to explore this too. Parasites pose a special problem for representation in a food web. Those with one host are presumably contained within the node of its host; but parasites with one or two intermediate hosts are hitchhiking along the links in the food web as a means of transport to their definitive host. This implies a kind of congruence between a parasite's life cycle and a section of the food web in which its hosts are embedded.

While the geometry of foodwebs is an important descriptive study in its own, the need to explain causally what is described will continue, as pioneered by Pimm (1982). Yet, I suspect that explaining a food web in terms of a system of differential equations may not offer the best way to proceed, especially in view of the findings by Smale on the mathematics of nonlinear dynamical systems cited earlier. Differential equations, especially in complex systems, require too much in their specification, and predict too much to test; just observing the trajectory of an N-species system requires censusing N species, an impossible task for N large. So we may need a new conception of what a dynamical model is for a community.

3. Finally, the seemingly natural bond that should exist between community and ecosystem ecology has been surprisingly slow to form. The flow of material and energy through ecosystems is, in part, mediated by population interactions. Material does not diffuse from producers to consumers, and thence to predators according to Fick's law, but by the kinetics of species interactions. It is in everyone's interest to get together on this.

CONCLUSION

Community ecology today has traveled far from its origins fifty years ago. Its central questions have evolved and its models have met with encouraging, though limited, success, so that now the goings-on in particular communities are well understood. Although present-day success consists of finding genuine answers in special cases, years ago one might have dispaired as to whether an ecological community was even scientifically tractable. Moreover, as a peer group, community ecologists now have more experience with how to propose, solve, and test models than at any time in the past. But future theory will not be a simple extension of present models. We must remain alert for altogether new metaphors, especially from less traditional sciences such as meteorology and computer science, and from areas of mathematics dealing with path-dependent processes and multiple scaling methods. And I trust we will be dreaming up some of our own too.

ACKNOWLEDGMENTS

I thank Peter Chesson and Joel Cohen for valuable comments on an earlier draft of this manuscript, and the National Science Foundation for the support that made this conference possible.

REFERENCES

Abrams, P. 1980. Consumer functional response and competition in consumer-resource systems. *Theor. Pop. Biol.* 17:80–102.

Abrams, P. 1986. Character displacement and niche shift analyzed using consumer-resource models of competition. *Theor. Pop. Biol.* 29:107–60.

Bowers, M., and J. Brown. 1982. Body size and coexistence in desert rodents: Chance or community structure? *Ecology* 63:391–400.

Brown, J. 1975. Geographical ecology of desert rodents. In M. Cody and J. Diamond, eds., *Ecology and Evolution of Communities,* pp. 315–41. Harvard University Press, Cambridge, Mass.

Brown, W., Jr., and E. Wilson. 1956. Character displacement. *Syst. Zool.* 7:49–64.

Buss, L., and J. Jackson. 1979. Competitive networks: Nontransitive competitive relationships in cryptic coral reef environments. *Amer. Natur.* 113:223–34.

Case, T. 1979. Character displacement and coevolution in some *Cnemidophorus* lizards. *Fortschr. Zool.* 25, 2/3: 235–82. Gustav Fischer, Stuttgart.

Chesson, P. 1986. Environmental variation and the coexistence of species. In J. Diamond and T. Case, eds., *Community Ecology,* pp. 240–56. Harper and Row, New York.

Chesson, P., and R. Warner. 1981. Environmental variability promotes coexistence in lottery competitive systems. *Amer. Natur.* 117:923–43.

Cody, M. 1974. *Competition and the Structure of Bird Communities.* Princeton University Press, Princeton, N.J.

Cody, M. 1986. Structural niches in plant communities. In J. Diamond and T. Case, eds., *Community Ecology,* pp. 381–405. Harper and Row, New York.

Cohen, J. 1978. *Food Webs and Niche Space.* Princeton University Press, Princeton, N.J.

Cohen, J., and C. M. Newman. 1985. When will a large and complex system be stable? *J. Theor. Biol.* 113:153–56.

Colwell, R., and D. Winkler. 1984. A null model for null models in biogeography. In

D. Strong, D. Simberloff, L. Abele, and A. Thistle, eds., *Ecological Communities: Conceptual Issues and the Evidence,* pp. 344–59. Princeton University Press, Princeton, N.J.

Connell, J. 1961a. The influence of interspecific competition and other factors on the distribution of the barnacle *Chthamalus stellatus*. *Ecology* 42:710–13.

Connell, J. 1961b. Effects of competition, predation by *Thais lapillus,* and other factors on natural populations of the barnacle *Balanus bananoides*. *Ecol. Monogr.* 31:61–104.

Connell, J. 1983. On the prevalence and relative importance of interspecific competition: Evidence from field experiments. *Amer. Natur.* 122:661–96.

Dayton, P. 1971. Competition, disturbance and community organization: The provision and subsequent utilization of space in a rocky intertidal community. *Ecol. Monogr.* 41:351–89.

Diamond, J. 1973. Distributional ecology of New Guinea birds. *Science* 179:759–69.

Diamond, J. 1975. Assembly of species communities. In M. Cody and J. Diamond, eds., *Ecology and Evolution of Communities,* pp. 342–44. Harvard University Press, Cambridge, Mass.

Elton, C. 1933. *The Ecology of Animals.* Reprinted 1966 by Science Paperbacks and Methuen, London.

Fenchel, T. 1975. Character displacement and coexistence in mud snails (Hydrobiidae). *Oecologia* 20:19–32.

Fenchel, T., and F. Christiansen. 1977. Selection and interspecific competition. In F. Christiansen and T. Fenchel, eds., *Measuring Selection in Natural Populations,* pp. 477–98. Vol. 19, Lecture Notes in Biomathematics. Springer-Verlag, New York.

Gaines, S., and J. Roughgarden. 1985. Larval settlement rate: A leading determinant of structure in an ecological community of the marine intertidal zone. *Proc. Natl. Acad. Sci. USA* 82:3707–11.

Gilpin, M., and J. Diamond. 1976. Calculation of immigration and extinction curves from the species-area-distance relation. *Proc. Natl. Acad. Sci. USA* 73:4130–34.

Gleason, H. 1926. The individualistic concept of the plant association. *Bull. Torrey Bot. Club.* 53:1–20.

Gleason, H. 1953. Dr. H. A. Gleason, distinguished ecologist. *Bull. Ecol. Soc. Amer.* 34:40–42.

Goel, N., and N. Richter-Dyn. 1974. *Stochastic Models in Biology.* Academic Press, New York.

Grant, P., and I. Abbott. 1980. Interspecific competition, island biogeography, and null hypotheses. *Evolution* 34:332–41.

Hirsch, M., and S. Smale. 1974. *Differential Equations, Dynamical Systems and Linear Algebra.* Academic Press, New York.

Horn, H. S., and R. MacArthur. 1972. Competition among fugitive species in a harlequin environment. *Ecology* 53:749–52.

Hubbell, S., and R. Foster. 1986. Biology, chance, and history in the structure of tropical rain forest tree communities. In J. Diamond and T. Case, eds., *Community Ecology,* pp. 314–30. Harper and Row, New York.

Hutchinson, G. 1959. Homage to Santa Rosalia, or why are there so many kinds of animals? *Amer. Natur.* 93:145–59.

Inouye, D. 1977. Species structure of bumblebee communities in North America and Europe. In W. Mattson, ed., *The Role of Arthropods in Forest Ecosystems,* pp. 35–40. Springer-Verlag, New York.

Iwasa, Y., and J. Roughgarden. 1986. Interspecific competition among metapopulations with space-limited subpopulations. *Theor. Pop. Biol.* 30:194–214.

Jackson, J. 1979. Overgrowth competition between encrusting cheilostome ectoprocts in a Jamaican cryptic reef environment. *J. Anim. Ecol.* 48:805–23.

Lack, D. 1947. *Darwin's Finches.* Cambridge University Press, Cambridge, England.

Lack, D. 1976. *Island Biology Illustrated by the Land Birds of Jamaica.* University of California Press, Berkeley.

Lawlor, L., and J. Maynard Smith. 1976. The coevolution and stability of competing species. *Amer. Natur.* 110:79–99.

Lazell, J., Jr. 1983. Biogeography of the herpetofauna of the British Virgin Islands, with a description of a new anole (Sauria: Iguanidae). In A. Rhodin and K. Miyata, eds., *Advances in Herpetology and Evolutionary Biology,* pp. 99–117. Museum of Comparative Zoology, Harvard University, Cambridge, Mass.

Lemen, C. 1978. Seed size selection in heteromyids, a second look. *Oecologia* 35:13–19.

Levin, S., and R. Paine. 1974. Disturbance, patch formation, and community structure. *Proc. Natl. Acad. Sci. USA* 71:2744–47.

Lewin, R. 1986. Supply-side ecology. *Science* 234:25–27.

Lubchenco, J. 1978. Plant species diversity in a marine intertidal community: Importance of herbivore food preference and algal competitive ability. *Amer. Natur.* 112:23–39.

Lubchenco, J., and B. Menge. 1978. Community development and persistence in a low rocky intertidal zone. *Ecol. Monogr.* 59:67–94.

MacArthur, R. 1968. The theory of the niche. In R. C. Lewontin, ed., *Population Biology and Evolution,* pp. 159–76. Syracuse University Press, Syracuse, N.Y.

MacArthur, R., and E. Wilson. 1963. An equilibrium theory of insular zoogeography. *Evolution* 17:373–87.

MacArthur, R., and R. Levins. 1967. The limiting similarity, convergence, and divergence of coexisting species. *Amer. Natur.* 101:377–85.

MacArthur, R., and E. Wilson. 1967. *The Theory of Island Biogeography.* Princeton University Press, Princeton, N.J.

May, R. 1973. *Stability and Complexity in Model Ecosystems.* Princeton University Press, Princeton, N.J.

May, R. 1974. Biological populations with non-overlapping generations: Stable points, stable cycles, and chaos. *Science* 186:645–47.

May, R. and R. MacArthur. 1972. Niche overlap as a function of environmental variability. *Proc. Natl. Acad. Sci. USA* 69:1109–13.

Menge, B., and J. Lubchenco. 1981. Community organization in temperate and tropical rocky intertidal habitats: Prey refuges in relation to consumer pressure gradients. *Ecol. Monogr.* 51:429–50.

Menge, B., and J. Sutherland. 1976. Species diversity gradients: Synthesis of the roles of predation, competition, and temporal heterogeneity. *Amer. Natur.* 110:351–69.

Menge, B., and J. Sutherland. 1987. Community regulation: Variation in disturbance, competition, and predation in relation to environmental stress and recruitment. *Amer. Natur.* 130:730–57.

Moulton, M., and S. Pimm. 1986. The extent of competition in shaping an introduced avifauna. In J. Diamond and T. Case, eds., *Community Ecology,* pp. 80–97. Harper and Row, New York.

Osman, R. 1977. The establishment and development of a marine epifaunal community. *Ecol. Monogr.* 47:37–63.

Pacala, S., and J. Roughgarden. 1985. Population experiments with the *Anolis* lizards of St. Maarten and St. Eustatius. *Ecology* 66:128–41.

Paine, R. 1966. Food web complexity and species diversity. *Amer. Natur.* 100:65–75.

Parrish, R., C. Nelson, and A. Bakun. 1981. Transport mechanisms and reproductive success of fishes in the California current. *Biol. Oceanogr.* 1:175–203.

Pimm, S. 1982. *Food Webs.* Chapman and Hall, London.

Price, M. 1978. Seed dispersion preferences of coexisting desert rodent species. *J. Mammal.* 56:731–51.

Roughgarden, J. 1972. Evolution of niche width. *Amer. Natur.* 106:683–718.

Roughgarden, J. 1975. Population dynamics in a stochastic environment: Spectral theory for the linearized N-species Lotka-Volterra competition equations. *Theor. Pop. Biol.* 7:1–12.

Roughgarden, J. 1976. Resource partitioning among competing species—a coevolutionary approach. *Theor. Pop. Biol.* 9:388–424.

Roughgarden, J. 1977. Coevolution in ecological systems: Results from "loop analysis" for purely density-dependent coevolution. In F. Christiansen and T. Fenchel, eds., *Measuring Selection in Natural Populations,* pp. 499–518. Vol. 19, Lecture Notes in Biomathematics. Springer-Verlag, New York.

Roughgarden, J. 1978. Influence of competition on patchiness in a random environment. *Theor. Pop. Biol.* 14:185–203.

Roughgarden, J. 1979. *Theory of Population Genetics and Evolutionary Ecology.* Macmillan, New York (reprinted 1987).

Roughgarden, J. 1986. A comparison of food-limited and space-limited animal competition communities. In J. Diamond and T. Case, eds., *Community Ecology*, pp. 492–516. Harper and Row, New York.

Roughgarden, J., and Y. Iwasa. 1986. Dynamics of a metapopulation with space-limited subpopulations. *Theor. Pop. Biol.* 29:235–61.

Roughgarden, J., S. Gaines, and S. Pacala. 1987. Supply-side ecology: The role of physical transport processes. In P. Giller and J. Gee, eds., *Organization of Communities: Past and Present,* pp. 459–86. Blackwell Scientific, London.

Roughgarden, J., S. Gaines, and H. Possingham. 1988. Recruitment dynamics in complex life cycles. *Science* 241:1460–66.

Roughgarden, J., D. Heckel, and E. Fuentes. 1983. Coevolutionary theory and the biogeography and community structure of *Anolis*. In R. Huey, E. Pianka, and T. Schoener, eds., *Lizard Ecology: Studies on a Model Organism,* pp. 371–410. Harvard University Press, Cambridge, Mass.

Roughgarden, J., Y. Iwasa, and C. Baxter. 1985. Demographic theory for an open marine population with space-limited recruitment. *Ecology* 66:54–67.

Rummel, J., and J. Roughgarden. 1985. A theory of faunal buildup for competition communities. *Evolution* 39:1009–33.

Sale, P. 1977. Maintenance of high diversity in coral reef fish communities. *Amer. Natur.* 111:337–59.

Schoener, T. 1983. Field experiments on interspecific competition. *Amer. Natur.* 122:240–85.

Schoener, T. 1984. Size differences among sympatric bird-eating hawks: A worldwide survey. In D. Strong, D. Simberloff, L. Abele, and A. Thistle, eds., *Ecological Communities: Conceptual Issues and the Evidence,* pp. 254–81. Princeton University Press, Princeton, N.J.

Schoener, T. 1986. Patterns in terrestrial vertebrate versus arthropod communities: Do systematic differences in regularity exist? In J. Diamond and T. Case, eds., *Community Ecology*, pp. 556–86. Harper and Row, New York.

Schoener, T., and A. Schoener. 1983. Distribution of vertebrates on some very small islands. II: Patterns in species number. *J. Anim. Ecol.* 52:237–62.

Simberloff, D., and W. Boecklen. 1981. Santa Rosalia reconsidered: Size ratios and competition. *Evolution* 35:1206–28.

Simberloff, D., and E. Wilson. 1969. Experimental zoogeography of islands: The colonization of empty islands. *Ecology* 50:278–96.

Simberloff, D., and E. Wilson. 1970. Experimental zoogeography of islands: A two-year record of colonization. *Ecology* 51:934–37.

Slatkin, M. 1974. Competition and regional coexistence. *Ecology* 55:128–34.

Smale, S. 1976. On the differential equations of species in competition. *J. Math. Biol.* 3:5–7.

Sousa, W. 1979. Experimental investigations of disturbance and ecological succession in a rocky intertidal community. *Ecol. Monogr.* 49:227–54.

Stenseth, N. 1985. Darwinian evolution in ecosystems: The Red Queen view. In P. Greenwood, P. Harvey, and M. Slatkin, eds., *Evolution: Essays in Honour of John Maynard Smith,* pp. 55–72. Cambridge University Press, Cambridge, England.

Sutherland, J., and R. Karlson. 1977. Development and stability of the fouling community at Beaufort, North Carolina. *Ecol. Monogr.* 47:425–46.

Terborgh, J. 1985. The role of ecotones in the distribution of Andean birds. *Ecology* 66:1237–46.

Tilman, D. 1982. *Resource Competition and Community Structure.* Princeton University Press, Princeton, N.J.

Tilman, D. 1986. Evolution and differentiation in terrestrial plant communities: The importance of the soil resource:light gradient. In J. Diamond and T. Case, eds., *Community Ecology,* pp. 359–80. Harper and Row, New York.

Underwood, A., E. Denley, and M. Moran. 1983. Experimental analyses of the structure and dynamics of mid-shore rocky intertidal communities in New South Wales. *Oecologia* 56:202–19.

Van Valen, L. 1973. A new evolutionary law. *Evol. Theor.* 1:1–30.

Werner, E. 1986. Species interactions in freshwater fish communities. In J. Diamond and T. Case, eds., *Community Ecology,* pp. 344–58. Harper and Row, New York.

Wiens, J. 1982. On size ratios and sequences in ecological communities: Are there no rules? *Annales Zoologici Fennici* 19:297–308.

Williams, E. 1983. Ecomorphs, faunas, island size and diverse end points. In R. Huey, E. Pianka, and T. Schoener, eds., *Lizard Ecology: Studies on a Model Organism,* pp. 326–70. Harvard University Press, Cambridge, Mass.

Williamson, M. 1983. The land-bird community of Skokholm: Ordination and turnover. *Oikos* 41:378–84.

Wilson, E. 1961. The nature of the taxon cycle in the Melanesian ant fauna. *Amer. Natur.* 95:169–73.

Chapter 15

Discussion: Structure and Assembly of Communities

MARTIN L. CODY

The purpose of this chapter is threefold: (1) to provide a general overview of the relations between theoretical and empirical ecology at the community level; (2) to reflect the tenor of the discussion in the Community Ecology section of the Asilomar conference; and (3) to attempt to identify, from my own perspective, some of the concerns of empirical ecologists for what theory is relevant, necessary, and sufficient, and for what topics a theoretical basis is still lacking but necessary. I use as source material the papers by J. Roughgarden and J. Cohen (chapters 13 and 14) as well as transcripts of the discussions. In addition I introduce topics, as relevant, which were discussed only peripherally or not at all during the conference because of insufficient time.

The community ecologists in this section of the conference were not easily segregated into theoretical versus empirical or field researchers. The overwhelming impression was one of appreciation by each contributor for the efforts and goals of the others, with the more theoretically inclined understanding the needs and challenges of fieldwork, and the more empirical welcoming the input of the less agarophillic. I hope to convey this sense of unity here.

THEORY AT THE COMMUNITY LEVEL

Since ecological communities are higher-order phenomena, in the sense that it is unlikely they will be well understood by simply adding the properties of their component species and populations, there is a need for community-level theory that is not necessarily based on the theory that has served well at the level of organisms and populations. Most ecological theory can be placed into one of three categories: (1)Natural Selection, or N-S, models, which deal with the properties of variable individuals in varying environments and operate generally through some optimization algorithm that, for example, maximizes fitness, or identifies optimal body size or the most efficient foraging strategy; these models are unlikely to be useful at the community level. (2) Lotka-Volterra, or L-V, models, which use differential equations of population growth and generally investigate the equilibrium solution, its stability, and the system dynamics around the equilibrium. L-V models have been important in the initial phases of community modeling and have led to a significant number of insights. These include especially the notion of a stable coexistence of particular species, the propensity of stable communities to return to equilibrial composition after local perturbations, the idea that some communities might be invasion-resistant to potential colonists, and the result that different, alternative species sets might be nearly equally likely, and persistent, in a given environment.

But for community-level theory the models need to diversify further, and a promising third class of models can be termed (3) Extinction-Colonization, or E-C, models. Good examples of E-C models are the MacArthur-Wilson (1967) equilibrium theory of island species numbers, and the theories based on the stochastic aspects of dispersal to and persistence in sites within patchy habitats (Horn and MacArthur 1972; Lande 1987; Levins and Culver 1971; Yodzis 1978). Given that environments are variable, that habitats are patchily distributed, and that most communties are not self-contained nor well buffered from extrinsic factors, including the biotic (epidemics, irruptions) as well as the physical (droughts, unseasonal frosts), then their status quo is not at all assured. For example, a certain organism cannot be expected to maintain the same competitive dominance at a given site over all ranges of conditions, and thus local population fluctuations and their extremes—colonization and extinction—will be imposed on community size and composition. To understand the resulting patterns of community change through space and time, both deterministic and probabilistic elements are needed. Such stochastic elements as dispersal and demographic extinction of E-C models will be central, but the elements of biologically deterministic growth, competition, and predation, as in the more classical L-V models, would be retained in the hybrids.

THE MAIN ISSUES IN COMMUNITY ECOLOGY

Species Packing and Community Membership

Although, as Roughgarden suggests in Chapter 14, there is now ample evidence that many communities are structured into predictable species sets with limited membership according to certain species-packing or assembly rules, there is also evidence that other communities are not. Communities might rank along a

"Clements-Gleason" axis, with respect to their degree of internal organization, broadly according to taxonomic criteria, in much the same way, and for similar reasons, as did species in the density-dependence versus density-independence debates of population regulation some thirty to forty years ago. Thus studies of community structure appear to be important to ecologists concerned with, for example, insectivorous birds or lizards, or with suites of predatory mammals such as mustelids, or with insects such as pompilid wasps; more equivocal to those studying, say, reef fish or perennial plants; and less relevant to those studying herbivorous insects, marine pelagics, or annual plants. The latter might even be communities without organizational rules—the "anomic" communities of Rosenzweig (1987).

Even in communities that might be clearly regulated by density-dependent competition or predation at some particular time or place, there may be strong density-independent components that make them difficult to model and to understand in general. Thus filter feeders or grazers might segregate ecologically on intertidal rocks along well-defined niche axes, yet owe certain aspects of their community structure, such as shifts in relative abundance of component species, to the vagaries of survival in and settlement from the plankton. Similarly, woodland or grassland birds might have reduced density dependence in overwintering sites, and contend with even less predictable influences during long-distance migration.

One of the stronger areas of interaction between theory and field data in community ecology is in studies of density compensation, where a given base-line (e.g., "mainland") community can be used to generate the most likely viable subsets of that community. Such tests of the theory (and of course of the relevance of the field data) are powerful tools in predicting the composition of successive species reductions or levels in stadial decomposition (see Diamond 1975) to smaller size. But here also the simple L-V theory appears only marginally sufficient, for it appears that observations deviate from predictions because fewer species than potentially expected are retained in the reduced community (Cody 1983a). Two possible explanations are (1) higher incidence of demographic extinction at low density, such that some species will not be retained at low density even though they might predictably occur, and (2) the evolution among some of the remaining species to broader niches and expanded resource use, hastening the exclusion of the less resourceful and further reducing the community size. It would seem feasible to expand the L-V models to encompass these possibilities.

The question of whether results from small communities can be generalized to larger communities is equivocal, since larger communities might include, with increasing likelihood, different kinds of species as well as more of similar kind. The likelihood of intransitivities will increase with larger species sets, and the chances, for example, of the coexistence of competitor species being mediated by predation or of predator species mediated by alternative prey would seem to enhance qualitatively the behavior of larger communities; extrapolations from simple theory should take such factors into consideration.

A still underexploited way to evaluate equilibrial aspects of community composition and organization may be through studies of convergent evolution: the extent to which the same community structure is reached and maintained in similar environments in different locations. This is especially valuable in well-separated areas in which the taxonomic similarities of community members are low. Similarly, community convergence can also be tested in the successive regeneration of structure after catastrophic community overhaul, as in vegetation records preserved in London clay after repeated Pleistocene glaciations. Overall the evidence is equivocal, with some consumer communities providing excellent matches across continents (e.g., insectivorous birds in Mediterranean-climate scrub on several continents), and others deviating for reasons strongly associated with inexact matches in the resource base (e.g., more species and higher densities of nectarivorous birds in southern African and southwest Australian Mediterranean-type scrub; Cody and Mooney 1978, and Cody unpublished). Yet other consumer communities show little or no correspondence among locations (e.g., bracken herbivores worldwide; Lawton 1983); or they show communities convergent in some characteristics such as the important niche axes, but not in others, such as species numbers. Clearly, the specifics of consumer-resource coevolution can critically affect convergence in community structure. With interactions less general than those of filter feeders or insectivores, such as in host-parasite interactions, or with the herbivores of plants with specific defense chemistries, these coevolutionary effects may override any tendency for convergent community structure. Furthermore, the historical effects of speciation patterns and adaptive radiations, and the locally pertinent influences of topographic isolation, restricted dispersal, and chance, provide powerful constraints on the possibilities for convergence. With regard to the many notable nonconvergences, such as the remarkable lack of succulent-stemmed plants in arid Australia, we can merely cite weak arguments about historical accident, and speculate vaguely about the chances of success should stem-succulents fortuitously invade that continent.

Invasibility

Community noninvasibility is often considered to be a criterion of equilibrium and stability. The requirements of a species being able to reach the site and then being able to increase its numbers will of course involve very different aspects of its ecology so that both need to be considered in terms of invasion potential. A species might arrive and invade, yet never reach any substantial population size. It might be found consistently only because of high arrival rates, it might be found in substantial numbers only occasionally, or it might in fact never achieve any significant ecological role in the community. Some threshold of density, importance, biomass, or incidence might be required to satisfy a criterion for invasibility.

The role of disturbance, and of environmental variability in space and time, appears to be critical to the invasibility of many communities, but a theoretical treatment of this aspect is not well developed. Certainly many communities owe their existence to repeated disturbances (e.g., periodic brush fires, intertidal log bashing, freshwater flooding of saline marshes, etc.), but more subtle disturbances might well leave more permanent communities open to intermittent invasions, with results that could then persist for a long time. Different taxa have very different persistence times in unsuitable environments: the fortuitous arrival of one seed per century might ensure the continuous presence of a particular tree in habitat where it can never reproduce, whereas small mammals would be required to demonstrate their fitness in the same habitat by reproducing many times a year in order to maintain a population.

Some anomalies with respect to invasion occur because alien plants are most successful, yet most troublesome from management and aesthetic perspectives, in the most species-rich vegetation, just the opposite of what one would expect from the standpoint of resource partitioning and species packing. Examples are the species of Australian shrubs that now dominate parts of the South African fynbos, a part of the Cape flora that may be the richest, in species by area, in the world. Even the mechanics of island species-area curves are not well understood from the point of view of colonization success. With increasing size, islands appear to be invasible by increasingly large numbers of species, but in most cases the alternative explanations of (1) larger targets; (2) larger area per se, and thus reduced chances of stochastic extinction; or (3) greater variety of habitats, and of resources in general with increasing area, have rarely been distinguished.

Rare Events, Past Events, Keystone Species

There is a nagging doubt among both theoretical and empirical community ecologists that rare events may be important far beyond their qualitative or numerical visibility. Rare events are difficult to observe, study, and understand, and hard to incorporate into theory; yet it seems a real possibility that their influence on communities, in makeup and structure, is significant. In the same way, key mutations might dramatically alter the course of subsequent evolution in a lineage and cause many ramifications. They may, for example, produce a plant, and subsequently a plant family, with herbivore-retardant cardiac glycosides; then butterfly families may evolve, with the enzymes to handle these poisons and incorporate them into their own defense, and later there may be mimetic butterflies in unrelated families which come to resemble the distasteful models, and so on. Such mutations (new genes, recombinations, or linkage groups)—for specific plant toxins, for key digestive enzymes, for aposematic and mimetic coloration—must be rare events in the evolutionary histories of plant and butterfly families; yet their results dominate—in fact, generate—the ecology and biogeography of large sets of coevolutionarily related species.

A wide range of phenomena is included under the banner of the "rare but probably important." Those events least tractable are occurrences with rates measurable only in geologic time ($>10^6$ years; e.g., the opening and closing of the Isthmus of Panama and the resulting faunal interchange, or the rafting north of the Baja California peninsula with its riders representing the Mexican highlands fauna and flora), or measurable in evolutionary time ($>10^4$ years), such as key but chancy mutations (as illustrated above). The chance crossing of barriers to dispersal, or the accumulation of the various facets of reproductive isolation, are similarly critical incidents in the evolution of new species, in the demise of old ones, and in fluctuations in geographic ranges, but they are incompletely recorded and rarely modeled (see, e.g., Rosenzweig 1975).

More amenable to evaluation are rare events on an "ecological" time scale (10^1–10^3 years), such as the widespread introductions of alien species worldwide with mankind's global traveling, and the major alterations of landscapes and habitats due to our agriculture and other land use and misuse. Globe-trotting ants in Polynesia, Mediterranean weeds in the California deserts, and the critical destruction of tropical forests are typical of such events. The hundred-year droughts or fifty-year floods, the mass fruiting of some bamboos, the reorganization of communities following the arrival of a novel and major pest (Dutch elm disease in the northeastern United States, avian malaria in Hawaii, perhaps AIDS in man), the removal of a key predator (e.g., sea otters from much of the Pacific coast; see Kitching 1986), or the startling and destructive rise of an erstwhile minor predator, such as the boom in sunstars on the Great Barrier Reef, are all examples of rare and seemingly isolated events that can have a major and lasting impact on ecological communities.

A major concern is not with events that are documented as rarities, but with those that have occurred unrecorded and with undetermined effects on present-day community structure. The various sorts of events mentioned above have the capacity to "ghost" the present, and without their documentation it might be that the present-day ecologist cannot correctly interpret the structure of present-day communities. Not only might community-level characteristics bear the stamp of events long past except for their current ramifications, but so also might the various properties of component species. Connell (1980) suggested that patterns of ecological segregation among species might be attributable to the "ghost of competition past," since ongoing competition is often hard to observe or prove. Yet such "ghost" events are surely no less real for their having taken place before the ecologist arrives to observe their consequences; they are only less easily studied, and surely not limited to competition.

In a similar sense, predation might be a rare event, in that the chances of an organism falling prey to a particular predator might be extremely low, yet the evolution of traits in potential prey for avoiding predation might be overwhelmingly obvious. Perhaps a dominant constraint on foraging behavior in oak-woodland warblers is the threat of predation by sharp-shinned hawks, yet the hawk is scarce and the event is almost never observed; such constraints are hard to quantify, but must nevertheless exist.

There appear to be various sorts of "keystone" species that can dramatically affect communities and component species even though they may appear numerically unimportant either in time or space. Rare predators, disease organisms and tapeworms, brood parasites such as cuckoos that irrupt only infrequently (perhaps when host densities cross certain thresholds), and certain prey species that, though rare, provide a predictable fallback for predators in times of food shortage (e.g., the fungi that are a critical slug food in dry summers in northwestern forests) share a certain inconspicuousness and numerical insignificance. Community ecologists must distinguish rarity from unimportance, and theoreticians must model more realistically the infrequent with the routine.

Community Perturbations, Catastrophes, and Extinctions

One sure consequence of an expanding and demanding human population is that harmonious natural communities at equilibrium with their surroundings will become increasingly difficult to find and study. Community ecology will eventually become the study of species' extinction in fragmenting and vanishing habitats. Conservation biology needs both empirical and theoretical work to try to offset the current and predicted trends to higher extinction rates, and as yet only a modest start has been made (see Soulé 1986). Brown (1984) finds strong correlations between the density of a species and proximity to the center of its geographical range, and between its peak density and the breadth of the geographical distribution. Given that extinction is the simultaneous reduction of both population density and geographic range to zero, the various possible paths to extinction (see fig. 15.1) need to be examined. Are any of these paths reversible, and if so, all to the same extent? Are the rates of progression along them similar, and are they taxon- and habitat-specific?

Woody plants, salamanders, and mammals will respond in quite different ways to reduced ranges and reduced numbers, given that habitat fragmentation is a function of species vagility, that persistence times in islandlike fragments are very species-specific (Case and Cody 1987), and that tolerable inbreeding and genetic homozygosity levels vary widely among taxa (reviews in Soulé 1986). The path to extinction might be much more rapidly traversed in birds and mammals than, say, in tropical-forest trees. Some taxa, such as scallops, appear from the geological record to have "innately" high turnover, with both high extinction rates and high generation rates of new species. The positive feedback loop from range

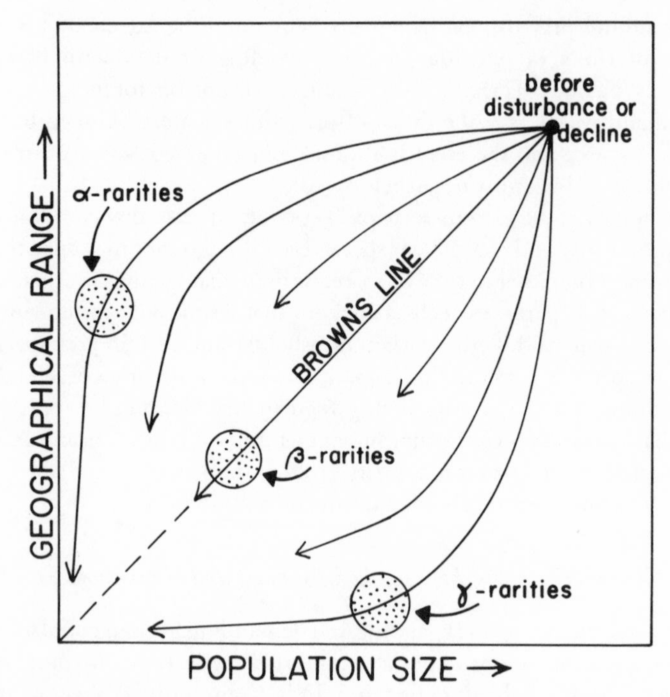

FIGURE 15.1. Possible tracks of declining abundance and shrinking range, exhibited collectively by the species in a taxon (genus, family), or by a single species over the course of time.

fragmentation to the generation of new species in new biotic environments might be much more significant in some taxa than in others.

Some marine communities, such as the fish in the English Channel or assemblages of benthic animals, seem to undergo wholesale changes, perhaps linked to subtle changes in their physical environment, in a relatively short time span (one or a few decades). The reconstitution of plant communities in postglacial times might represent a similar phenomenon on a time scale of about one hundred times longer. The interplay of local extinction, environmental changes over the longer term, species range expansions and contractions, etc., contribute to these vague patterns in poorly understood ways. The theory needs to make a smooth transition between events operating at the local community level and those determining biogeography. This work will require a major effort, especially where potential geographic ranges are patchily and ephemerally occupied because of biotic as well as physical limiting factors.

"Supply-Side" Ecology, Mass Effects, and Open Communities

A theme that surfaced several times during the discussions at the conference is that often only to a limited degree can the communities studied by ecologists be realistically regarded, measured, or modeled as self-contained. Such reservations are clearly appropriate in intertidal communities, in which many residents have pelagic larval stages. The problems associated with onshore and offshore transport are considerable, but at least (in Roughgarden's models, chapter 14) they are being considered. Matters of sampling effort and scale are especially challenging with regard to the potentially wide movements of larvae within ocean currents, and the potential problems of generalizing from small-scale models to those for larger systems were already mentioned. Evaluating the extent of density dependence in these transport processes will be difficult, and even then the capacity of predators to regulate prey, given some form of density dependence, is a complex issue (see Erlinge et al. 1983; Kidd and Lewis 1987).

The same problems are inherent in other systems, such as those of well-studied terrestrial bird communities. Fretwell (1972) discussed the possibilities for breeding bird communities of temperate-zone migrants being as much or more determined by processes in wintering grounds than by resource limitation in the summer. In North American grasslands, the bird community composition and organization seem more predictable in the eastern plains where the members are resident or make short (within-plains) migrations, but become less predictable in the western plains and in shrub-steppe, from which species make longer migrations to unpredictable overwintering sites in the southwestern deserts (Cody 1985). Long-distance bird migration, like the offshore transport of intertidal larvae, stand a good chance of being affected, even dominated, by physical factors such as the vagaries of climate in transit or the shifts of ocean currents that are wholly extrinsic to the community of immediate interest. It is likely that even small variations in these extrinsic factors could have large effects on the community and the makeup and relative abundance of its constituents at the breeding grounds. And in all likelihood the variations in such extrinsic factors will be larger than a few percentages.

To date, the common, and reasonable, approach to this difficulty has been to ask how the assemblage resolves itself into a set of consumers, given that we can expect variations in species and numbers of individuals jostling for community membership. Is the actual composition relatively invariant considering the variations in the potential, and how does the resulting assemblage match the resources available? Currently there is a lot of interest in the more challenging alternative of trying to understand the factors that generate the variations in potential community membership, and in bringing more of the "extrinsic" events into direct consideration as an integral part of the community ecology.

Most terrestrial landscapes are a rather complicated mix of different habitat types of different and variously intergrading areal extents, shapes, and distributions. Studies of a community of some particular taxonomically or trophically delimited species subset in one of these habitats are open to the risk that some community attributes are due to positional effects of the study site. The circumstances of type and abundance of surrounding habitat, for instance, might in fact predominate in the species number or composition at the selected site. For example, many bird species typically found in karoo vegetation breed in South African fynbos at sites where this habitat is largely surrounded by karoo, and some woodland birds extend into the fynbos as breeding members even when woodlands are common nearby (Cody 1983b). In plants as well, the species composition of a vegetation type that fingers into another bears the stamp of this second sort of vegetation (Auerbach and Shmida 1987). These influences are twofold: they might be due largely to the copious production of propagules from neighboring nabitat, in which case they are more appropriately termed "mass effects," or they might be better understood by the increased opportunities for invasion of smaller and isolated habitat patches that are species- or numbers-poor in terms of their normal residents, in which case they are better regarded as "areal-" or "isolation-effects." Such effects depend on a variety of ecologically interesting factors, such as the plasticity of different species in habitat use or their flexibility in germination sites; on the determinants of opportunities for living in nontypical habitat, as indicated by the slopes of species-area relations and consequent species impoverishment of isolated habitat patches; on the dispersal of propagules across variously alien habitat; and on the efficacy of nonpreferred habitat as barriers to dispersal, to name just a few.

Overall, there may be many occasions for the study of relatively self-contained communities within geographically large tracts of homogeneous habitat. But it seems that the consideration of mass and areal effects in opening up communities to extrinsic influences will be increasingly necessary, as habitats become ever more fragmented into small and isolated patches.

Food Webs and Community Organization

As summarized in Chapter 13 by Cohen, several recent avenues of theoretical study on various aspects of food webs promise to contribute to our understanding of community organization in novel and promising ways. The analogy of food webs to road maps of community organization might seem facile but brings out the possibilities for study at different scales and resolution. Food webs are quite commonly used to report descriptive facts about communities of several trophic levels, while they skirt the difficulties inherent in this concept. Food webs exhibit various sorts of structural regularities of chain length, diet breadth, and overall connectance, properties that are apparently not those expected to obtain in, for example, randomly assembled webs; the new theories appear robust, with a low ratio of assumptions to results.

While the theoreticians call for more food-web data that is more carefully collated and in standardized format, the empiricists caution that much of the extant food-web data may be of limited value for the sorts of derivations undertaken. Published food webs might be little more than superficial summaries of the system, drawn up with artistic license, omitting rare species, and stressing the author's own biases as to where the important interactions lie. Food-web details might subsequently be submerged with such devices as "trophic species" as nodes rather than separate biological species; the condensation of different species with widely differing properties into a single artificial unit seems to be risky at best.

There appear to be interesting correspondences between the food webs and their more orthodox representation as general dynamical systems; the recently developed loop analysis (Puccia and Levins 1985) could provide the bridge here, stressing as it does the qualitative aspects of these systems (as in food webs), rather than quantitative details that are usually nonexistent. Some effort seems worthwhile to strengthen this bridge and to determine whether the results of food-web theory can be obtained through alternative systems models.

COMMUNITY-LEVEL PATTERNS IN SEARCH OF A THEORY

Some areas of empirical work in community ecology, as hinted above, have progressed to the point of identifying patterns, but as yet have received little or no theoretical development. In fact, the larger part of the efforts in community ecology over the last decade has appeared to retract from communities per se to concentrate on limited subsets of its members, and to emphasize the properties of component species. But in the other direction, community ecology leads naturally to studies of species diversity at the continent-wide scale, and to the elucidation of patterns of species densities, diversities, and distribution over gradients of varied habitats and diverse topography that characterize continental land masses. The patterns are inadequately documented, their controls poorly understood, and their evolution rarely speculated; particularly in view of its potential as a conservation and management tool, it would seem that community ecology should progress rapidly in this direction.

Our theoretical understanding of communities, as limited-membership species sets with a degree of stability and invasion resistance and a certain capacity for density compensation, is useful for predicting community size within habitats of fixed and limited resources. But this aspect of species diversity, α-diversity, is just one component of diversity. Another is β-diversity, species turnover rate between habitats (see, e.g., Cody 1986 for an overview of diversity components), and here both theory and data are less well developed. It appears that β-diversity has a lot to do with the relative and absolute extents of different habitat types and the distances and barriers among them. Rare habitat will seldom retain species endemic to it, and common habitat will support species that extend their ambits into neighboring habitat along the gradient, as described above for bird species in southern Africa. To generate predictors for β-diversity, the ecologist will have to quantify and interrelate the mass, areal, and isolation effects discussed above.

The third diversity component is γ-diversity, species turnover within habitats between different parts of the habitat range; γ-diversity measures the extent to which ecological counterparts occur as allopatric replacements throughout the habitat type. For example, widely and contiguously distributed habitat, such as the mulga in southwest Australia, has a large and predictable component of core bird species (for a relatively low 36% species turnover among census sites within the habitat), whereas the more fragmented woodlands in the same region have higher species turnover, 48%, among sites. Heaths occur naturally in patches, often small and well isolated from each other, on the most nutrient-poor soils; bird species turnover between sites, γ-diversity, is much higher (69%) in these heath habitats (Cody 1986). Different taxa differ widely in their diversity components (see table 15.1), in as yet poorly understood ways.

TABLE 15.1

Variability among diversity components in some southwest Australian taxa.

Diversity component:	α	β	γ
Acacia			
Cum. spp. # (and % of taxon):	12 (1%)	≈40 (5%)	850 (100%)
Banksia			
Cum. spp. # (and % of taxon):	4 (5%)	≈30 (41%)	73 (100%)
Lichenostomus			
Cum. spp. # (and % of taxon):	2–3 (13%)	6–7 (36%)	18 (100%)
Acanthiza			
Cum. spp. # (and % of taxon):	4 (33%)	6 (50%)	12 (100%)
Artamus			
Cum. spp. # (and % of taxon):	3–4 (50%)	4–5 (75%)	6 (100%)

NOTES: *Acacia* is the largest Australian plant genus; *Banksia* is the second largest genus in the family Proteaceae; *Lichenostomus* is the largest genus in the bird family Meliphagidae (honeyeaters); *Acanthiza* (thornbills) is the largest genus in the family Acanthizidae (scrub wrens, "warblers"); *Artamus* is the single genus in the wood swallow family Artamidae. Figures in the table give cumulative species numbers and percentage of taxon total, beginning with α-diversity within southwest Australian sites, and then including additional species on local habitat gradients (β-diversity), and finally species in other geographical areas in Australia (γ-diversity.

SOURCE: Cody, unpublished data and general sources.

It is noteworthy that in the development of a theory of island biogeography, little attention has been paid to the variety of habitat types within islands (generally subsumed under island size) and the distribution of taxa over or dispersal across habitat gradients (β-diversity factors); and most islands are small enough to preclude geographic replacements (γ-diversity) within habitats. Thus the island theory constitutes a particular and simplified subset of a more general class of E-C models. The extension of this sort of theory to the more complex "mainland" land masses seems to constitute a timely challenge for community ecologists and might serve as an umbrella under which many of the themes in community ecology could be united. The emerging concepts of mass effects and areal effects are likely to contribute to this sort of theory, and the connections to the recently synthesized notions of "landscape ecology" (Forman and Godron 1986) are clearly worth pursuing. The common interests and common goals of the community ecologists at the Asilomar meeting indicate that these goals are quite realistic.

ACKNOWLEDGMENTS

Thanks are due to the organizers and sponsors of these very useful meetings, and especially to the participants in the discussions on community ecology, whose many ideas, observations, and insights I have tried to summarize in this chapter. These participants included R. Anderson, P. Chesson, J. Cohen, S. Gaines, M. Hassell, P. Kareiva, S. Pacala, R. Paine, S. Pimm, T. Powell, H. R. Pulliam, L. Real, M. Rosenzweig, J. Roughgarden, D. Simberloff, W. Sousa, S. Stanley, J. Steele, G. Sugihara, D. Tilman, D. Urban, and E. O. Wilson.

REFERENCES

Auerbach, M., and A. Shmida. 1987. Spatial scale and the determinants of plant species richness. *Trends in Ecol. & Evol.* 2:238–42.

Brown, J. H. 1984. On the relationship between abundance and distribution of species. *Amer. Natur.* 124:255–79.

Case, T. J., and M. L. Cody. 1987. Testing theories of island biogeography. *Amer. Sci.* 75:402–11.

Cody, M. L. 1975. Towards a theory of continental species diversities. In M. L. Cody and J. M. Diamond, eds., *Ecology and Evolution of Communities,* pp. 214–57. Belknap Press of Harvard University, Cambridge, Mass.

Cody, M. L. 1983a. Bird species diversity and density in South African forests. *Oecologia* 59:210–15.

Cody, M. L. 1983b. Continental diversity patterns and convergent evolution in bird communities. In F. Kruger, D. T. Mitchell, and J.U.M. Jarvis, eds., *Ecological Studies #43,* pp. 347–402. Springer-Verlag, Vienna and Berlin.

Cody, M. L. 1985. Habitat selection is grassland and open country birds. In M. L. Cody, ed., *Habitat Selection in Birds,* pp. 191–226. Academic Press, Orlando, Fla.

Cody, M. L. 1986. Diversity and rarity in Mediterranean ecosystems. In M. Soulé, ed., *Conservation Biology,* pp. 122–52. Sinauer, Sunderland, Mass.

Cody, M. L., and H. A. Mooney. 1978. Convergence versus nonconvergence in Mediterranean climate ecosystems. *Ann. Rev. Ecol. Syst.* 9:265–321.

Connell, J. H. 1980. Diversity and the coevolution of competitors, or the ghost of competition past. *Oikos* 35:131–38.

Diamond, J. M. 1975. Assembly of species communities. In M. L. Cody and J. M. Diamond, eds., *Ecology and Evolution of Communities,* pp. 342–444. Belknap Press of Harvard University, Cambridge, Mass.

Erlinge, S., G. Gòransson, G. Hògstedt, O. Liberg, J. Loma, I. Nilsson, T. von Schanz, and M. Sylvén. 1983. Predation as a regulating factor in small rodent populations in southern Sweden. *Oikos* 40:36–52.

Forman, R.T.T., and M. Godron. 1986. *Landscape Ecology.* Wiley, New York.

Fretwell, S. H. 1972. *Populations in a Seasonal Environment.* Princeton University Press, Princeton, N.J.

Horn, H. S., and R. H. MacArthur. 1972. Competition among species in a harlequin environment. *Ecology* 53:749–52.

Kidd, N.A.C., and G. B. Lewis. 1987. Can vertebrate predators regulate their prey? A reply. *Amer. Natur.* 130:448–53.

Kitching, R. L. 1986. Predator-prey interactions. In J. Kikkawa and D. J. Anderson, eds., *Community Ecology: Pattern and Process,* pp. 214–39. Blackwell Scientific, Oxford.

Lande, R. 1987. Extinction thresholds in demographic models of territorial populations. *Amer. Natur.* 130:624–35.

Lawton, J. 1983. Herbivore community organization: General models and specific tests with phytophagous insects. In P. W. Price, C. N. Slobodchikoff, and W. Gaud, eds., *A New Ecology: Novel Approaches to Interactive Systems,* pp. 206–27. Wiley, New York.

Levins, R., and D. Culver. 1971. Regional coexistence of species and competition between rare species. *Proc. Natl. Acad. Sci. USA* 68:246–48.

MacArthur, R. H., and E. O. Wilson. 1967. *The Theory of Island Biogeography.* Princeton University Press, Princeton, N.J.

Puccia, C. J., and R. Levins. 1985. *Qualitative Modeling of Complex Systems.* Harvard University Press, Cambridge, Mass.

Rosenzweig, M. R. 1975. On continental steady states of species diversity. In M. L. Cody and J. M. Diamond, eds., *Ecology and Evolution of Communities,* pp. 121–40. Belknap Press of Harvard University, Cambridge, Mass.

Rosenzweig, M. R. 1987. Community organization from the point of view of habitat selectors. In P. Giller and J. Lee, eds., *Organization of Communities: Past and*

Present. British Ecological Society Symposium #27, Blackwell Scientific, Oxford.

Soulé, M., ed. 1986. *Conservation Biology.* Sinauer, Sunderland, Mass.

Yodzis, P. 1978. Competition for space and the structure of ecological communities. Lecture Notes in Biomathematics #25. Springer-Verlag, Berlin and New York.

VI ECOSYSTEM STRUCTURE AND FUNCTION

Chapter 16

Challenges in the Development of a Theory of Community and Ecosystem Structure and Function

SIMON A. LEVIN

The impetus for improved mathematical approaches to the study of the structure and function of ecosystems has its foundations both in basic and in applied research. In the United States, the importance of such investigations is reflected in the research priorities of federal agencies as diverse as the National Science Foundation (NSF), Department of Energy (DOE), Environmental Protection Agency (EPA), and National Oceanic and Atmospheric Administration (NOAA). The funding by NSF of the International Biological Program (IBP) was in recognition of these needs, and those efforts were paralleled by the even earlier commitment by the Atomic Energy Commission (AEC), the predecessor of DOE, to the development of mathematical approaches to ecosystem problems (Mooney et al. 1987). EPA's involvement with the area is more recent, but was in direct response to the Congress's mandate that the Agency develop long-range and anticipatory research programs. The primary stimulus came from the need to evaluate the effects of chemicals on ecosystem structure and function, but the concerns are generic ones that apply to the diversity of problems confronted by EPA and other regulatory agencies, as well as to foundations and agencies concerned with basic research: What are the natural patterns and dynamics of ecosystems, how are they regulated, and how robust are they to perturbations?

The need for a scientific basis for environmental management has led to the development of increasingly sophisticated models that address the responses to stress of community and ecosystem structure and function, and of patterns of succession, productivity, and nutrient cycling. It is clear that, as scientific attention turns to regional and global environmental problems, it will be necessary to interface such models with those for the physical environment, including atmospheric models of climate change, and this is going to involve the coupling of biotic and abiotic systems operating on vastly different temporal and spatial scales. The rapid increase in supercomputing ability will facilitate study of such systems; but unless numerical work is soundly based in an understanding of the mathematical properties of large-scale dynamical systems, egregious errors are inevitable. Thus, fundamental mathematical analyses of the dynamics of such systems, of their hierarchical relationships, and of the techniques available for simplifying them, are indispensable.

What is clear is that the solutions to the environmental problems we face are going to require the integration of experimental and theoretical approaches at a variety of levels of specialization. No single model, and no single level of description, will suffice. We will require multiple models, directed at different levels of detail, and ranging from the generic to the site- and situation-specific. That is, we will need to be concerned with the development both of special and of general theory. For example, in the modeling of the distribution of chemicals in the environment, whether this refers to those chemicals discharged from pipes into aquatic systems or from smokestacks into the terrestrial environment, the interdependence of specific and generic approaches has been highly successful. Generic models incorporate the basic mechanisms—diffusion, advection, and reaction—and parameterize these phenomenologically, and often over very broad scales. These approaches are complemented by site-specific models of particular bodies of water or complex terrains, incorporating the effects of local geometries and topographies. The generic models are absolutely essential in providing the framework within which the site-specific studies are set, but the site-specific studies provide the test cases that make the general theory useful. Similar comments apply to the interplay between generic and specific approaches to the movement of organisms, discussed elsewhere in this volume by Kareiva.

For the analysis of ecosystem structure and function, the complementarity between specific and general approaches is less well developed, but no less essential. As Kimball and Levin (1985) argue, "We must develop a theory for the response patterns of different ecosystems" to stresses. "We must develop standards of comparisons among ecosystems, based on the identification of common, functionally important processes and properties. . . . Such understanding can emerge only from theoretical syntheses based on a comprehensive program of microcosm research and experimental manipulation coupled with retrospective studies." Furthermore, we need to couple system-level testing that allows identification of emergent phenomena with mechanistic studies designed to provide understanding and the basis for extrapolation, and we need to develop the generic theoretical structure that makes extrapolation possible. Theory without data is sterile, while data without theory is uninterpretable.

SYSTEM PROCESSES

It would be easier to begin with a discussion of community structure rather than ecosystem function, since historically a much larger body of mathematical theory has been developed for dealing with community properties. In speculating on the reasons for this imbalance, it is useful to point out that the separation between population and community studies, on the one hand, and ecosystem studies, on the other, is not unique to mathematical investigations. In part, the explanation for the gulf may lie in the fact that the community tradition was a natural outgrowth of basic studies in ecology, whereas many of the modeling approaches to ecosystems were inspired by applied problems, involving the perturbation of system dynamics by extrinsic abiotic factors. This explains, but does not excuse, the fact that the preponderance of community models ignore extrinsic factors and the abiotic environment, and typically involve only a few of the interactions and factors known to be important. The pathway to understanding is simplification, the separation of the signal from the noise, and this requires isolating pieces of the ecosystem in order to examine their importance and properties. On the other hand, it also explains, but does not excuse, the almost mindless inclination to include in many ecosystem models the full complexity of the biotic and abiotic environment, on the mistaken notion that highly detailed and reductionistic approaches make the best tools for prediction and management. Such an approach ignores the practical problems of parameter estimation and error propagation, and the importance of understanding in guiding experimentation, prediction, and management. The time has come to merge these traditions, to build on the strengths of each, and to explore the interface between population and community biology and ecosystem science.

Although theory can suggest and thereby guide empirical studies, it is virtually impossible to develop theories without benefit of some data base. Thus, a critical precursor to the development of ecosystems theory is the systematization of data, and the elucidation of patterns. For example, a wide data base related to the external loading of nitrogen and phosphorus into ecosystems exists, but encompasses a tremendous diversity of systems and several orders of magnitude (e.g., fig. 16.1).

Kelly and Levin (1986) organize this information, and show that some striking patterns emerge if one relates net primary production (NPP) of a variety of aquatic ecosystems to nutrient input (fig. 16.2). In general, NPP responds strongly to nutrient input at low levels of input, but the relative responsiveness appears

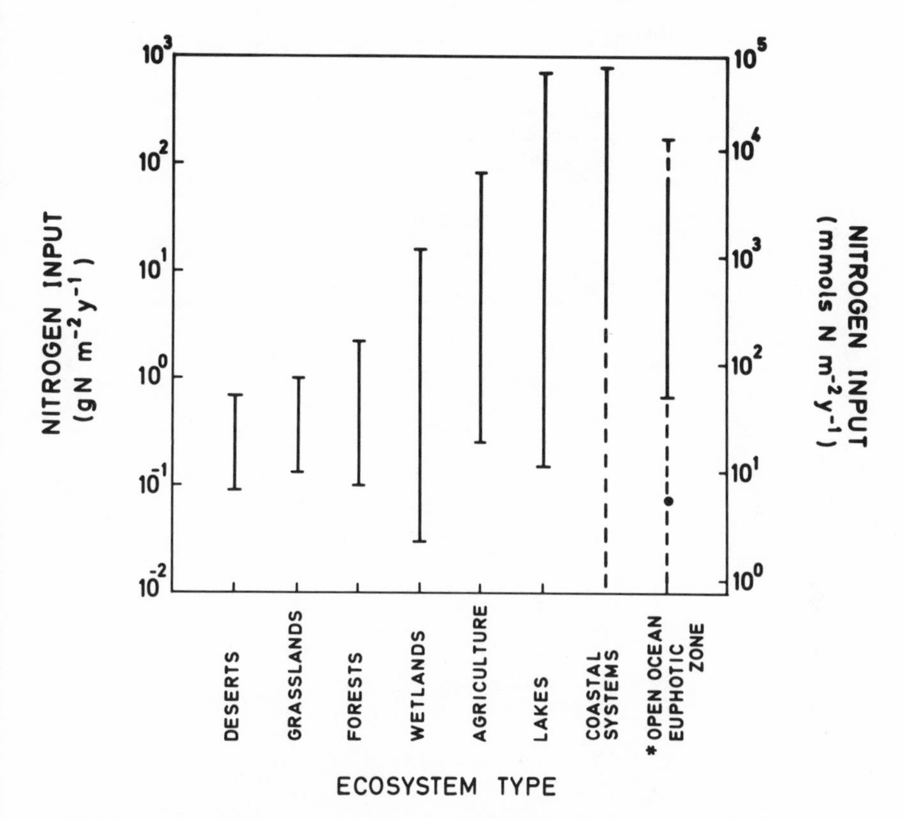

FIGURE 16.1 The range of nitrogen inputs to ecosystems. Data are from values reported in Kelly and Levin (1986, Table 1, Appendix 1). Values do not include nitrogen fixation. For coastal systems (which here include salt marshes), dotted lines indicate that values will extend lower where land runoff does not contribute (e.g., tropical embayments). Net import from adjacent surface systems has been measured in several such instances (Smith 1984), but its relationship to gross influx is uncertain. For open ocean surface water systems, the asterisk denotes that values have been calculated, not measured. The dotted line at the upper end of the range indicates the Peru upwelling based on either a 6-month or 1-year period of intense upwelling. The solid point represents an average for the oceans on a global basis, considering the inputs to be precipitation and land runoff. The dotted line at the lower end of open ocean range is to suggest the possibility that inputs are lower in areas where there is little precipitation, very weak upwelling, or even downwelling (which imports surface waters with indetectable *N*). It is likely that lower input terrestrial systems also occur. (From Kelly and Levin 1986).

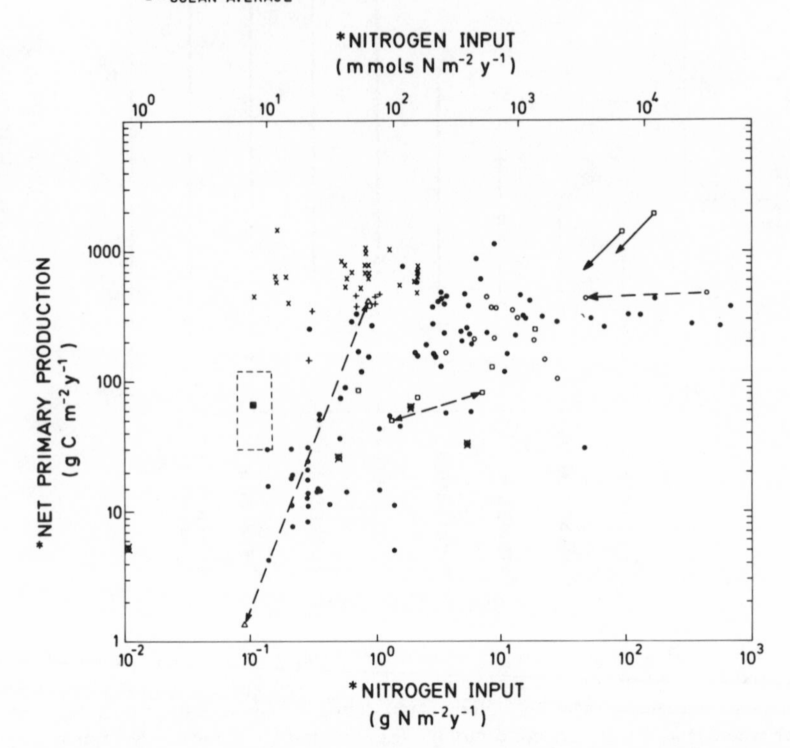

FIGURE 16.2 Nutrient inputs and primary productivity in terrestrial and aquatic ecosystems. The asterisks denote where a value has been calculated from data or is dissimilar to the rest of the data set. Four estuarine systems did not report a standard primary production estimate (see Kelly and Levin 1986, Appendix), other manipulations and definitions of NPP are given in the text. The double-headed arrows show range estimated for deserts and for Sargasso Sea. Open ocean points with single-headed arrows show range, assuming upwelling (Peru and Costa Rica Dome) occurs for entire year or only 6 months. The dotted line with single-headed arrow connects points moving from New York Harbor (higher) to New York Bight Apex. The box around the average ocean value gives a range assuming plus or minus 100% around the estimate. (From Kelly and Levin 1986).

to dissipate at high input levels. Coastal marine areas in general seem near saturation, suggesting that they will show a low efficiency of autotrophic use of available nutrients. Such comparisons have immediate applicability to applied issues, such as the evaluation of waste disposal options; but they also demonstrate the need for theoretical analysis. Kelly and Levin (1986) fit the observed data to a hyperbolic (Michaelis-Menten) relationship, but this is a purely phenomenological approach and not based on any mechanistic understanding of the relationship between nutrient input and system productivity. Thus, the relationship between nutrient input and ecosystem productivity, the most basic of issues in ecosystems science, is virtually untouched by theory aimed at elucidating mechanisms, although it is a concept that has its roots in the work of Liebig and Lotka. Also untouched by theory are freshwater Vollenweider curves, used for prediction of chlorophyll or productivity from phosphorus loading.

Similar comments apply to the regulation of element cycles. One of the first steps in describing the dynamics of an ecosystem is the construction of budgets for energy and materials. A variety of very useful, and relatively site-specific, compartmental models describe these flows; but there is a notable absence of general theory addressed to the question of how these budgets are regulated, and how biogeochemical cycles affect and are affected by the biotic community. Recent work (e.g., Agren and Bosatta 1987) makes strides in this direction; but in general, it is an area that has been neglected by theoreticians for too long. The absence of a general nonequilibrium theory frustrates attempts to extrapolate from one ecosystem study to another, or even to predict how a particular ecosystem will behave when exposed to stresses or conditions that previously have not been observed.

COMMUNITY PATTERN AND DYNAMICS

Our view of any system depends on our scale of investigation, and thus the choice of scale can fundamentally affect our perspective. In *Poetics,* Aristotle attributes to Agathon the remark that "It is probable that the improbable will sometimes happen." Indeed, it is the arrangement of unique or rare phenomena into a broader conceptual framework that converts curiosity into science, changing the scale of investigation from that of the isolated event to that of the ensemble. A case in point is provided by community theory.

Classical views of ecological communities were dominated by the Clementsian approach and treated the community as an integral unit, a superorganism, essentially homogeneous and equilibrial in the sense of tending to a unique, locally defined climax state. Mathematical theory was even more orthodox in its orientation, and the weakness of Lotka-Volterra and similar model systems is not so much in their reliance on particular model forms, but rather on their adherence to the notions of homogeneity and equilibrium, and on the concept of the community as a well-defined unit. Indeed, in some cases, there even has been a willingness to adopt the extreme group selectionist approach, and to view the ecosystem as an evolutionary unit, selected for its particular features.

In reality, we know that the situation is much more complex. Whittaker (1975), by studying the distribution of species along environmental gradients, showed that those distributions were broadly overlapping. The community, rather than being a crisply defined entity, is a concept of convenience that changes in space and time, reflecting the individualistic properties of its component species. This view, emphasizing the importance of stochastic phenomena, was much closer to that advanced by Gleason than by Clements, and was mirrored by A. S. Watt's elucidation of the importance of gap phase phenomena in forest dynamics.

Thus, ecological communities and ecosystems are heterogeneous in space and time, and this heterogeneity affects diversity and the evolution of life histories. Such biotic heterogeneity reflects underlying heterogeneity in the abiotic environment, frequency-dependent habitat and niche partitioning, and the stochastic phenomena associated with disturbance and colonization. Mathematical models of communities have contributed greatly to our understanding of how these factors, individually and in concert, contribute to the patterns we observe. In recent years (Levin 1976), a large literature has developed for treating the dynamics of populations and materials interacting in complex and heterogeneous environments, and these approaches have provided the theoretical basis for the emerging discipline of landscape ecology. However, as landscape ecology has developed, the need has become clear for new approaches that can deal with the irregular, fractal patterns exhibited by fragmented landscapes, and with the fluctuating spatio-temporal mosaics that are characteristic of ecosystems and landscapes (see, e.g., Forman and Godron 1986).

The approaches of the past, emphasizing equilibrium, constancy, homogeneity, stability, and predictability have served their purpose in defining the null model, but they are inadequate for dealing with real landscapes. Equilibrium and its related concepts are not absolutes but depend critically on the scale of investigation. Furthermore, it should be clear that there is no single correct scale of investigation, and that description of the dynamics of ecosystems and landscapes must look across scales and examine how system description depends on the scale of investigation. Just as the measurement of coastline length is meaningless without reference to the scale of measurement (Mandelbrot 1983), so too the measurement of dynamic system properties is dependent on the chosen spatial, temporal, and hierarchical scale. Furthermore, typically this dependence involves slow gradation across a continuum of scales rather than simply sharp discontinuity, although the existence of discontinuities is of considerable importance. Oceanographers long have been aware of this continuum of scales, through the recognition that plankton exhibit patchiness on almost every scale of investigation, and that particular oceanographic studies give us a biased picture of system properties by focusing on a small range of spatial and temporal scales (Steele 1978).

The temporal variability of community composition was the stimulus for the development of the theory of island biogeography (MacArthur and Wilson 1967), which was predicated on the concept that island communities exhibit high turnover rates in terms of the specific species assemblages, but might achieve "equilibrium" in terms of such macroscopic descriptors as the number of species. The tremendous interest generated by this theory was not limited to its relevance to true island situations, but raised the hope that it would be applicable to insular habitats in terrestrial systems, in which one type of habitat was virtually surrounded by a sea of contrasting habitat.

The most natural extension of the theory of island biogeography was to nonequilibrium situations, in which temporary islands form and disappear. The forest gaps discussed by Watt (1947) and later by numerous other authors (Levin and Paine 1974; Pickett and Thompson 1978; Pickett and White 1985); the intertidal gaps studied by Levin and Paine (1974) and Paine and Levin (1981); and the gopher mounds and badger mounds that disturb grassland vegetation (Platt 1975; Hobbs, Mooney, and Hobbs, ms.; Hobbs and Hobbs 1988) all provide examples of such systems. In each of these, disturbance or a comparable event upsets or prevents an equilibrium situation, reinitiating a temporal or successional dynamic and providing colonization opportunities for species that would be eliminated from the climax or equilibrium state. Levin and Paine (1974) modeled such systems by focusing attention on the individual patch, and on the distinct temporal and spatial scales associated with within-patch versus among-patch events. In their approach, one develops first a demographic accounting of the distribution of patches, relating the age and size frequency distribution of patches to descriptors of patch birth, growth, and death. Complementing this are descriptions of the possible colonization sequences for patches, in relation to properties such as size, location, and time of formation. The model as described was a generic one, and its usefulness came from specific applications to particular systems, and from the in-depth studies of recolonization and competition in gaps. Such studies now have been carried out for a wide variety of terrestrial and intertidal systems (Pickett and White 1985).

Such approaches are a beginning, but they leave unresolved the issue of the determination of pattern on a continuum of scales. In general, disturbances come in a wide variety of shapes and sizes, and these and other events introduce complicated patterns of spatial and temporal correlation. To address these issues, and to develop a methodology for the analysis of such disturbance-controlled systems, Linda Buttel and I (Levin and Buttel, ms.) have developed a general model of disturbance and succession, with the view of applying it to forests, grasslands, and the intertidal. Our investigation, being carried out at the Cornell

National Supercomputer Facility (CNSF), imposes disturbance of various sizes on a grid composed of 10,000 cells, according to a set of rules that are determined by stochastic functions of the local state variables. Disturbances are centered in particular cells, and their size and frequency distributions are conditional upon the current status of the cell; disturbances are allowed to radiate outward from their centers to adjacent cells.

Any cell, whether or not it is newly disturbed, is continually bathed in propagules from neighboring cells and from a dispersal pool, and the local dynamics are then determined by a particular set of colonization rules, mimicking the dynamics of the particular system of interest. The resultant spatial and temporal patterns are then analyzed in a variety of ways. In the succeeding paragraphs, we discuss the application of such analyses to particular forms of the disturbance model.

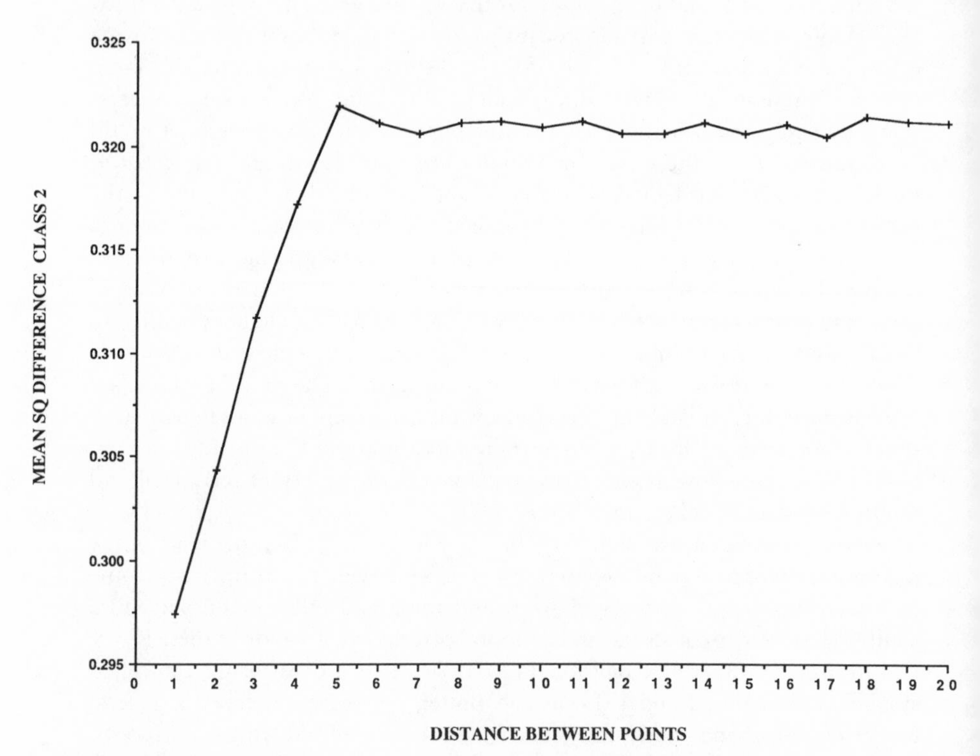

FIGURE 16.3.

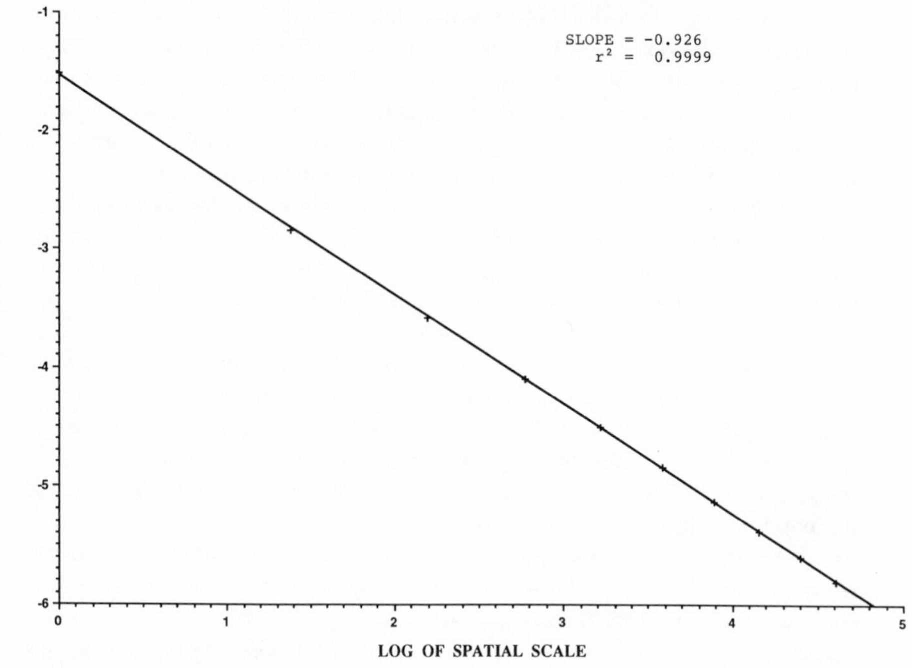

FIGURE 16.4.

One of the simplest methodologies for analyzing spatial pattern is the semivariogram, which relates, for a particular measure, the average squared deviation between two points as a function of the linear distance separating them. A typical such relationship is shown in figure 16.3, in which the variable of interest is the presence or absence of a selected species. The squared deviation rises with distance, leveling off when the size of the largest patch is attained.

Alternatively, one can choose a particular descriptor, such as the species density of a particular species, or species diversity, or average successional state, and consider how that descriptor changes if measured at different scales. Consider a nested sequence of quadrats centered at a particular point. For any given descriptor and quadrat size (*aggregation level*), one may compute a mean value taken over the quadrat, which may be thought of as the window through which the system is viewed. Intuitively, at least for homogeneous systems, one expects both the temporal and spatial variability of the measure to decrease as the window is increased, and this is borne out by our analyses. In figure 16.4 the spatial variability of one descriptor (density of successional class 3) is shown; the relationship between variability and aggregation level (on a log-log plot) is remarkably linear, and this linearity is observed for virtually any descriptor chosen. It suggests a self-similarity across a broad range of scales, as observed for a variety of natural processes (see Mandelbrot 1983). Furthermore, consideration of such patterns at different points in time leads to the discovery of identical patterns, except for a very brief transient phase, and this suggests that the patterns are stationary and that the dynamics are ergodic. Ergodicity means that temporal sequences at a particular spatial location and spatial transects at a particular point in time are statistically equivalent; ergodicity can be confirmed directly by examination of the relationship of temporal variance to spatial scale. On much larger spatial scales, theory suggests that the slope of the curve should approach -1; but this does not diminish the importance of the self-similarity observed on the smaller scales.

The above conclusions were derived for simple and homogeneous systems, and do not reflect patterns that should be expected to hold under all conditions. For example, when we complicated the model by superimposing patterns of spatial heterogeneity, the linear relationship and ergodicity broke down, although there still was strong scale dependence (figure 16.5). Thus, spatial and temporal heterogeneity interact in complicated ways, and we are just beginning to explore those interrelationships. Furthermore, for more detailed dynamical behaviors, or when correlations arise from local dispersal or neighborhood interactions, deviations from constancy are to be expected. What is needed is the exploration of a variety of measures such as these, in relation to a family of models of disturbance and regrowth, and in relation to corresponding data sets from natural systems.

The general implications of such studies are diverse. First of all, it is clear that even for the simplest dynamical models, the measurement of the variance of virtually any descriptor is critically dependent on the scale of measurement. The regularity seen in figure 16.4 holds out the hope that comparisons can be made across systems and across scales, and that scaling relationships can be derived when the dynamics are simple. Curvilinearity, as seen in figure 16.5, reflects the operation of distinct mechanisms on different scales.

Such studies are purely theoretical, and are meant to provide a framework for the investigation of the dynamics of particular systems. Thus, with Kirk Moloney and Hal Mooney, we have begun to particularize our model to the serpentine grassland of Jasper Ridge, California, for which Mooney and his colleagues have developed an understanding of the dynamics of gopher disturbances, the nutrient dynamics of mounds, and the successional dynamics of mounds in relation to soil depth and water availability. Parallel theoretical studies are being carried out for deciduous forests, by implementation of standard forest growth simulators.

The availability of techniques for analyzing spatial pattern, for understanding its development, and for investigating its implications will be critical for the

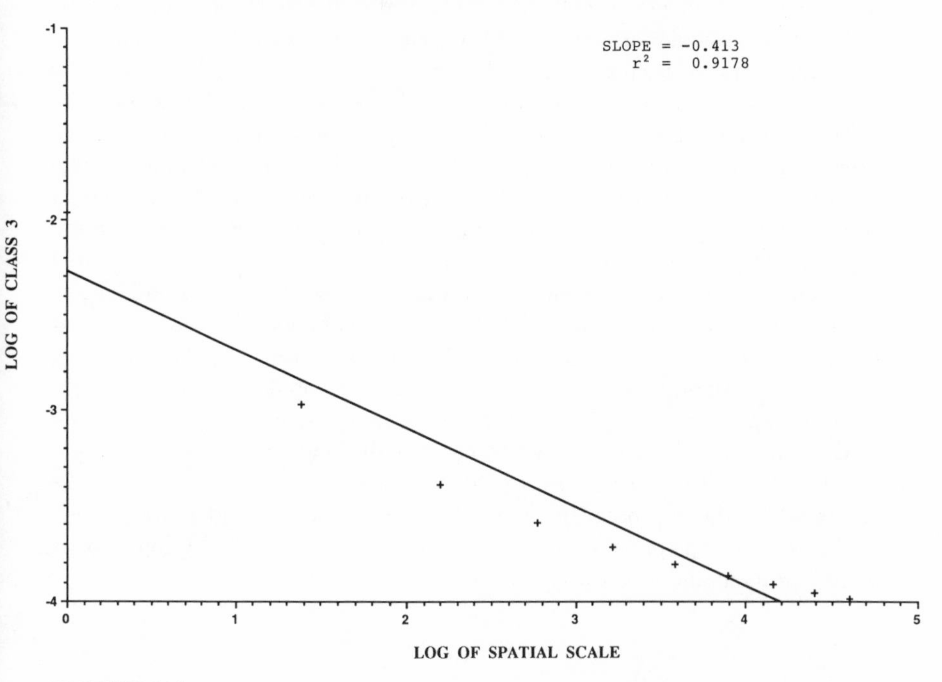

FIGURE 16.5.

exploration of ecosystem, landscape, and global dynamics. Such techniques are being developed in a variety of disciplines, ranging from soil science to meteorology, and their implementation is facilitated by the availability of supercomputers. Spatio-temporal patterning on a continuum of scales is evident not only in the distribution of species, but also in the measurement of pH and of nutrients and other materials. We cannot progress without recognition of that variability, and we must develop methodologies for describing and analyzing it. Here, the need for the coupling of generic and site-specific approaches is evident, as we must interface the development of general methodology with applications to particular systems.

CONCLUSIONS

Any mathematical model has its strengths and limitations, but many criticisms of community models have failed to understand the multiple purposes that models serve and the need to define the limits to applicability. In particular, a fundamental aspect of the development of science is the interfacing of generic and specific approaches. The development of general theory and methodologies seeks common features in diverse problems, and its goal is the development of generic approaches to a wide variety of situations. Such general theory is the indispensable framework for science, but must be complemented by a hierarchy of models of increasing specificity, tailored to the special features of particular systems.

In community theory, general models of population interactions have guided experimental investigations. The interplay between theory and experiment led to recognition of the importance of spatial heterogeneity and to the development of mathematical approaches to the dynamics of spatially distributed populations. More recently has come the recognition that patterns in the distribution both of the biotic and abiotic components of ecosystems are interrelated and can be detected on almost every temporal and spatial scale. Major advances in the next few years will involve the development of methods for detecting and analyzing spatial and temporal pattern in the distribution of populations and materials, understanding how such pattern arises, and elaborating upon its consequences.

Community ecologists are very familiar with the biogeographical method and the insights that are to be derived from it. Numerous papers in this volume make strong use of the approach. In ecosystems science, similar studies are needed, studies that draw patterns from comparisons across systems and develop theories to explain them and to allow extrapolation to new situations.

ACKNOWLEDGMENTS

This research was conducted at the Cornell National Supercomputer Facility, Center for Theory and Simulation in Science and Engineering, which is funded, in part, by the National Science Foundation, New York State, and IBM Corporation. Support was also provided by NSF Grant DMS-8406472 to the author. Additional support was provided by the Environmental Protection Agency Cooperative Agreement CR812685. This publication is ERC-150 of the Ecosystems Research Center, Cornell University. The views expressed herein are those of the author and do not necessarily represent those of the granting agencies.

I am pleased to acknowledge helpful comments by Richard Forman, Egbert Leigh, and Colleen Martin.

REFERENCES

Agren, G., and E. Bosatta. 1987. Theoretical analysis of the long-term dynamics of carbon and nitrogen in soils. *J. Ecol.* 68:1181–89.

Forman, R.T.T., and M. Godron. 1986. *Landscape Ecology.* Wiley, New York.

Hobbs, R. J., and V. J. Hobbs. 1987. Gophers and grassland: A model of vegetation response to patchy soil disturbance. *Vegetatio* 69:14–46.

Kelly, J. R., and S. A. Levin. 1986. A comparison of aquatic and terrestrial nutrient cycling and production processes in natural ecosystems, with reference to ecological concepts of relevance to some waste disposal issues. In G. Kullenberg, ed., *The Role of the Oceans as a Waste Disposal Option,* pp. 165–203. D. Reidel, Dordrecht, Holland.

Kimball, K. D., and S. A. Levin. 1985. Limitations of laboratory bioassays and the need for ecosystem level testing. *BioScience* 35(3):165–71.

Levin, S. A. 1976. Population dynamic models in heterogeneous environments. *Ann. Rev. Ecol. Syst.* 7:287–311.

Levin, S. A., and R. T. Paine. 1974. Disturbance, patch formation, and community structure. *Proc. Natl. Acad. Sci. USA* 72(7):2744–47.

MacArthur, R., and E. Wilson. 1967. *The Theory of Island Biogeography.* Princeton University Press, Princeton, N.J.

Mandelbrot, B. B. 1983. *The Fractal Geometry of Nature.* Freeman, San Francisco.

Mooney, H. A., F. A. Bazzaz, J. Berry, J. H. Cushman, W. F. Harris, S. A. Levin, J. J. Magnuson, P. L. Parker, W. P. Porter, and P. Risser. 1987. *Review of the Office of Health and Environmental Research Program: Ecology.* U.S. Department of Energy, Washington, D.C.

Paine, R. T., and S. A. Levin. 1981. Intertidal landscapes: Disturbance and the dynamics of pattern. *Ecol. Monogr.* 51:145–78.

Pickett, S.T.A., and J. N. Thompson. 1978. Patch dynamics and the design of nature reserves. *Biol. Conserv.* 13:27–37.

Pickett, S.T.A., and P. S. White. 1985. *The Ecology of Natural Disturbance and Patch Dynamics.* Academic Press, Orlando, Fla.

Platt, W. J. 1975. The colonization and formation of equilibrium plant species associations on badger disturbances in a tall-grass prairie. *Ecol. Monogr.* 45:285–305.

Steele, J. H., ed. 1978. *Spatial Pattern in Plankton Communities.* NATO Conference Series, Series IV: *Marine Sciences,* Vol. 3. Plenum Press. New York and London.

Watt, A. S. 1947. Pattern and process in the plant community. *J. Ecol.* 35:1–22.

Whittaker, R. H. 1975. *Communities and Ecosystems.* 2d ed. Macmillan, New York.

Chapter 17

Simulators as Models of Forest Dynamics

HENRY S. HORN, HERMAN H. SHUGART,
AND DEAN L. URBAN

The diversity of modeling approaches in ecology can be conveniently partioned into two brand categories that we shall refer to as analytical models and simulators. By "analytical models" we mean mathematical models that can potentially be solved in closed form. These include differential equations, Markov models, and other such formulations. Most current analytical models of forest dynamics are either based on systems of linear equations or solved by linear expansion about points of interest. In contrast, "simulators" typically incorporate richer biological detail, including explicit nonlinearities, at the expense of mathematical intractability. They are generally solved by computer using Monte Carlo techniques or other numerical methods.

The bulk of ecological theory to date has been derived from analytical models, while simulators have contributed mainly to the understanding of particular cases. But simulators have matured over the past two decades to the point of making important contributions to ecological theory in general and to theories of forest dynamics in particular.

In this chapter we argue that simulators, used in concert with more abstract and more analytically tractable models, can considerably advance ecological theory. In particular we believe that simulators of forest dynamics provide a way to convert quantitative details of the natural history and physiology of trees into parameters for either analytical models or further simulation at community and landscape levels. If this belief is true, we have a particular and successful instance of theoretical relations between hierarchical scales of space and time such as O'Neill (chapter 10) envisions. A further and as yet unexploited role for simulators is exploring spatial models of forest dynamics, models that include nonlinearities and variance in seedling establishment, competitive environment, and mortality.

Our view is toward the future rather than the past. After introducing some general principles and results of simulators of forest dynamics, we shall describe a representative simulator in which regeneration depends on whether or not the death of a particular tree leaves a distinct gap in the canopy. We shall then use qualitative analysis of this model as an excuse to describe a very general result of an analytical model of forest dynamics that can be applied, at least in principle, over a wide range of scales of space and time, from tree-by-tree replacement over a generation to regional shifts in landscape over many centuries. This result could also be extended to qualitative analysis of many of the box-and-arrow models that are current in studies of the flow of energy and materials through ecosystems. Finally, we shall outline how we expect simulators to contribute to models of forest dynamics in the near future.

GENERAL PRINCIPLES AND RESULTS OF FOREST SIMULATORS

A class of simulators that has been especially successful in ecological application is the individual-based model, which takes the life histories of each individual in a community, and integrates these individual behaviors to simulate the behavior of the whole community or ecosystem. Such models tend to be dominated by structure rather than by detail, in that model behavior may depend more on the manner in which elements are linked together than on the detailed form of state equations. Nevertheless, devotees of simulation have been reluctant to jettison ecological details when building models. This is partly because the complexity and mathematical difficulties of simulators arise either from a multiplicity of individually trivial details or from a few significant nonlinearities that combine into an intractable mess. Removing enough of the details to make the model tractable might remove most of its interest. Consequently, as Gross (chapter 1) laments, we have yet to see an analytical model, or even a simulator, that translates the well-known details of photosynthetic physiology of individual leaves into the dynamics of growth of any natural plant community.

The penchant for detail in simulators has produced a number of models that are individually fitted to particular circumstances. As a holdover from the early days of such modeling, when computers were only capable of printing upper-case letters, simulators have their own taxonomy which overemphasizes differences of detail. Names like BRIND, FORAR, FOREST, FORET, FORICO, FORMIS, FORNUT, FORTNITE, JABOWA, KIAMBRAM, and SWAMP (Shugart 1984) give a false impression of inflexibility. In fact, these models have an impressive record of varied accomplishments, which are listed in table 17.1. They have been subjected to two levels of tests, which have been christened "verification" and "validation" (Mankin et al. 1977; Cale, O'Neill, and Shugart 1983). A simulator is verified if its output can be matched to a given set of data by using those data to estimate its parameters. It is validated if its output is consistent with a new set of observations that are independent of the data used to frame the model and to estimate its parameters.

Verification not only demonstrates the potential utility of a model in a particular instance, but it also assesses our conceptual understanding of a system by testing our ability to specify appropriate ingredients for the model. Verified simulators can be used for a number of purposes for which their validity is moot. Examples of such applications include (1) outlining priorities for research; (2) environmental hazard assessment, where the concern is not so much accuracy

TABLE 17.1

Some successful tests and applications of forest models based on the dynamics of individual trees and gaps. See Shugart (1984) for details and references.

Verification = fitting of parameters to match data

Models can be made to represent these known features of forests:

1. Forestry yield tables for loblolly pine in Arkansas
2. Succession of forest types at middle altitudes in Australian Alps and Smoky Mountains
3. Response to clear-cut in Arkansas wetlands
4. Forest types changing in response to flood frequency in Arkansas and Mississippi
5. Structure and composition of forests in New Hampshire, Tennessee, Puerto Rico, and flood plain of the Mississippi River
6. Age-specific structure in a subtropical rain forest
7. Arkansas upland forests based on 1859 survey

Validation = tests with independently gathered data

Model predicts or is consistent with these independently known features of forests:

8. Response of Eucalyptus forests to fire
9. Effects of chestnut blight on forest dynamics in southern Appalachians
10. Yield tables for *Eucalyptus delegatensis* in New South Wales
11. Average annual increment in diameter at four sites in Tennessee
12. Distribution of tree diameters in Arkansas upland and Puerto Rican rain forest
13. Elevational gradients of forest composition and structure in New Hampshire and Australian Alps
14. Effects of hurricanes on diversity of Puerto Rican rain forest

as "worst case" scenarios; and (3) developing "null hypotheses," where purposefully inadequate models are used to create artificial standards against which to measure the statistical behavior of the real world. The examples of verification in table 17.1 have the potential for all of these applications, and in addition they mimic reality in a wide range of circumstances.

Validation greatly increases our interest in exploring the theoretical content of a model, because a valid model is an adequate caricature of reality, at least in the domain of the observations used to test it. The more often a model is tested and validated, the more clearly defined is its domain of applicability, and the more reliable are its predictions. Validated models like those in table 17.1 are ready for tentative predictive use in practical forest management (Shugart 1984).

SPECIES' ROLES IN A GAP MODEL OF FOREST DYNAMICS

The models represented in table 17.1 form a general class of forest simulators called "gap models," and they have been produced by several collaborators over a bit more than a decade (see Shugart 1984 for a review). Gap models simulate forest dynamics by accounting the establishment, annual growth in height and diameter, and mortality of every tree on a small model plot corresponding to the zone of influence of a single dominant tree in the canopy. The models have passed tests at time scales from decades to centuries, and at spatial scales typically in the range of 0.1 to 10 hectares. Gap models have successfully predicted qualitative patterns of species' composition and structure, and have even had some quantitative success (table 17.1). They are particularly useful for simulating the dynamics of forests with mixed ages and/or mixed species of trees, and for exploring theories about patterns of forest dynamics at time scales that are long enough to prohibit direct observation. Unfortunately, such time scales are the rule in forest dynamics because the life spans of most canopy trees exceed those of their investigators.

The long-term behavior of several gap models has been generalized by Shugart (1984). His general model couples the mode of death and the mode of regeneration by dividing all tree species among four categories based on a pair of loose dichotomies. One dichotomy contrasts trees that can attain a large size and that leave a distinct gap in the canopy versus those without that potential. The second contrasts species that require a large gap overhead for successful establishment versus species that do not require a gap in the canopy to regenerate. Shugart (1984) referred to the resultant categories as four numbered functional "roles" that different species play in different forests (fig. 17.1).

Of course, other categories and corresponding roles could be defined. For example, several authors have discussed attributes of species that give them differ-

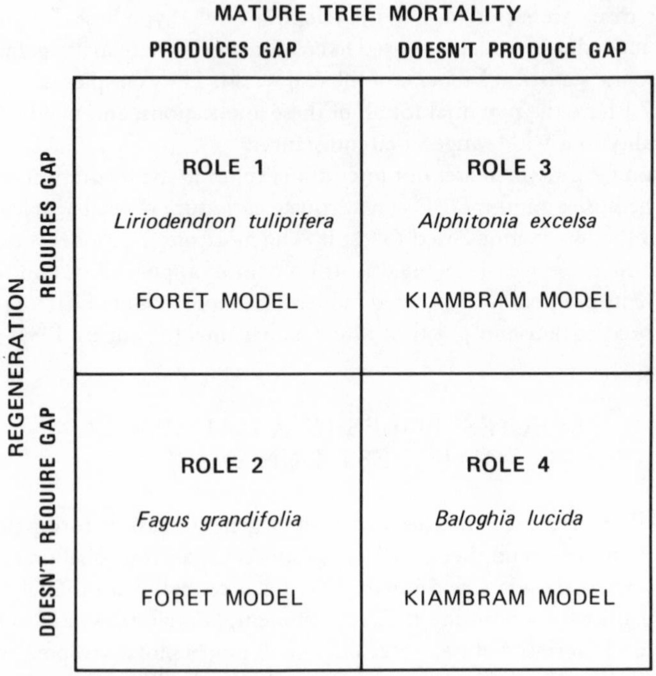

REGENERATION \ MATURE TREE MORTALITY	PRODUCES GAP	DOESN'T PRODUCE GAP
REQUIRES GAP	ROLE 1 *Liriodendron tulipifera* FORET MODEL	ROLE 3 *Alphitonia excelsa* KIAMBRAM MODEL
DOESN'T REQUIRE GAP	ROLE 2 *Fagus grandifolia* FORET MODEL	ROLE 4 *Baloghia lucida* KIAMBRAM MODEL

FIGURE 17.1. Categorization of four ecological roles of trees based on whether death of a mature tree does or does not produce a distinct gap in the canopy, and on whether or not a gap is required for regeneration. Representative species names and particular MODELS are given for each role as discussed in the text.

ential success in regenerating in gaps of various sizes (van der Pijl 1972; Whitmore 1975; Grubb 1977; Bazzaz and Pickett 1980; Denslow 1980). Nevertheless, the simple four-role categorization provides a rich range of forest behaviors. This range is illustrated in figure 17.2, which gives the simulated dynamics of pure stands of representative species that play each of the four roles. The examples are denizens of either the lush forests of the southern Appalachians or the subtropical rain forests of Australia, and they are simulated, respectively, by the FORET model (Shugart and West 1977) or the KIAMBRAM model (Shugart et al. 1980).

Trees of Role 1 often attain large size before they die, and they require a canopy gap to regenerate. The tulip tree (*Liriodendron tulipifera*) of temperate forests is shade-intolerant, grows rapidly, and can reach 55 meters in height. Large trees typically die from windthrow, leaving a huge gap. The simulation of a small (ca. 0.1 ha) pure stand of a Role 1 species produces oscillations in biomass caused by the growth and thinning of trees in a given cohort (fig. 17.2, Role 1). Copious

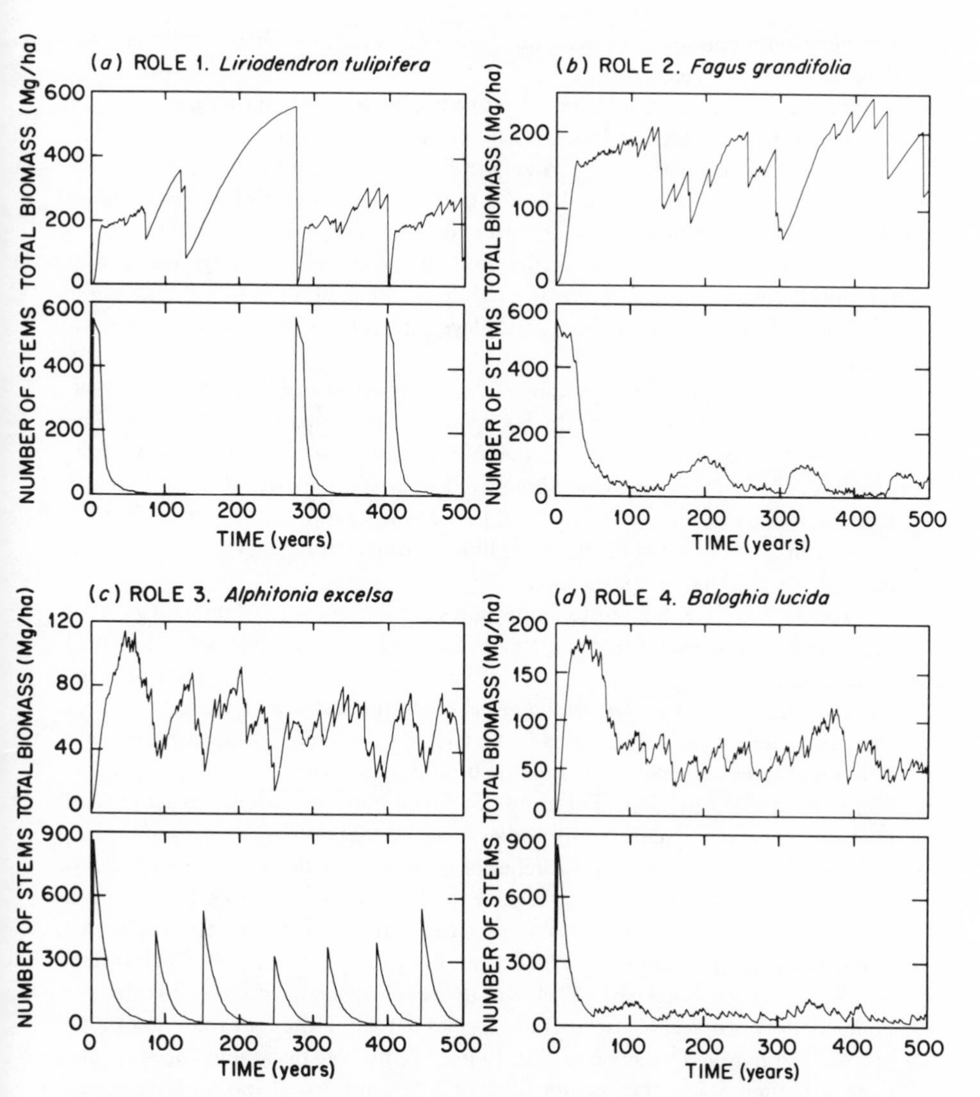

FIGURE 17.2. Dynamics of biomass (at top in each diagram) and of numers (below) in hypothetical monospecific stands for each of the four roles shown in figure 17.1. Simulations are for 500 years on 0.1 to 0.05 hectares of land, approximately the zone of influence of a large tree.

regeneration is episodic following the death of a canopy dominant, and numbers drop as the regenerated cohort thins.

Trees of Role 2 attain sufficient size to generate gaps when they die, but their regeneration is not restricted to such canopy gaps. The example, American beech (*Fagus grandifolia*), reaches large size and leaves a gap, and is tolerant of shade throughout life. Like the Role 1 forest, the simulation of a Role 2 forest shows strong oscillation in biomass as large trees die, but the peaks in numbers are less dramatic because regeneration occurs both after the death of a large tree and in the slightly increased light following the deaths of small trees (fig. 17.2, Role 2). The Role 2 forest has a mixed-age, multilayered structure, compared to the even-aged cohorts of a Role 1 forest.

Trees of Role 3 require gaps to regenerate, but because of their relatively small size and diffuse foliage, they do not leave distinct gaps when they die. The example, *Alphitonia excelsia,* an element of the Australian rain forest, is frail, fast-growing, shade-intolerant, and short-lived. The biomass dynamic of a Role 3 forest is oscillatory, but damped relative to the Role 1 forest because single trees do not dominate stand biomass (fig. 17.2, Role 3). Bursts of regeneration must await multiple deaths that break up the canopy.

Trees of Role 4 neither create canopy gaps nor do they depend on such gaps for regeneration. The example, *Baloghia lucida,* is small, shade-tolerant, and abundant in the subcanopy of some Australian rain forests. The creation of large canopy gaps frequently stresses *Baloghia* seedlings and causes mortality rather than increased growth. The Role 4 forest has a damped biomass dynamic because individual trees are small, and a damped numerical dynamic because regeneration is not pulsed by gaps. The resultant forest has a mixed-age structure and relatively constant dynamics of numbers and biomass.

Thus the dynamics of small patches of monospecific forests differ in predictable ways that depend on the life-history attributes of their species. Even though this generalized gap model is a purely theoretical construct, it is useful for marshalling observations about real forests. For example, we can observe interesting, though largely unexplored, patterns in the role diversity of forests in different biomes. Moist tropical forests include species in each of the four roles, but much of their taxonomic diversity is within Role 4. Temperate deciduous forests in the eastern United States also include all four roles, but Role 4 species are uncommon. In boreal forests at high latitudes, sunlight always penetrates at such an oblique angle that the death of a single tree in an intact stand is not likely to create a canopy gap of sufficient size to allow direct sunlight to fall on the forest floor. Therefore, boreal species should be mainly shade-intolerant Role 3 or shade-tolerant Role 4, though Role 2 would be expected wherever extensive openings are regularly caused by some combination of wind, fire, disease, and/or plagues of insects.

The output of the generalized gap simulator (fig. 17.2) would remind electrical engineers of a noisy relaxation oscillator, and there may be some hope that the engineer's machinery for complex analysis can be used to obtain analytical solutions for gap models (for inspiration, see Acevedo 1981, and Antonovsky and Korzukhin 1986). An analytical gap model would allow many of the details of natural history to be condensed into a few significant parameters of oscillation, damping, and noise filtration. The gap simulators could then be economically run for larger spatial scales, and the spatial propagation of the oscillations could be explored.

QUALITATIVE ANALYSIS OF THE GENERALIZED GAP MODEL

A qualitative analysis of the gap model of figure 17.1 provides an example of the sort of interaction between simulators and analytical models that we envision in the near future. Imagine a forest composed of four species, each of which plays one of the four roles. Knowing which produce gaps and which require gaps, we can draw a diagram with arrows that show which species can regenerate themselves locally and which species can replace one another (fig. 17.3). The arrows may be condensed into a matrix of transition probabilities, which can be subjected to analytical techniques after the fashion of Horn (1975), or to a mixture of analysis and simulation like that of Acevedo (1981). Horn and Acevedo have analyzed linear models and a particular form of nonlinearity in which local regeneration is proportional to the regional density of mature trees. The model with density-dependent regeneration has complex dynamics, featuring alternative stationary states that depend on the initial proportions of the species. The behavior of the linear models is straightforward, as argued below.

If the arrows of figure 17.3 represent fixed transition probabilities, some simple and intuitively sensible results of finite Markov chains can be applied (Kemeny and Snell 1960). In particular, since the system is closed and it is possible for any given species to be replaced by any other in only two steps, all of the species will persist indefinitely in a large enough sample. There may be some local three-species cycles, but these will be damped and will drift out of phase with one another because of asynchronous self-replacements within Roles 1, 2, and 4, as well as asynchrony among instances of the four reciprocal replacements represented by the double-headed arrows (fig. 17.3). The relative proportions of species in each role at equilibrium will of course depend on the exact values of the transition probabilities represented by the arrows of figure 17.3.

The above result is so simple and so intuitively obvious that by itself it is hardly worth noting. However, some intriguing results appear when we change the structure of the model. For example, Role 2, the shade-tolerant tree that leaves a distinct gap, is crucial to the persistence of the other species at equilibrium. When

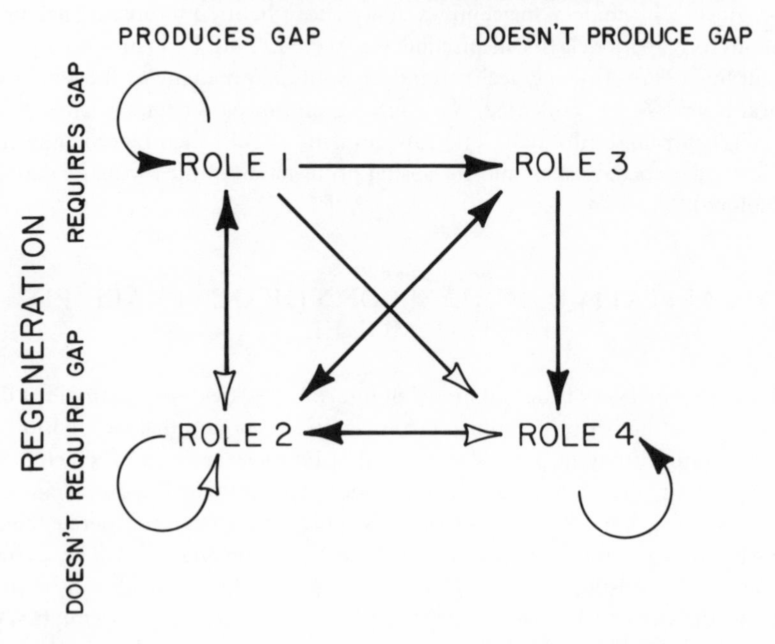

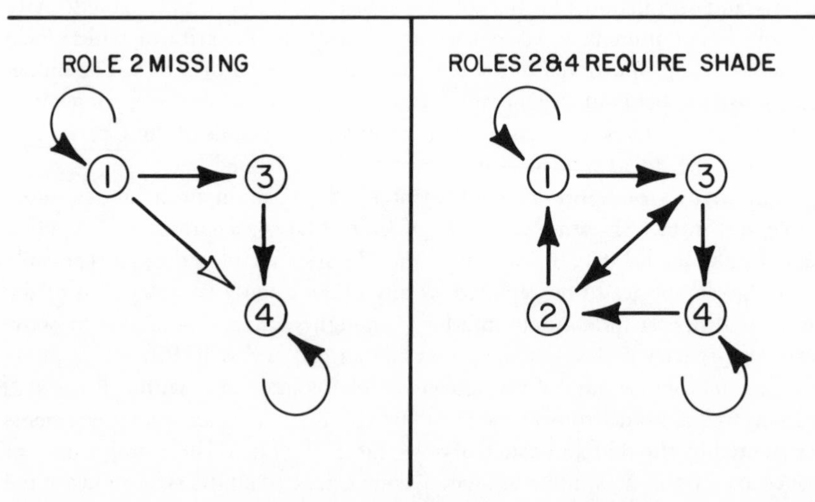

FIGURE 17.3. Top: Possible replacements of individual trees playing the roles of figure 17.1. Solid arrowheads represent replacements that are always possible. Open arrowheads represent regeneration in a gap by a species that does not require a gap. *Lower left:* The same but with Role 2 missing. *Lower right:* Possible replacements if Role 2 and Role 4 require shade, that is, require the absence of a gap. See text for a discussion of the dramatically different dynamics of these three models.

Role 2 is removed (fig. 17.3, lower left), the dominant dynamic is a relentless succession from Role 1, to Role 3, to an equilibrium in which Role 4 regenerates itself as the sole survivor. The succession to Role 4 may be slowed by local self-regeneration of Role 1, but it cannot be prevented. Removing any species other than Role 2 leaves a three-species system in which all three persist in equilibrium. Role 2 is therefore what Paine (1966) has called a "keystone" species, one on whose presence the whole structure and dynamics of a community depends. Indeed, the Role 2 keystone in this generalized gap model is the dominant competitor for space, just as is the mussel in Paine's beloved intertidal community (Paine 1974).

We can make another interesting change in the structure of the model if we specify that shade-tolerant species actually require shade for regeneration and cannot become established in a gap (fig. 17.3, lower right). This system would have a strong propensity to cycle from Role 1, to Role 3, to Role 4, to Role 2, and back to Role 1. The self-replacements of Role 1 and Role 4, and the reciprocal replacement between Role 2 and Role 3, would cause adjacent cycles to drift out of phase with one another if regeneration were linear and globally homogeneous. However, the four-species cycle could theoretically be either reinforced or stalled either by simple nonlinearities of regeneration (Horn 1975; Acevedo 1981), or by spatial restrictions on where young trees may be established relative to their female parent.

Horn (1981a) has generalized this form of qualitative analysis of forest dynamics, and has presented a mildly sleazy proof that in a closed, linear, box-and-arrow system nearly all of the significant details of the dynamics are bound up in the topology of the arrows. Fortunately, technical details of his proof support some intuitively satisfying interpretations. Double-headed arrows speed convergence on an equilibrial configuration because any two species whose abundances depart from equilibrium are able to reassort themselves in as little as one generation. Self-replacements slow convergence and may even tend to stall the system in a configuration that is far from the equilibrium; that is, self-replacement tends to perpetuate whatever exists at a given time. These qualitative results apply not only to linear models of tree-by-tree replacement from one generation to the next, but also to linear shifts in regional landscapes over many centuries, and to linear flows of energy and materials through closed ecosystems. They underscore our notion that qualitative behavior of linear systems is relatively easy to intuit, but serious departures from simple intuition arise from the simplest of nonlinearities.

PROSPECT

We believe that simulators have recently come of age in practical studies of forest dynamics on a small spatial scale, and that they serve a useful function by incorporating realistic details of the natural history of individual trees into a form that allows extension to temporal scales of many generations. Shugart (1984) provides massive support for this view. Extensions to wider spatial scales may be made by more massive simulations as the computational and storage machinery of computers continues to increase. However, we expect that simulators will also play a valuable role in converting details of natural history at a small scale into parameters for modeling at larger scales of both space and time, and that the resulting models, be they simulators or analytical models, need not be massive or unduly complex. Simulation has also been used by Acevedo (1981) to prompt further exploration of analytical models, and we expect this usage to increase, particularly for models that incorporate nonlinearities that are analytically intractable but crucial in the real world.

Throughout our review of representative simulators, we have been led to the expectation that intriguing complications will accompany the addition of explicit spatial patterns either to simulators or to analytical models of forest dynamics. For example, Horn (1981b) has presented a rudimentary spatial model for regeneration of two species in which the quantitative value of a single parameter determines whether forest dynamics will appear to be a uniform and unidirectional succession from the less to the more shade-tolerant species, or a history-dependent spatial patchwork of the two species.

We joint Pacala (chapter 4) in a plea for more attention to explicit spatial patterns in models of forest dynamics. We expect simulators to be prominent among such models. The models will necessarily be complex and analytically intractable, because they will simultaneously involve nonlinearities, variance in parameters, and spatial bookkeeping. Even if these features can be incorporated into analytical models, we expect simulators to be involved in their development.

Ideally, we would like to see a coherent model of forest dynamics that combines natural history, spatial pattern, simulation, and analysis. We have yet to encounter all four features in a single model, but the number of pairwise combinations in recent models is encouraging.

ACKNOWLEDGMENTS

Research of H.H.S. and D.L.U. has been supported by the National Science Foundation's Ecosystem Studies Program (grant no. BSR-85-100099) to the University of Virginia.

REFERENCES

Acevedo, M. F. 1981. On Horn's Markovian model of forest dynamics with particular reference to tropical forests. *Theor. Pop. Biol.* 19:230–50.

Antonovsky, M. Ya., and M. D. Korzukhin. 1986. Predictive forest ecosystem models and implications for integrated monitoring. Working Paper 86-36, Int. Inst. for Appl. Systems Analysis, Laxemburg, Austria.

Bazzaz, F. A., and S.T.A. Pickett. 1980. Physiological ecology of tropical succession: A comparative review. *Ann. Rev. Ecol. Syst.* 11:287–310.

Cale, W. G., R. V. O'Neill, and H. H. Shugart. 1983. Development and application of desirable ecological models. *Ecol. Modelling* 18:171–86.

Denslow, J. S. 1980. Gap partitioning among tropical rain forest trees. *Biotropica* 12 (suppl.):47–55.

Grubb, P. J. 1977. The maintenance of species-richness in plant communities: The importance of the regeneration niche. *Biol. Rev.* 52:107–45.

Horn, H. S. 1975. Markovian properties of forest successions. In M. L. Cody and J. M. Diamond, eds., *Ecology and Evolution of Communities,* pp. 196–211. Harvard University Press, Cambridge, Mass.

Horn, 1981a. Succession. In R. M. May, ed., Theoretical Ecology: Principles and Applications, pp. 253–71. 2nd ed. Blackwell, Oxford.

Horn, H. S. 1981b. Some causes of variety in patterns of secondary succession. In D. C. West, H. H. Shugart, and D. B. Botkin, eds., *Forest Succession: Concepts and Application,* pp. 24–35. Springer-Verlag, New York.

Kemeny, J. G., and J. L. Snell. 1960. *Finite Markov Chains.* Van Nostrand, New York.

Mankin, J. B., R. V. O'Neill, H. H. Shugart, and B. W. Rust. 1977. The importance of validation in ecosystem analysis. In G. S. Innis, ed., *New Directions in the Analysis of Ecological Systems,* part 1, pp. 63–71. Simulation Councils of America, La Jolla, Calif.

Paine, R. T. 1966. Food web complexity and species diversity. *Amer. Natur.* 100:65–75.

Paine, R. T. 1974. Intertidal community structure: Experimental studies on the relationship between a dominant competitor and its principal predator. *Oecologia* 15:93–120.

Shugart, H. H. 1984. *A Theory of Forest Dynamics.* Springer-Verlag, New York.

Shugart, H. H., and D. C. West. 1977. Development of an Appalachian deciduous forest succession model and its application to assessment of the impact of the chestnut blight. *J. Environ. Managemt.* 5:161–79.

Shugart, H. H., A. T. Mortlock, M. S. Hopkins, and I. P. Burgess. 1980. A computer simulation model of ecological succession in Australian sub-tropical rain forest. ORNL/TM-7929. Oak Ridge National Laboratory, Oak Ridge, Tenn.

van der Pijl, L. 1972. *Principles of Dispersal in Higher Plants.* 2d ed. Springer-Verlag, Berlin.

Whitmore, T. C. 1975. *Tropical Rain Forests of the Far East.* Clarendon Press, Oxford.

Chapter 18

Discussion: Ecosystem Structure and Function

WILLIAM H. SCHLESINGER

Traditionally, ecosystem studies have considered the movement of energy and materials through arbitrarily defined units of biotic communities. While Tansley (1935) first offered the ecosystem concept to ecology, the traditional focus was perhaps best enunciated by Evans (1956), who suggested that an ecosystem ecologist studies "the circulation, transformation, and accumulation of energy and matter through the medium of living things and their activities." Early examples of ecosystem analysis are best found in the literature of aquatic ecology, where the classic paper by Lindeman (1942) remains a benchmark in the understanding of energy flow through connected trophic levels. Progress in ecosystem analysis was more widespread and rapid after World War II, as the engineering concept of systems analysis matured with growing sophistication in electronic design (Boulding 1956). I would suggest that the 1950s comprise a golden age of energy-flow studies, with such classic papers as Odum's (1957) study of energy flow in Silver Springs, Florida. Since the boundaries of aquatic ecosystems are usually well defined, the ecosystem concept was readily applied to understanding energy flow in wetland habitats.

With the improvement of analytical capabilities—first the atomic absorption spectrophotometer and later the autoanalyzer—ecosystem ecologists could concentrate on the movement of materials through ecosystems. About the same time, recognition of the watershed concept for ecosystems (Bormann and Likens 1967) gave terrestrial ecosystem ecologists a useful boundary definition. Thus, during the 1960s terrestrial ecosystem studies flourished, and ecosystem studies became biogeochemically oriented. Indeed, a major goal of the International Biological Program was to compare energy and material flow through a diversity of natural ecosystems (e.g., Webb et al. 1983). A vast bank of data was accumulated describing the pools and annual transfers in terrestrial (e.g., Reichle 1981) and aquatic ecosystems. During this period, elaborate, large-scale ecosystem models were developed in collaboration with the empirical studies in the field. Most of these models were elegant descriptions of the system for which they were developed, but they had limited predictive power. Little effort had been addressed to the factors that *control* ecosystem processes, in contrast to a vast store of accumulated knowledge on pools and flux (Miller 1976; Innis 1976).

The Estes Park conference in 1975 reflected a growing interest among ecosystem ecologists in "process-level" studies (Sollins et al. 1976). Well-known centers of ecological research shifted their emphasis to studies that might have been recognized as soil microbiology, plant physiology, or atmospheric sciences in an earlier era. This change led to the strong interaction between physiological ecology and ecosystem science that is seen in recent theoretical papers (e.g., Chapin, Vitousek, and Van Cleve 1986) and textbooks (e.g., Waring and Schlesinger 1985). At the same time, modelers increasingly concentrated on subsystem components, which could include control parameters. With all this came a fragmentation of the traditional field of ecosystem science. Reiners (1986) suggested that the discipline had become incoherent, leading to a "lack of useful, theoretical development since approximately 1960 and the present digression of ecosystem research into a largely reductionist mode." Aside from the application of the laws of thermodynamics and of conservation of mass, one might ask if there is a paradigm for modern ecosystem science. At the Asilomar conference, a small subgroup attempted to identify a number of avenues that would lead to a productive and healthy development of ecosystem science and theory for the rest of this century.

Certainly one area that will include an important role for ecosystem science, and perhaps lead to new theoretical developments, is in the growing interest in global biogeochemistry. The major, current initiative in the International Geosphere Biosphere Program (IGBP) offers an important forum for this development. This program is motivated by the recognition that man is profoundly affecting the environment at a global scale, and that we have only the beginnings of a theory for how the Earth functions as a unified system.

Global ecology will treat the Earth as a single ecosystem, in which energy flow and material transfers are studied. Global studies of energy flow address questions of global heat balance and albedo, as these are affected by the presence of biota on the surface of the Earth. Global changes in the level of net primary productivity wrought by humans (e.g., Houghton et al. 1983; Vitousek et al. 1986) are likely to affect the energy balance of the Earth by changes in atmospheric CO_2 and other gases. Studies of the exchange of oxidized and reduced gases (e.g., N_2O) between the biosphere and the atmosphere will enhance our knowledge of the global circulation of essential elements (e.g., nitrogen) through biota. Indeed, studies of various atmospheric trace gases may offer the best index of the "health" of the biosphere since the atmosphere is well mixed and many of the biogenic components exist in small pools with relatively short residence times. While such measurements are likely to be reported from an international network of cooperative field studies, their interpretation will depend on the effective development of global ecosystem theory.

Studies of global biogeochemistry build on theories of planetary evolution that compare the conditions of the Earth today to those that might have developed on a lifeless Earth or on the adjacent planets, Mars and Venus (Walker 1977, 1984; Lovelock 1979). Most of the present-day atmospheric composition is derived or influenced by biotic activity—nitrogen through denitrification and oxygen through photosynthesis. A large pool of reduced carbon—the biosphere—subtending an atmosphere of 21% O_2 allows the spatial partitioning of oxidative and reducing reactions that control the global cycles of other biochemical elements, such as nitrogen, phosphorus, and sulfur. The uniform, abiotic environment on Mars is much less interesting. Oxidative weathering of exposed sediments and carbonation weathering of minerals in the plant rooting zone speak for the importance of biota to most geochemical reactions that occur on the surface of the Earth. In fact, the study of the chemistry of the surface of the Earth defines the science of *bio*geochemistry, since it is difficult to conceive of many reactions in this arena that are not influenced by biota. The stable conditions that we regard as the normal environment on Earth are in many cases maintained by the biosphere (Lovelock, 1979; Reiners 1986).

In its earliest developments, biogeochemistry was dominated by studies of microbial transformations. Indeed, the field was often recognized as geomicrobiology. Global biogeochemistry will unify a diversity of scientific investigations, including the tendency to address energy flow and biogeochemical flux separately. One new, unifying paradigm that will link studies from microbial to global levels is found in the linkage between all biochemical elements through oxidation and reduction reactions. Studies of energy flow are best addressed through measurements of the biogeochemical flux of carbon in oxidized or reduced form. Reduced carbon is measured as a proxy for the energy available for biotic transformations of other elements. Thus, carbon links traditional studies of energy flow to studies of biogeochemical cycling at various scales. Changes in the global pool of reduced carbon—the biosphere—are directly linked to changes in the form and amount of other elements of biochemical interest in a connected reservoir model developed by Garrels and Lerman (1981). This thinking can be expanded to include transformations of most elements of primary biogeochemical interest (e.g., fig. 18.1). For example, the global rate of denitrification is linked to the availability of reduced carbon, though the process has traditionally been studied at the microbial level. This theory builds on the ideas of Reiners (1986), who recognized the "stoichiometry of life" as a major principle organizing

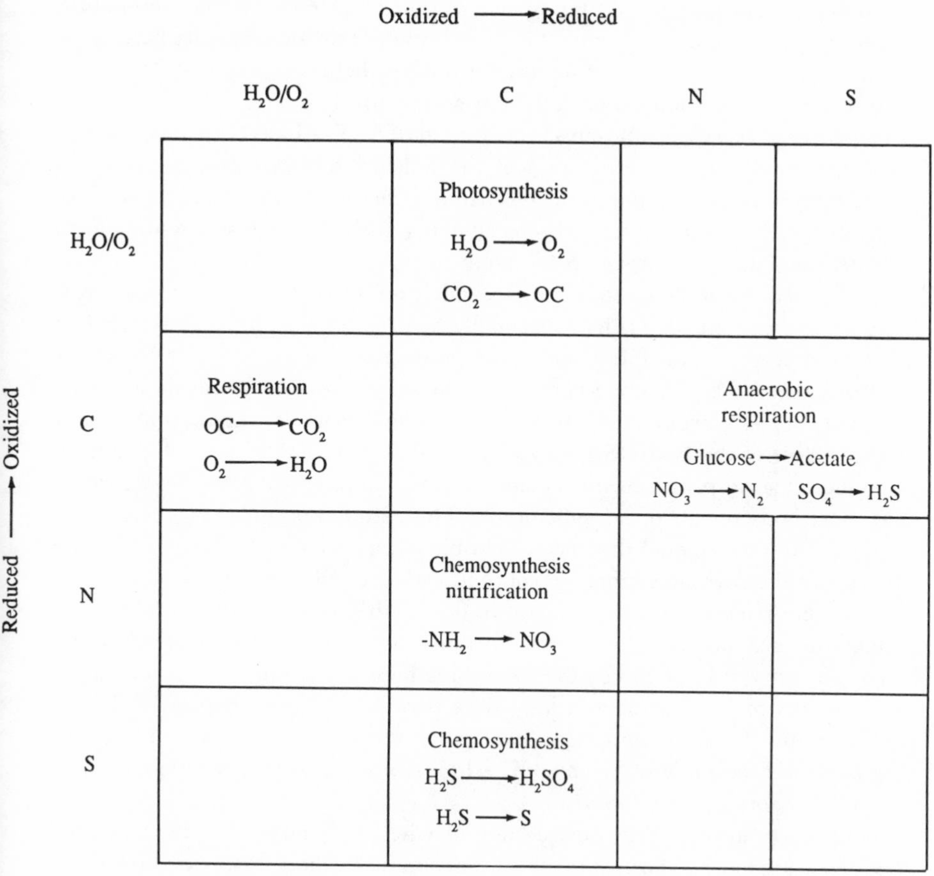

Oxidized $\rightarrow$ Reduced

Reduced $\rightarrow$ Oxidized

	H_2O/O_2	C	N	S
H_2O/O_2		Photosynthesis $H_2O \rightarrow O_2$ $CO_2 \rightarrow OC$		
C	Respiration $OC \rightarrow CO_2$ $O_2 \rightarrow H_2O$		Anaerobic respiration Glucose $\rightarrow$ Acetate $NO_3 \rightarrow N_2$	$SO_4 \rightarrow H_2S$
N		Chemosynthesis nitrification $-NH_2 \rightarrow NO_3$		
S		Chemosynthesis $H_2S \rightarrow H_2SO_4$ $H_2S \rightarrow S$		

FIGURE 18.1 Coupling of biogeochemical reactions through oxidation and reduction.

ecosystem function, but I stress reactions rather than final accumulations in biotic pools.

Linkages in biogeochemistry are seen at scales smaller than the global level. When potential key linkages are posed as hypotheses, ecosystem science can test the existence and importance of these linkages using strong inference and large-scale field experiments. For example, it is often difficult for ecosystem scientists to make useful statements about the mechanisms of forest dieback due to acid rain; usually there are no convenient control forests! In some areas forest dieback is thought to be due to excessive deposition of nitrate-nitrogen in precipitation. Plant uptake of NO_3 may exceed the capacity of the nitrate-reductase system, inhibiting root growth and the uptake of other nutrients, such as phosphorus (Waring, pers. comm.). We might test a hypothesis that an imbalance between *N* and *P* nutrition may be affecting forest growth by field application of *P* fertilizer. If forest dieback is ameliorated, there would be strong inference for an underlying effect of excessive atmospheric deposition of *N*. The basis of this approach is, of course, through a recognition of the linkage of these elements in redox reactions with carbon and in the biochemical composition of living tissue. Ecosystem science must proceed by recognizing such linkages and testing them experimentally (Waring and Schlesinger 1985).

The activity of the biosphere varies widely on the surface of the Earth. Areas of ice and extreme desert have essentially no net primary productivity, whereas areas of tropical rain forest may show productivity in excess of 2000 $g/m^2/yr$ (Whittaker 1975). Organic accumulations in soils and marine basins show the net activity of biota through time; these also show an enormous range in content over the surface of the Earth (Schlesinger 1977; Post et al. 1982; Berner 1982). Even studies of atmospheric chemistry show regional differences in the concentration of trace components of biogenic origin. Thus, studies of global processes must appreciate the regional diversity of biosphere and its activity. Moreover, while traditional ecosystem science has concentrated on equilibrium concepts, modern ecosystem science must be prepared to deal with nonequilibrium conditions at both the local and the global level. This diversity in space and time leads to the current interest in hierarchy theory and methods for scaling process-level measurements to large areas and long time periods (O'Neill, chapter 10; Levin, chapter 16). These are important developments in ecosystem science, although it is unclear if their role will be merely technical or whether they will lead to a new level of theoretical development.

Diversity in biota begs the question of whether some species, or some ecosystems, are more important to global function than others. Is species diversity of interest to ecosystem ecology and biogeochemistry? Certainly it is easy to think of examples in which "keystone" species control much of the biotic function in particular systems. However, the much sought-after link between community and ecosystem approaches must address the question of species diversity more broadly. Community ecologists usually concentrate on the loss of species diversity in response to ecosystem stress, but often changes in species populations are seen without obvious changes in ecosystem-level function (Rapport, Regier, and Hutchinson 1985). Similarly, Vitousek (1986) analyzed the ecosystem-level changes that have resulted following well-documented cases of species invasion in North America and Hawaii. Properties such as net primary productivity and nutrient cycling were affected significantly only in those cases in which there was a change in the major physiognomy of the community. Ecosystem scientists will appreciate the diversity of nature when community studies can demonstrate that such diversity frequently affects well-established measurements of ecosystem function, such as net primary productivity or streamwater losses of nitrogen.

A productive avenue for examining the interface between community and ecosystem studies is likely to be found in models that link the life processes of particular species to properties of entire systems (e.g., Botkin, Janak, and Wallis 1972; Horn, Shugart, and Urban, chapter 17). I suspect that the importance of diversity will be seen most strongly in models that include nonequilibrium conditions, in which diversity leads to resilience and long-term persistence of ecosystem function at relatively stable levels. Such work may rekindle interest in the minimal constraints to the development of closed systems for space exploration (Botkin, Janak, and Wallis 1979) and in predicting the effects of genetically engineered organisms that escape to natural communities.

A quantum leap in progress could be made if evolutionary and population biologists mellowed in their widespread animosity for ecosystem science. Their help and insight are needed to develop the kind of productive interactions that are now found between physiological and ecosystem approaches. I have, in the past, made the prediction that "there are no emergent properties of ecosystems, such as nutrient conservation, that cannot be predicted from a thorough knowledge of the components of the ecosystem and their interactions" (Waring and Schlesinger 1985). Given the complexity of nature, a test of this hypothesis will require the integrated knowledge of many fields.

REFERENCES

Berner, R. A. 1982. Burial of organic carbon and pyrite sulfur in the modern ocean: Its geochemical and environmental significance. *Amer. J. Sci.* 282:451–73.

Bormann, F. H., and G. E. Likens. 1967. Nutrient cycling. *Science* 155:424–29.

Botkin, D. B., J. F. Janak, and J. R. Wallis. 1972. Some ecological consequences of a computer model of forest growth. *J. Ecol.* 60:849–72.

Botkin, D. B., B. Maguire, B. Moore, H. J. Morowitz, and L. B. Slobodkin. 1979. A foundation for ecological theory. *Mem. Inst. Italian Idrobiol.* (suppl.) 37:13–31.

Boulding, K. 1956. General systems theory. *General Systems* 1:11–17.

Chapin, F. S., P. M. Vitousek, and K. Van Cleve. 1986. The nature of nutrient limitation in plant communities. *Amer. Natur.* 127:48–58.

Evans, F. C. 1956. Ecosystem as the basic unit in ecology. *Science* 123:1127–28.

Garrels, R. M., and A. Lerman. 1981. Phanerozoic cycles of sedimentary carbon and sulfur. *Proc. Natl. Acad. Sci. USA.* 78:4652–56.

Houghton, R. A., J. E. Hobbie, J. M. Melillo, B. Moore, B. J. Peterson, G. R. Shaver, and G. M. Woodwell. 1983. Changes in the carbon content of terrestrial biota and soils between 1860 and 1980: A net release of CO_2 to the atmosphere. *Ecol. Monogr.* 53:235–62.

Innis, G. S. 1976. Is there a paradigm for nutrient cycling? In D. C. Adriano and I. L. Brisbin, eds., *Environmental Chemistry and Cycling Processes,* pp. 113–120. U.S. Department of Energy, Washington, D.C.

Lindeman, R. L. 1942. The trophic-dynamic aspect of ecology. *Ecology* 23:399–418.

Lovelock, J. E. 1979. *Gaia: A New Look at Life on Earth.* Oxford University Press, Oxford.

Miller, P. C. 1976. Problems of synthesis in mineral cycling studies: The tundra as an example. In D. C. Adriano and I. L. Brisbin, eds., *Environmental Chemistry and Cycling Processes,* pp. 59–71. U.S. Department of Energy, Washington, D.C.

Odum, H. T. 1957. Trophic structure and productivity of Silver Springs, Florida. *Ecol. Monogr.* 27:55–112.

Post, W. M., W. R. Emanuel, P. J. Zinke, and A. G. Stangenberger. 1982. Soil carbon pools and world life zones. *Nature* 298:156–59.

Rapport, D. J., H. A. Regier, and T. C. Hutchinson. 1985. Ecosystem behavior under stress. *Amer. Natur.* 125:617–40.

Reichle, D. E., ed. 1981. *Dynamic Properties of Forest Ecosystems.* Cambridge University Press, Cambridge, England.

Reiners, W. A. 1986. Complementary models for ecosystems. *Amer. Natur.* 127:59–73.

Schlesinger, W. H. 1977. Carbon balance in terrestrial detritus. *Ann. Rev. Ecol. Syst.* 8:51–81.

Sollins, P., D. Coleman, B. S. Ausmus, and K. Cromack. 1976. A new ecology? A view from within. *Ecology* 57:1101–1103.

Tansley, A. G. 1935. The use and abuse of vegetational concepts and terms. *Ecology* 16:284–307.

Vitousek, P. M. 1986. Biological invasions and ecosystem properties: Can species make a difference? In H. A. Mooney and J. A. Drake, eds., *Ecology of Biological Invasions of North America and Hawaii,* pp. 163–76. Springer-Verlag, New York.

Vitousek, P. M., P. R. Ehrlich, A. H. Ehrlich, and P. A. Matson. 1986. Human appropriation of the products of photosynthesis. *Bioscience* 36:368–73.

Walker, J.C.G. 1977. *Evolution of the Atmosphere.* Macmillan, New York.

Walker, J.C.G. 1984. How life affects the atmosphere. *BioScience* 34:486–91.

Waring, R. H., and W. H. Schlesinger. 1985. *Forest Ecosystems.* Academic Press, Orlando, Fla.

Webb, W. L., W. K. Lauenroth, S. R. Szarek, and R. S. Kinerson. 1983. Primary production and abiotic controls in forests, grasslands, and desert ecosystems in the United States. *Ecology* 64:134–51.

Whittaker, R. H. 1975. *Communities and Ecosystems.* Macmillan, New York.

ECOLOGY AND RESOURCE MANAGEMENT VII

Chapter 19

Bioeconomics

COLIN W. CLARK

The word "bioeconomics" (sometimes "bionomics") has been employed with two quite different meanings. First, "bioeconomics" has been used to describe the ways in which biological organisms allocate resources to optimize reproductive success. For example, the *r-K* selection dichotomy has been characterized as a bionomic question (Southwood 1981).

In this article I employ the word "bioeconomics" to refer instead to the gamut of interactions between biological systems on the one hand and human economic systems on the other. It is manifestly obvious that biological resources are crucial to man's welfare—we depend on them exclusively for food, and to a major degree for clothing, housing, medicines, and recreational benefits. Yet until recently mainstream economics has tended by and large to ignore the fundamental role of biology, leaving bioeconomics in the hands of specialized disciplines such as agricultural economics—the "economics of the barn" to many professional economists. Biologists have displayed greater concern for the health and persistence of ecosystems as a foundation for human well-being, but have usually tended to oversimplify the economic side of the relationship.

It is hardly surprising to find this professional specialization on one side or the other of the field of bioeconomics. Both biological and economic systems are themselves complex almost beyond comprehension. A bioeconomic system is all the more complex, being much more than the sum of its biological and economic parts.

Forest management may be considered as one example of this complexity. A forest is a highly involved biological system, typically containing thousands of interacting species ranging from soil micro-organisms to the largest of trees. Timber production may be only one of many economic uses of the forest, others including recreation and wildlife preservation, regulation of water flow, protection of stream habitats for fish, and on a global scale, modulation of climate and atmosphere. The economic demands of the primary exploiters—the logging industry—may be in sharp conflict with alternative uses. The literature in forest economics, however, has been almost exclusively addressed to the problem of optimal harvesting, with limited attention paid to multiple uses (see Reed 1986 for a good review).

The question of conflicting demands on resources is one of the leading themes in bioeconomics. Other important themes include time discounting and the effects of uncertainty. Bioeconomic modeling, based on these themes, has primarily been addressed to local or regional problems, which can in principle be dealt with by local and federal management agencies. A subsidiary theme concerns the comparative effectiveness of alternative management policies. Each of these themes is discussed briefly in the following pages.

In recent years the importance of global bioeconomic problems has become increasingly apparent. Examples include the potential biological and economic implications of acid rain, climatic modification, desertification, and the reduction of genetic diversity. The classical themes of conflicting uses, time discounting, and uncertainty apply in full force to global environmental and resource issues, but at present there exist few if any institutions capable of influencing these vital developments.

THE ROLE OF MODELING

Over the past two or three decades there has been an increasing trend toward the use of models as a basis for addressing bioeconomic problems. These models can be divided into two general classes: (1) management models, and (2) descriptive models. Management models are specific models of particular resource systems, designed to assist resource managers in determining optimal management regimes. Such models are commonly used in fisheries (which have a voluminous literature, going back at least to Beverton and Holt 1957), and in forestry, a field in which large-scale computer models have become common (see Reed 1986). Quantitative management models have also been employed in areas such as range and wildlife management. Many existing management models have little or no economic component, the usually assumed objective being simply the maximization of sustained yield (MSY) from the resource stock. Some economists have argued that this one-sided approach to resource management is responsible for the failure of many resource industries to contribute positively to national wealth, but the problem involves many additional social and institutional factors (see, e.g., Pontecorvo 1986).

Descriptive bioeconomic models are an attempt to improve our understanding of how resource systems work. Such models lead at least to qualitative (but ideally also quantitative) predictions of how economic and biological systems interact, and how the advent of various management institutions and policies may affect the operation of resource systems (Clark 1985).

CONFLICTS AMONG RESOURCE USERS

In elementary economic theory, two extreme or ideal forms of market control are modeled—monopoly and pure competition. A similar (but not identical) dichotomy pertains to resource exploitation, in which private resource ownership contrasts with open, competitive access to the resource. A simple graphic model of this dichotomy for the case of a fishery resource is due to Gordon (1954) (fig. 19.1). The so-called "bionomic equilibrium" of the open-access fishery occurs at effort level E_∞, where total revenue just balances opportunity cost. Net profits (more accurately, rents) would be maximized at a lower level of effort E_0. Gordon's model predicts that if unregulated, fishing effort will increase until net revenues become zero—because the existence of any positive net revenues will always attract additional fishermen. Subsequent increases in fish price, or decrease in effort costs, both of which would result in increased profits at the optimal effort

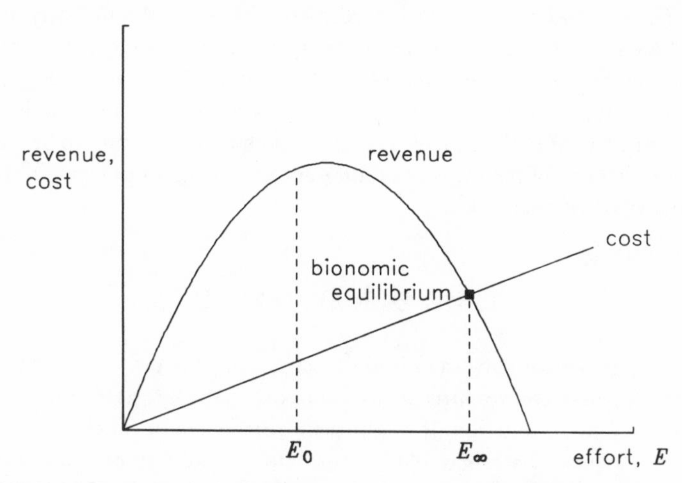

FIGURE 19.1. H. S. Gordon's (1954) model of bionomic equilibrium in an open-access fishery.

level, simply lead to further increases in fishing effort and the consequent further depletion of the fish stock.

Gordon's very simple model reveals a fundamental difficulty in renewable resource exploitation: unregulated, open-access resources will become overexploited, leading to the impoverishment of the associated resource industry. In extreme cases, the resource stock may be driven to the point of extinction.

It is important to stress that this "Tragedy of the Commons" (to use the term of Hardin 1968), resulting from open-access resource use, does not result from irrational behavior on the part of individual exploiters. Quite the opposite is true—individual rationality leads to a socially undesirable result because of the way that bioeconomic systems work. The current fad for governments to withdraw from "interference" in the private sector, if applied to common-property resource systems, can only make matters worse. The responsible approach is one of continually monitoring and improving management regulations and institutions, with a view toward achieving long-term sustainability of the resource base, and long-term viability of the associated resource industry.

Although simplistic to the point of caricature, Gordon's model is at least implicitly a bioeconomic model. The fact that the yield, or total revenue curve, reaches a peak and declines with further increases of the input variable, effort, is a consequence of the biological limitations of the fishery resource. Thus the bionomic equilibrium is in effect both a biological and an economic equilibrium.

Although Gordon's model was directed specifically to the case of the commercial fishery, it is clear that the principle of overexploitation of common property resources is a universal one. The nature and scope of the problem is now widely recognized among resource scientists and managers, if not politicians. A Common Property Resource Network has been established, with currently over one thousand members from seventy countries. The central office is at the University of Minnesota–St. Paul, Center for Natural Resource Policy and Management.

The practical policy and management implications of Gordon's model are far from obvious; basically the model is just too much of a caricature to be used as a basis for predictions of the impacts of different management policies. This question is discussed further below.

TIME DISCOUNTING

It is tempting to conclude from Gordon's analysis that the difficulties inherent in competitive, open-access resource use could be overcome by establishing exclusive rights to resource exploitation. It might seem reasonable to conclude, for example, that the owner of a renewable resource stock would automatically wish to conserve the stock so as to maximize his annual net profits. In 1973 I pointed out in a *Science* article (Clark 1973) that this is not necessarily true for K-selected species such as whales and redwood trees, which possess low intrinsic rates of growth. The owner of any such resource may be able to earn a greater income by depleting the resource and investing the profits at a rate of interest greater than the "internal rate of return" of the resource stock. There are many provisos to this rule—for example, the market price of the resource product should not grow at a superinflationary rate. Nevertheless, there can be little doubt that future discounting (an equivalent way of looking at interest-rate calculations) is often a primary force mitigating against resource conservation (cf. Ciriacy-Wantrup 1968). In negotiations with governmental management authorities, resource exploiters regularly use "current economic need" arguments to plead for increased quotas, regardless of projected impacts on future resource productivity. This holds true in fisheries, forestry, water resources, and many other areas.

Once again it is important to realize that the motivations for resource depletion that derive from discounting are not necessarily irrational from the viewpoint of resource users. Present survival is a logical prerequisite to future existence—a fundamental biological fact that influences the behavior of every organism and every organization. The economic reflection of this principle is time discounting, by individuals and by society at large. Some social philosophers have argued that it is the duty of the state to ensure that discounting does not affect the welfare of future generations, but how this is to be accomplished by a nontotalitarian state remains unclear.

The discounting and common-property influences on conservation are not unrelated; nor are their effects by any means limited to the depletion of currently commercially valuable resource stocks. For example, the urgent need for foreign capital may lead tropical nations to undertake the rapid exploitation of forests. A side effect may be a significant loss of endemic forest-dwelling species, and a corresponding loss of genetic diversity (Oldfield 1984), which may ultimately have untold economic implications for the future production of pharmaceutical and other chemicals, and for horticulture and the viability of agricultural crops worldwide. These future potential losses are largely incalculable, and are unlikely to enter into any computation of the profits of the forest industry.

UNCERTAINTY

Conservation by definition implies a concern for the future. Unfortunately the future is highly uncertain. Responding to future uncertainty is a rational act, which in the case of resource exploitation can seriously affect management decisions. An interesting example is the forest rotation problem: what is the optimal age at which trees should be harvested? The answer depends critically on the discount rate (Clark 1976), and also on the risk of forest fire (Reed 1984). Indeed, as Reed shows, forest fire risk simply has the effect of imposing an additive increase to the discount rate. In economics, this is called "risk discounting"; it is a separate effect from time preference. Risks of disease or pest outbreaks in the forest have effects on harvest policy which are similar to those of the risk of fire.

Many other sources of uncertainty are also important in bioeconomics. Economic uncertainties pertaining to future prices and markets can be important in renewable resource industries, but such risks are fairly normal in *all* resource industries, and not uncommon in the economy generally. Uncertainties that have a purely biological basis are perhaps more important, and are certainly often more difficult to deal with.

In fishery systems, uncertainty is particularly pervasive because of the simple fact that wild fish populations cannot be observed directly. The size of a fish stock can only be inferred, either from sample surveys or from fishery data. In either case, mathematical-statistical models of the survey or fishery process, as well as of the demography of fish populations, are inescapable. It is no accident that fishery modeling has long been dominated by mathematically literate scientists.

An interesting and important but theoretically difficult question is how to calculate, even roughly, the optimal level of expenditure on stock assessment in fisheries (Clark and Kirkwood 1986). Management agencies now spend millions of dollars estimating fish stock abundance, but my spies tell me the Management Councils are still not doing a very good job of conserving fish stocks. The economic realities of the fishing industry force the Councils implicitly to adopt high discount rates.

Although the case is not perhaps as one-sided as those for common property and discounting, uncertainty also has a primarily anticonservationist bias. Government expenditures on biological research in the resource field are justified by this fact—if the government does not support the research it will not get done.

The bioeconomics of uncertainty have only recently begun to be worked out (Mangel 1985; Clark 1985, chapter 6; Walters 1986), and I will say more about this later.

THE THEORY OF RESOURCE REGULATION

Most resource industries are subject to intensive regulation by government agencies. Regulations are presumably introduced with some set of objectives in mind. The results of a new system of regulations, however, often turn out to be very different from the apparent original objectives. Bioeconomic modeling has the potential for predicting the resource industry's reaction to alternative management policies, and may thus lead to more effective management strategies.

It must be emphasized that regulations are necessary precisely because rational individual behavior leads to socially undesirable results. It follows that there is always a natural tendency to resist and thwart regulations. Even if nothing is done illegally, individual resource users will always be searching for ways to get around any regulations that hinder their natural proclivities.

For example, consider the fishery model depicted in figure 19.1. Suppose that, bionomic equilibrium at $E = E_\infty$ having been reached, the government decides to intervene and reduce effort to E_0. There are various types of regulations that can be employed for this purpose, one of the most common being simply to shorten the annual fishing season. This method is currently employed in the Pacific fisheries for salmon, herring, halibut, and other species. An alternative but essentially equivalent method is to set an annual catch quota, and then to close the fishery as soon as the quota has been caught. This method was used in the 1960s and 1970s by the International Whaling Commission, and has also been used for tuna, cod, and many other species.

Suppose that some such method is successful in reducing effort to E_0. What will happen? A dynamic model of the fishery (in contrast to the purely equilibrium model of fig. 19.1) will predict that catches will first be reduced but will ultimately increase as the fish stock is rebuilt. This essentially dynamic scenario was surprisingly overlooked by many resource economists following the publication of Gordon's (1954) paper, although it was analyzed very clearly by Beverton and Holt (1957).

The end of the story is still to come, however. Assume that effort has been reduced to E_0, that the fish stock has recovered from overexploitation, and that revenues now exceed costs, as in figure 19.1. The fishermen are making good money, and additional fishermen will therefore be attracted to the fishery. Effort will tend to increase again; remember that rational behavior always attempts to overcome regulations. To counter this expansion in fleet size, the government must further reduce the length of the fishing season, thereby engendering further expansion, and so on. We reach the paradoxical conclusion that the original regulations, which were deemed necessary because there were too many fishermen, have had the effect of *increasing* the number of fishermen to an even higher level.

Ultimately, a "regulated bionomic equilibrium" may be reached when the costs of fishing have increased sufficiently to offset revenues (fig. 19.2). Costs will have increased simply because expensive fishing vessels can now be used only for a few weeks per year. Individual fishermen are as badly off under the regulated equilibrium as under the original unregulated equilibrium. (In practice, fishermen may actually be worse off, if the temporary period of high net revenues convinced them to go into debt to purchase new vessels and equipment.)

For the benefit of readers not familiar with fishery management, I hasten to point out that the scenario just described is not an academic abstraction. It is hard to think of any regulated fishery on earth that is not suffering simultaneously from overexpansion of fishing capacity and impoverishment of fishermen. The

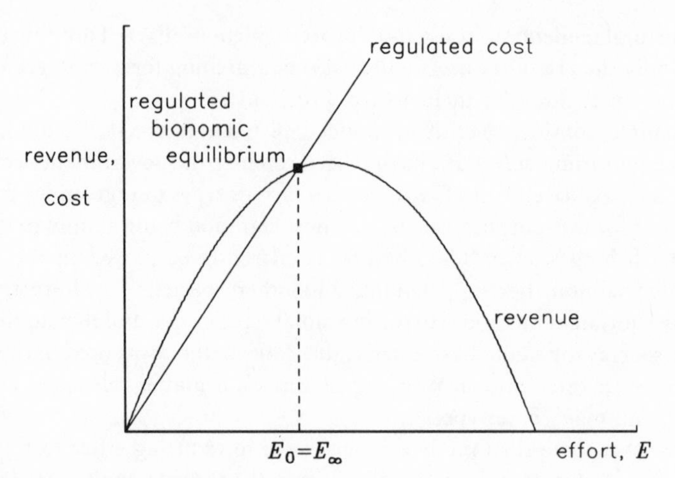

FIGURE 19.2. Regulated bionomic equilibrium.

only exceptions would be cases where prices have recently increased, so that there is again a temporary differential between revenue and cost. Such differentials can occasionally be spectacular (witness a one-day catch of $9 million worth of herring by four seiners in British Columbia!). If so, they naturally tend to attract yet further entry into the fishery.

Because of the failure of these traditional approaches to fishery management, many countries have begun to experiment with alternative methods that recognize the social and economic dimensions of the problem. The modeling literature is still underdeveloped in this area, although several authors have addressed some of the main issues theoretically (Clark 1980, 1985; McKelvey 1985; Anderson 1986; Rosenman and Whiteman 1987).

A first step toward controlling fishery overexpansion is a limitation on the number of fishermen and vessels. Many countries have introduced limited licensing systems into commercial fisheries. Such limited entry programs have met with varied degrees of success, as in many instances licensed fishermen have been able to replace smaller vessels with larger, more powerful, and better-equipped vessels, again resulting in fishery power far in excess of that needed to catch the available fish. A simple game-theoretic analysis (Clark 1980) indicates that competition between as few as two exploiting firms will still lead to resource depletion and overcapacity. Explicit agreement and cooperation in sharing the resource are required if the problem is to be resolved.

A current trend in fisheries management is the introduction of allocated catch quotas. This approach has the potential for removing the natural tendency toward overexpansion of capacity—a fisherman who owns a certain quota has no incentive to expand his fishing capacity, unless he thinks he can cheat on his quota (thus strict enforcement of the quotas is essential). Interestingly, quota allocations (by country) were introduced in Antarctic whaling as early as 1961 (Clark and Lamberson 1982). Apparently the quota system increased the profitability of whaling, although it failed to save the whales. Resource conservation often requires a control authority with more than a short-term profit motive. No such authorities exist on the international level.

It is an elementary result of welfare economics that allocated transferable quotas are theoretically equivalent, in their influence on the activities of resource users, to taxes or other charges. This result has been extended to the dynamic case by Clark (1980); see also McKelvey (1985). However, quotas and taxes are obviously far from equivalent in their effect on incomes. Allocated catch quotas in renewable resources place the resource rents in the hands of the quota holders, while taxes reap the rents for the government. Resource taxes are understandably unpopular with producers; on the other hand, allocated quotas granted freely to favored individuals have the potential of turning the recipients into rich men. The fair alternative is the auctioning of quotas to the highest bidder. Quota auctions are in fact common in various resource industries, from forestry to offshore oil drilling.

The management of common-property biological resources is no easy task. Mathematical theories have helped to disentangle some of the bioeconomic complexities, and these theories are gradually being put to the test in various countries.

RECENT DEVELOPMENTS

Thirty years ago the use of mathematical modeling approaches in resource management was unheard of. Gordon's fishery model was graphical rather than mathematical, but it led eventually to the development of relatively sophisticated mathematical bioeconomic models (control theoretic, game theoretic, stochastic programming, etc.). Research activity in this area, covering exhaustible and renewable resources as well as environmental problems generally, is now being pursued at a high level. Important recent books include Herfindahl and Kneese (1974), Clark (1976), Dasgupta and Heal (1979), Mirman and Spulber (1982), May (1984), Mangel (1985), Scott (1985), Clark (1985), Walters (1986), Starfield and Blelock (1986), Workman (1986), Conrad and Clark (1987).

The journals devoted exclusively to resource and environmental modeling include the *Journal of Environmental Economics and Management, Marine Resource Economics,* and *Natural Resource Modeling*. A new journal, *Ecological Economics* came to my attention during the writing of this article. While several other journals in the fields of economics, biology, management science, and applied mathematics publish occasional papers devoted to resource issues, the growing list of specialized journals attests to the approaching maturity of bioeconomics as a recognized field of specialization.

This is not the place to discuss the practical implementation of the insights obtained from bioeconomic modeling. In most cases, the main impediments to a more rational long-term approach to resource and environmental management seem to be primarily political. The political system is by no means exempt from the realities of future discounting, although the standard aphorism that politicians look forward only to the next election is perhaps a superficial view—we all discount the future, and if not we would throw the shortsighted politicians out of office.

Political events such as the proclamation of 200-mile zones and the passage of environmental protection and conservation legislation call for an increasingly professional approach to the understanding of bioeconomic systems. The future may see a progressive unification of biology, economics, and political science, culminating in a grand view of man and his vital resource systems.

GLOBAL BIOECONOMICS

As the world population grows, and technology increases, the demand on biological systems expands, and an entirely new class of global bioeconomic problems is coming into existence. These problems display the three main motives of bioeconomics—multiple users, future discounting, and uncertainty—often in extreme and complex forms. The true scope of these problems is only now beginning to become apparent. Acid rain may be mentioned as an example of a recently recognized environmental phenomenon, which crosses international boundaries, has vast biological and economic consequences, and remains poorly understood scientifically.

As yet there are no specific bioeconomic models of global systems. Given the lack of any management agency with international scope, it is not clear what the role of such models would be, in any case. Some important international agencies, however, have recently begun to address global bioeconomic problems.

In May 1987, for example, the president of the World Bank announced a new environmentally oriented policy centered around sustainable rather than exhaustible development of third-world resource systems (Wolf 1987). Also, in April 1987 a small group of Soviet and Western scientists met in Leningrad to advise the Soviet environment minister on "Criteria for the Sustainable Development of the Biosphere." Such events may indicate an increasing awareness by the world's leading decision makers of the fragility of biological systems induced by man's economic activities.

REFERENCES

Anderson, E. E. 1986. Taxes vs. quotas for regulating fisheries under uncertainty: A hybrid discrete-time continuous-time model. *Mar. Res. Econ.* 3:183–207.

Beverton, R.J.H., and S. J. Holt. 1957. *On the Dynamics of Exploited Fish Populations.* Ministry of Agriculture, Fisheries and Food (London), Fish. Invest. Ser. 2(19).

Ciriacy-Wantrup, S. V. 1968. *Resource Conservation: Economics and Policies,* 3d ed. University of California Press, Berkeley.

Clark, C. W. 1973. The economics of overexploitation. *Science* 181:630–34.

Clark, C. W. 1976. *Mathematical Bioeconomics: The Optimal Management of Renewable Resources.* Wiley-Interscience, New York.

Clark, C. W. 1980. Restricted access to common-property fishery resources: A game-theoretic analysis. In P. T. Liu, ed., *Dynamic Optimization and Mathematical Economics,* pp. 117–32. Plenum Press, New York.

Clark, C. W. 1985. *Bioeconomic Modelling and Fisheries Management.* Wiley-Interscience, New York.

Clark, C. W., and G. P. Kirkwood. 1986. Optimal harvesting of an uncertain resource stock and the value of stock surveys. *J. Environ. Econ. Manag.* 13:235–44.

Clark, C. W., and R. H. Lamberson. 1982. An economic history and analysis of pelagic whaling. *Marine Policy* 6:103–20.

Conrad, J. M., and C. W. Clark, 1987. *Resource Economics: Notes and Problems.* Cambridge University Press, New York.

Dasgupta, P., and Heal, G. 1979. *Economic Theory and Exhaustible Resources.* Cambridge University Press, Cambridge, England.

Gordon, H. S. 1954. The economic theory of a common-property resource: The fishery. *J. Polit. Econ.* 62:124–42.

Hardin, G. 1968. The tragedy of the commons. *Science* 162:1243–47.

Herfindahl, O. C., and A. V. Kneese. 1974. *Economic Theory of Natural Resources.* Merrill, Columbus, Ohio.

McKelvey, R. W. 1985. Decentralized regulation of a common property renewable resource industry with irreversible investment. *J. Envir. Econ. Manag.* 12:287–307.

Mangel, M. 1985. *Decision and Control in Uncertain Resource Systems.* Academic Press, New York.

May, R. M., ed. 1984. *Exploitation of Marine Communities.* Dahlem Conference. Springer-Verlag, Berlin.

Mirman, L. J., and D. F. Spulber, eds. 1982. *Essays in the Economics of Renewable Resources.* North Holland, Amsterdam.

Oldfield, M. L. 1984. *The Value of Conserving Genetic Resources.* U.S. National Park Service, Washington, D.C.

Pontecorvo, G., ed. 1986. *The New Order of the Oceans: The Advent of a Managed Regime.* Columbia University Press, New York.

Reed, W. J. 1984. The effects of risk of fire on the optimal rotation of a forest. *J. Envir. Econ. Manag.* 11:180–90.

Reed, W. J. 1986. Optimal harvesting models in forestry management—a survey. *Nat. Res. Modeling* 1:55–80.

Rosenman, R. E., and C. H. Whiteman. 1987. Fishery regulation under rational expectations and costly dynamic adjustment. *Nat. Res. Modeling* 1:297–320.

Scott, A. D., ed. 1985. *Progress in Natural Resource Economics.* Clarendon Press, Oxford.

Southwood, T.R.E. 1981. Bionomic strategies and population parameters. In R. M. May, ed., *Theoretical Ecology,* 2d ed. Blackwell Scientific, Oxford.

Starfield, A. M., and A. L. Blelock. 1986. *Building Models for Conservation and Wildlife Management.* Macmillan, New York.

Walters, C. J. 1986. *Adaptive Management of Renewable Resources.* Macmillan, New York.

Wolf, E. C. 1987. *On the Brink of Extinction: Conserving the Diversity of Life.* Worldwatch Paper No. 78. Worldwatch Institute, Washington, D.C.

Workman, J. P. 1986. *Range Economics.* Macmillan, New York.

Chapter 20

Theoretical Issues in Conservation Biology

STUART L. PIMM AND MICHAEL E. GILPIN

The world is on the threshold of a catastrophic loss of biological diversity. Hundreds of thousands, possibly even millions, of plant and animal species will be lost if current trends continue. There are four major causes of this loss of diversity—the so-called "evil quartet" (Diamond 1984): (1) destruction and fragmentation of some habitats and the pollution and degradation of others; (2) overkilling of plants and animals by man; (3) introduction of alien animals and plants; and (4) secondary effects of extinctions—the extinction of one species caused by the extinction of another. While each of these evils strikes differently at the three levels of biological organization—ecosystem, community, or population—each can reduce the amount of species diversity, and each challenges the science of conservation biology for an appropriate response.

First, habitat loss is clearly the most important member of the quartet. Even when apparently large amounts of land are protected, these areas may be spatially isolated from each other, and each fragment, or habitat patch, may be too small to maintain viable populations; that is, in small ecological systems normal ecological processes may fail and species may be lost as the result of scale. An understanding of the dynamics of very rare species, on single patches and also over a network of such patches, is essential both to predicting which species are prone to extinction and to shaping the options for managing the habitat on which they occur.

Second, many animal species have been hunted to extinction—the passenger pigeon in North America is a familiar example. These species losses may strike us as examples of gross incompetence. But there are many species whose continued survival should have been in our own economic interests, but which, nonetheless, have been lost. The fisheries literature is replete with such examples. Since the last individual in the population is probably not killed in the harvest or the hunt, this suggests that "overkilling" is not as simple an issue as it might seem. That is, overhunting may overwhelm the natural self-stabilizing mechanisms of the species, which may thus go extinct for unrelated reasons.

Third, man has moved many species around the world, sometimes deliberately, though often accidentally. Some species, like the English garden birds introduced by New Zealand settlers, are merely quaint curiosities. But other introductions have been devastating. Goats, pigs, and rabbits on oceanic islands have sometimes removed nearly all the vegetation. A snake introduced to Guam has caused the extinction of most of the island's bird species (Savage 1987). A fungus introduced to North America has all but eliminated the chestnut trees—once one of the commonest trees in the forest. Understanding which species are likely to invade which communities and what effects they will have is another pressing need for conservation biology.

Fourth, there are secondary effects. The loss of a species that is the major food source of a second species can be expected to cause the loss of that second species. So can the loss of a mutualistic species. The demise of a particular species may even cause the extinction of many other species, and theory may help to predict which and how many species these will be.

Thus, each member of the "evil quartet" poses scientific questions for conservation biology, and there are theoretical aspects to all these challenges. In fact, the theoretical side to these challenges is greater than might first be appreciated. Conservation biology is a crisis discipline (Soulé et al. 1986). Since we must make a number of our decisions in the absence of good data, complete knowledge, and appropriate models, our theories, adduced from the full range of our population biology experience, will be especially important. It is incumbent on us, then, that we be familiar with the strengths and weaknesses of these theories and that we be prepared to modify our theory as our experience in conservation biology dictates.

HABITAT LOSS

Habitat may be reduced, fragmented, degraded, or even lost altogether. For the case of habitat reduction, we need to be able to predict, from measurable dynamical indices of species, what the effects will be on the system and its constituent species. In particular, we need a way to relate area to population size, and this, in turn, to extinction probability. With fragmentation, we not only need to know the extinction probabilities of species on the small habitat patches, but we also need to know the species movement dynamics over the entire network of patches, so that the balance point, if there is one, between local extinction and recolonization can be assessed.

If we know the extinction probabilities of species on various spatial configurations of habitat, we will be able to come to more rational decisions concerning reserve design. For example, we may be able to assess the relative merits of a singl larger reserve compared to several small reserves.

It may turn out that a species actually loses its habitat—as, for example, tl California condor—and that we will need to know how to manage the species captivity until the day its habitat is again suitable and the species can be introduced. During the period of captive breeding, genetic considerations will paramount for the preservation of the species. And for reintroduction, again dynamics of rare species will suggest efficient strategies to accomplish the g . We consider each of these topics in the following five sections.

Population Vulnerability Analysis

Conservation efforts are usually reactive rather than preventative. As such ey place their emphasis on managing already very rare species. In general ire species are the most likely to become extinct, but there are considerab lif-ferences in the likelihood of extinction for a species of a given populatio ize. Can we anticipate which species are extinction prone? If we understa the dynamics of very rare species, can we alter them in order to preserve the s ies? We may fail and lose some species from the wild, probably retaining usua very small populations in captivity. Captive populations have special genetic p lems that we will discuss in a later section. If we can overcome these, and i can reverse whatever caused the species to become extinct in the wild, ca e re-introduce the species to the wild, either to its original habitat or elsewhe To be successful, this reintroduction must be based on an understanding of t actors predisposing very rare species to become extinct.

We first examine the factors that cause population change and the amine how species with different life histories will respond to these factors. W iall see that there are two kinds of processes that cause extinction: those asso d with demographic accidents, which depend very strongly on populatio ze, and those associated with environmental variation, which depend less ngly on population size. The population characteristics—how fast a popul grows, how long the individuals live, etc.—interact with these two factor give the resultant patterns of which species are extinction prone.

Consider two extreme cases. In the first, extinction is caused sol y demo-graphic accidents in an unvarying environment, with constant pe pita birth and death probabilities up to some population ceiling, *K,* above w the birth probabilities are assumed to be zero. These models predict very rap icreases in times to extinction (T^{ext}) as *K* increases (Leigh 1981; Richter-Dyn Goel 1972; Goodman 1987).

This very rapid increase in T^{ext} with *K* suggests that there minimum viable population size (MVP) below which a species is almost ce ily doomed to extinction and above which the population's survival is assure iaffer 1981).

Certain other processes besides dynamics can lead to this rapid transition (Gilpin and Soule 1986), but there are other dynamical processes that lead to much more modest increases in T^{ext} with K.

For the opposite extreme case, consider the ultimate form of external environmental disturbance—the total, catastrophic destruction of the habitat, such as might result from logging of a forest, a massive flood, or an asteroid collision. The time to extinction will be the same for all species, irrespective of population size, provided that this time is less than the time to be expected in the absence of the disturbance.

These extreme cases may not be very realistic, but they demonstrate a point. For demographic accidents alone, time to extinction increases rapidly with population size. For a given population size, environmental variation reduces the time to extinction below that expected from demographic accidents alone in an unvarying environment. As environmental variability increases, the increase in time to extinction with population size becomes more modest, reaching our extreme case (of no increase at all) when environmental disasters are extreme. Leigh (1981) suggests that, under some kinds of environmental disturbance, the time to extinction may increase as the logarithm of K. Figure 20.1 distinguishes the time to extinction for demographic stochasticity, environmental stochasticity, and catastrophes.

Three intuitively obvious theoretical results show how extinction times should vary among species (Leigh 1981; Goodman 1987). Extinction should on average take longer for species with

> 1. High intrinsic rates of increase (r); such species can recover more quickly from environmental disasters.

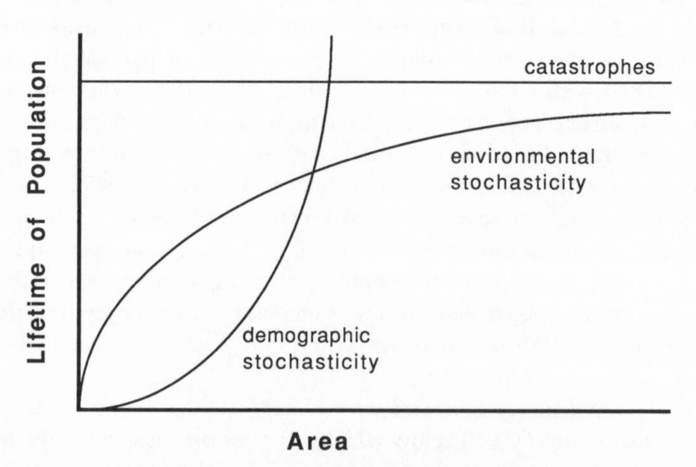

FIGURE 20.1. Functional relationships between area of patch and the lifetime of a population on it for three classes of stochasticities.

2. Long lifetimes; such species are less likely to go extinct in any year through demographic accidents because a smaller proportion of the population is likely to die.

3. Low temporal coefficients of variation in density (CV).

The difficulty is that all three parameters are linked.

Across animal species from protozoa to elephants, a one-order-of-magnitude increase in r corresponds to a one-order-of-magnitude decrease in generation time. But both r and longevity are closely related to body size (Bonner 1965), in ways that affect extinction rates oppositely: large body size is associated with low r and with long life times. What is the net effect of body size on time to extinction?

A summary of a quantitative argument (Pimm, Jones, and Diamond 1988) can be made by considering two extremes. First, consider single individuals of a large-bodied and a small-bodied species. Both are doomed to die, but the former will probably live longer; its yearly extinction rate is lower, even if the per generation rates of extinction for the two species are the same. Second, consider identically sized large populations of the two species. The large-bodied species is at a disadvantage because, following some severe reduction in numbers, its lower intrinsic rate of increase makes it take longer to climb to higher numbers than the small-bodied species. In most models of extinction it is this effect that causes times to extinction to decrease so rapidly with increases in r, for species with the same equilibrium numbers. These two extreme cases show that large-bodied species have an advantage over small-bodied species at low densities and that the reverse is true at high densities. However, these verbal arguments do not tell us at what density the advantage switches.

If we look at time to extinction measured in generations, not years, small-bodied species are always at an advantage. Measuring extinction times in generations may be sensible for some applications, but conservationists' plans for managing species will be measured in years, not in the different generation times of different species. Nor is the year a biologically arbitrary time unit. For many species, dispersal—and hence movement between isolated populations—will occur on a yearly cycle. Thus years, not generations, will be the appropriate measure of the time until a floundering population may be rescued by immigrants.

The chance of extinction also depends on the temporal variability of population size, which, in turn, also depends on body size, r, and generation time. We might expect that large-bodied species would, despite some counterexamples, tend to survive an environmental disturbance better than small-bodied species. Yet we have also argued that large, slowly growing species might recover more slowly from reductions in density than small species with short reproductive cycles. Thus, body size affects not only time to extinction but also CV in two opposite ways. Should the net result be that large-bodied or small-bodied species have higher CVs?

For British bird species, there are both advantages and disadvantages of large body size. Hard winters are a major cause of abrupt population declines in British birds, and small-bodied species suffer greater numerical reductions in hard winters than do large-bodied species (Cawthorne and Marchant 1980). However, much of the temporal variation in population densities comes from the slowness of recovery from these reductions. Pimm (1984) has shown that CV correlates negatively with effective rate of increase: slowly growing populations have higher CVs, even given the more modest declines of slowly growing species following a severe disturbance. While there might thus be an overall negative correlation between CV and body size, this is not a strong effect: large-bodied species show much the same range of CVs as small-bodied species (Pimm 1984). However, over a larger range of body sizes (insects to large mammals), CVs tend to be larger for smaller-bodied species (Pimm and Redfearn 1988).

Pimm et al. (1988) show that the suggested tradeoffs in body size and population size explain much of the variation in extinction rates, (once corrected for population size) of populations of birds on small islands off the coast of Britain. At low population densities, less than seven breeding pairs, large-bodied species are less likely to become extinct than small-bodied species. For populations greater than seven breeding pairs the reverse is true. Given these patterns, populations with higher CVs are more likely to become extinct than populations with small CVs. In short, theory predicts which species are extinction prone, and this limited data set supports those predictions.

Fragmentation

The preceding discussion has focused on a single population, that is, on a single habitat patch. Yet many populations are composed of a number of partially isolated subpopulations (Gilpin 1988). Now, small populations have a much greater chance of extinction because of random demographic accidents and local environmental variation. And these problems may be particularly acute if the population has been fragmented, because each fragment is so much smaller. On the other hand, isolated populations are likely to be less susceptible to certain forms of environmental disasters, such as those limited in time and space, for example, a local fire or a fast-spreading disease, which would therefore affect only some of the subpopulations, leaving others as the source areas for recolonization. Thus whether the existence of a population structure increases or decreases, the likelihood of total extinction depends, yet again, on the interplay between demographic and environmental causes of extinction and how the population responds in both time and space (Gilpin 1988).

Fragmentation has a qualitative effect on a population's probability of extinction. A fragment will likely be different from the continuous habitat of which it earlier was a part. Edge effects (Wilcove, McLellan, and Dobson 1986; Lovejoy et al. 1986) are an important consideration. An edge may allow the penetration of biotic factors that would not otherwise be felt within the system. Predators from the surrounding habitat are one clear example (Wilcove, McLellan, and Dobson 1986). There may even be abiotic changes in the edge zone. Lovejoy et al. (1986) found that changes in light penetration, wind, and humidity extend hundreds of meters into the interior of a patch.

Diseases are an excellent example of how population structure may be essential for the long-term survival of a species. Diseases will spread only if the population is dense enough (Hassell and May, chapter 22). Thus large population size can increase the chance of extinction from disease. Some populations may survive only if they are spatially fragmented. The history of the black-footed ferret in the United States is an example of not appreciating this point: the one known population was lost through disease. No other populations were known and, despite pleas to the contrary, none were established (May 1986). Only a last-minute captive breeding program offers the species any hope.

One management option, then, is to maximize the long-term chances of the population by moving individuals from the larger subpopulations to ones where local extinction has occurred or is imminent. Such a procedure has been carried out for the whooping crane in North America. Theoretical insights may be valuable here in deciding when extinction is, indeed, imminent, and how large a population can be, before we can remove individuals from it.

Reserve Design

We may have choices about which habitat fragments we can protect: money to buy land may be limited but we may have some flexibility in how the areas are chosen. This has been the topic of an extensive debate: whether we should put our efforts (fixed land resources) into one large or several small reserves. Species diversity (S) increases rather slowly with increasing area (A); approximately $S = cA^{Z}$, where Z is about ¼. Thus small reserves contain more species, per unit area, than large ones. But large areas often contain species not found in smaller areas—particularly large-bodied species that may require very large areas to maintain viable population sizes. These large species may, coincidentally, be keystone species (see the last two sections in this chapter), on whose survival the nature of the community depends. Conversely, a single large reserve may only sample a small part of the full range of communities to be found within the region. Many reserves, carefully selected to encompass a wide range of habitats, may be an effective way to preserve species for given, limited, resources. But these reserves must be of sufficient size to preserve viable populations of the species they were designed to protect.

A fragmented reserve design may be imposed by outside factors: not all suitable habitat may be left and what is left may not be managed. In such cases, special effort might be put into creating corridors of habitat, joining the habitat fragments. This will help reduce the isolation of the populations and so minimize the effects of low population size.

Since different species will react differently to habitat fragmentation, habitat isolation, and the character of ecological corridors, a general solution to the management problem for an arbitrary species may not be forthcoming. Nonetheless, since communities and ecological systems follow certain statistical regularities, there may be some hope for determining an optimal solution for raw species diversity. Higgs and Usher (1980) and Gilpin and Diamond (1980) have made a start in this direction. The work of McLellan et al. (1986) on British birds in woodlots of different sizes is sufficiently specific to be the basis of management action.

Species Reintroductions

If an introduction or reintroduction is to succeed, the species must survive for a time at very low population densities and increase despite the presence of potential competitors, predators, and diseases. Moreover, the species must not have lost, in captivity, those behavioral skills or genetic adaptations which are essential for its survival (May and Lyles 1987).

Experience with introduced species shows that populations that are eventually successful often fail initially (Long 1981). Thus, even when predators, parasites, competitors, and the physical environment will not prevent success, success is not inevitable. The problem we face in designing an introduction can be stated quite simply. Suppose we have a limited number of individuals, X. Should we introduce all X in one place at one time, or break the X into several groups? If we do the latter, what is the smallest size Z that each group can be? The advantage of introducing all X at one place is that we minimize the problems encountered by very small populations. The disadvantage is that such a strategy places "all one's eggs in one basket"—we may, unfortunately, introduce all X individuals into the wrong place at the wrong time. Again, the answer appears to be how species respond to environmental and demographic factors: some species are more susceptible to one than the other and this will alter the allocation of individuals.

Genetics

Small populations often lose genetic variability: they become inbred. This loss can cripple a population in one of several ways. First, inbred populations often show inbreeding depression: fertility decreases among genetically related individuals (Ralls, Brugger, and Ballou 1979). Second, inbred populations have a reduced chance at exploiting new opportunities, so that continual changes in the environment pose an increased threat. Third, the genetic uniformity means that diseases that might only affect a proportion of a genetically variable population may now totally destroy a less variable population. These three effects are often most visible in captive populations, which for a variety of reasons may be descended from a very small number of individuals. How then can we maintain genetic variability?

A species that is bred in captivity can be viewed as an information link between a formerly free-ranging species and what could become a restored population maintained in some type of biological reserve or park. The Arabian oryx is one of the first species to have completed this cycle, though the California condor should follow soon. From this standpoint of genetic information, the manager of such a captive species needs to pass the clearest and most accurate signal of the species' past history to those who will be in charge of its eventual ecological restoration, for this genetically coded signal contains the species' answers to the environmental problems it has faced and surmounted in the past, problems that are likely to be met again in the future. The tools such a manager has at his disposal are the theories and knowledge of population genetics and population ecology.

The best way to pass genetic information from generation to generation is over a broad-band, noiseless channel. This implies a large population in which equal contributions of offspring from all members of each generation are maintained for every generation. In the terms of population genetic theory, this implies no drift and no selection. For plant species in which long-term seed storage is possible, such an information channel may be economically feasible. For large-bodied zoo animals, this is not possible, and the manager must submit his or her program to certain economic constraints involving, among other things, limited cage space.

At the level of the DNA molecule or the chromosome, genetic information is partitioned between two classes: fixed loci and variable loci. Using the Shannon-Weaver measure of information, I, where p_{jk} is the frequency of the kth allele at the jth locus, one can see that only the variable loci actually contribute to this measure of genetic information. This makes a certain degree of sense. Fixed loci tell about the permanent identify and character of the species. They contain information that concerns unchanging form and behavior.

$$I = -\sum_{jk} p_{jk} \log p_{jk}, \tag{20.1}$$

The variable loci are different. Some fraction of them represent solutions to problems of environmental variability, whether of space or of time. If the frequency of one of the alleles were to change during captive management, the population would not be as fit if faced with the same variability in its new, postcaptivity environment. And the total loss or variability at a locus could actually compromise the continued existence of the species if, say, the lost allele coded for the resistance to a rare disease. Other variable loci represent the historical record of mutations that have not yet been lost from the population. To the extent a locus harbors a deleterious allele, any increase in this allele frequency will be harmful.

Rather than use the information formulation (20.1), which is difficult to interpret, most managers prefer some more traditional measure, such as the mean level of heterozygosity or the average inbreeding (the probability of autozygosity) of an individual. These are measures that apply to mean properties of the population. A different measure involves allele diversity, the number of loci still segregating for originally polymorphic loci. This measure tells something about the variance of the decay of genetic information; it is, however, sensitive to the original distribution of allele frequencies, about which we are incompletely informed.

For certain kinds of management effort, the mathematical theory is adequate. Some number of founders establish the population. Some additional genetic founders may be added as migrants at a later date. The population goes through an exponential growth phase and then reaches (probably in nonlogistic fashion) a ceiling, at which it is maintained until a reserve population is established. Using discrete relationships for population size, N, and loss of heterozygosity, H,

$$N_{t+1} = MIN\ (\lambda N_t, K) \tag{20.2a}$$

and

$$H_{t+1} = H_t[1 - 1/(2EN_t)], \tag{20.2b}$$

where λ is the discrete growth rate per time step (generation), MIN takes the minimum value of the postgrowth population and its ceiling level K, and E is a correction factor to scale the census population size to the genetically effective size, and where special techniques are required to account for the later-arriving migrants. Relationships (20.2) allow the manager, among other things, to make "what if" calculations to see the effects that changes in growth parameters have on the final level of retained heterozygosity (Soule et al. 1986).

If the population size remains constant, it is possible to calculate the variance in the retained heterozygosity. For more realistic scenarios of population growth, simulation techniques are currently required (Fuerst and Maruyama 1986).

The above techniques are ways to treat "noise" due to genetic drift in the information communication process. They are powerless to treat distortion, that is, selection, in the channel. Selection will be locus-specific, and a single scalar variable such as mean heterozygosity is an insufficient characterization of state.

Analyses over the pedigree of the captive species are more powerful, for they permit state identification of individual animals, and thus they give the manager greater control over reproduction. Although some pedigree relationships can be expressed in analytic form, they invariably require computers to carry out the required calculations, e.g., path analysis. And some closed-form relationships, especially those involving joint probabilities, "blow up" so fast with increasing animal and allele numbers that they cannot be computed in reasonable times with existing computer hardware. Often, however, these relationships can be estimated through Monte Carlo simulation techniques, so-called gene dropping (MacClure et al. 1986), wherein unique founder alleles are passed from parents to offspring according to the standard Mendelian rules.

There are two standard relationships of pedigree analysis (Thompson 1986). The first is the kinship coefficient between any two animals, which gives the probability that a gene picked in one animal is identical by descent ("idb") with a gene picked at random from the second animal. This coefficient is denoted $\psi(A, B)$, and is defined by the relationship

$$\psi(A, B) = (\psi(M_B, A) + \psi(F_B, A))/2, \qquad (20.3)$$

where A and B are individuals and where F_B is the father of B and M_B is the mother of B. This relationship, together with the fact that kinship values between different founders are 0, allows recursive computation on computers. The inbreeding coefficient of a single animal is the kinship coefficient between that animal's parents.

The second, and similar, coefficient is the probability that an animal possesses either of the genes of a founder. From this second probability two different things can be calculated: (1) the contribution of founder alleles to an individual, and (2) the distribution of founder alleles over the extant population.

Maintaining an equal distribution of founder genes from generation to generation is a sure way to eliminate the possibility of selection. It is not, however, a way to reduce drift, which is not based on expected allele frequency values. For a population that is displaced from equality of founder representation in some generation, the effort to even the founder representation will necessarily increase inbreeding. Thus, these two goods—equal founder contribution and low inbreeding coefficients—can be in conflict. This is not a solved optimization problem in genetic management. The solution to this optimization problem depends in part on the "cost" of inbreeding, which in turn depends on the population structure the species had in its wild state, and on the distribution of the genetic load between single-locus and quantitative traits (Lande and Barrowclough 1987).

Speke's gazelle went through a founding population size of four (three females and one male) and consequently soon showed high inbreeding coefficients and an uneven representation of founder alleles. At the same time, inbred individuals exhibited higher rates of infant mortality. The managers of this population (Templeton and Read 1984) were in a quandary. They chose to even the representation of founder genes, even at the expense of rapidly increasing inbreeding coefficients. This strategy has proven successful, as the inbreeding depression, that is, the high infant mortality, was apparently due to a few genes of large effects that were selected against during the process.

The expected absolute variability of founder alleles over members of the population can also be calculated by pedigree analysis (Gilpin and Thompson, in prep.). This statistic is related to a manager's freedom to change allele frequencies at a locus. It may happen that, during the "bottleneck" phase of colony development, a deleterious allele acquires a high frequency through drift (e.g., hairlessness as exhibited in the red-ruffed lemur). Some selection in later generations will be required to correct this. This selection will impact both levels of inbreeding and the distribution of founder alleles, as correlations among founder alleles will be built up in the initial generations of the pedigrees.

To these purely genetic tradeoffs must be added tradeoffs related to demographic management. For example, a strategy for rapid colony growth might involve culling old animals and skewing the sex ratio toward females. This alters both the generation length and the effective population size of the colony.

It is thus an extremely difficult problem to project the optimal path that a manager should follow in creating a breeding plan, that is, a pedigree. A solution to this problem must involve greater information about the cost of demographic and pedigree variables, both for short-term fitness and also for long-term adaptability. It must also include a more integrated theory of population and genetic processes, together with computer techniques to treat the information simultaneously.

OVEREXPLOITATION OF RESOURCES

The above discussion might indicate that scientific issues of conservation stem from man's indifference or hostility toward natural resources. Yet, there are many examples of the loss of species when it was in our own best interests to preserve them. Humans have harvested the sea for millennia, and this exploitation provides some important examples of species losses, despite good intentions. Though we will restrict our discussion to fisheries, the lessons apply to other exploited resources.

Modeling exploitation of a fishery is simple in concept—man is another variable in the model—but difficult in practice. May (1984) provides a summary of these difficulties, from which we have selected a few of the major issues.

Even classical "single-species" fisheries (such as the cod or herring in the North Sea or anchovies off Peru) are members of complex communities harboring, and providing, food and shelter for predators and competitors. The modeling task is difficult for even these systems (May et al. 1979). It is even more complex for fisheries of tropical waters—off the Senegal-Mauritian coast, for example, where as many as 174 species are involved in the demersal fishery (May 1984).

Classical single-species models are of little use for these species-rich systems. But detailed simulation models involving all the commercial species and their food supplies are not a solution either. The sensitivity of very complex models to the errors inherent in estimating the plethora of parameters necessary for their construction makes them unreliable. Simplified models are less sensitive to parameter variation. The problem, however, is how to go about simplifying, and how to aggregate the variables (O'Neill and Rust 1979; Sugihara 1984).

Even for the simplest systems, model parameters may be difficult to estimate in an uncertain world where changes in environmental variables may more or less accidentally coincide with changes in fishing intensity. One imaginative solution is to manage by experiment: to use deliberate changes in fishing effort as a means of understanding the dynamics of the fishery (Walters 1986).

But managing exploited resources in an uncertain environment presents other difficulties that may well explain the frequent collapse of heavily exploited fish stocks. Unpredictable environmental events can lead to significant fluctuations in stocks and thus to uncertainty in catch from year to year. A constant level of fishing will impact a variable population much more at low densities than at high densities. In this, and in other ways, harvesting may reduce a population's ability to recover from environmental disasters. Such effects can lead to situations where it is better to fish less intensely, diminishing the fishing effort below that estimated to give the maximum average yield, in order to reduce the severity of the fluctuations in yield (Beddington and May 1977). Fluctuations in the processing and marketing of fish also impose fluctuations in the biological stock, forcing us to consider the interaction of ecological and economic factors (Clark, chapter 19).

THE PROBLEMS OF INTRODUCED SPECIES

Man-made habitats are often readily invaded by a variety of species that have become commensal with man worldwide. As fragmented natural areas come in ever closer proximity to these man-modified habitats and to the introduced species they house, these introduced species can have an increasingly deleterious effect on the native species. The control or removal of introduced species may be essential for the survival of many endangered species. For some introduced species, control may be an expensive technique, and for others it may be impossible, given current technology. In most cases, early detection of potentially harmful invasive species is required if control is going to be economically feasible. The obvious questions are: Can we predict which species are likely to invade which communities, and what effects they will have when they get there? (Mooney and Drake 1986).

To answer these questions, ecologists have devoted considerable effort recently to trying to understand what determines whether a community will be invaded or not, as well as what are the attributes of successful invaders. This is a central topic in community ecology (Roughgarden, chapter 14). Species introductions sometimes provide unique opportunities for testing community theory (Diamond and Case 1986; Moulton and Pimm 1986) as well as many patterns that are a challenge to theoreticians.

Multispecies models of community invasions may again be useful in suggesting which species introductions are likely to be the most destructive. Species introductions may succeed, and may have the greatest impact, when those species are introduced into habitats where their competitors and predators are absent. Thus, the literature on modeling species removals (Pimm 1982, 1987) is relevant here and anticipates our discussions in the final section of this chapter.

Community models typically start with systems of Lotka-Volterra equations:

$$\frac{dX_c}{dt} = X_i(b_i + \sum_{j=1}^{n} a_{ij}X_j),$$

subject to the equilibrium densities X_i^* (where all $dX_i/dt = 0$) being positive and the equilibrium being locally stable. Species then can be removed from this system one at a time, and the resulting $n - 1$ species system can be analyzed either for the feasibility ($X_i^* > 0$) and stability of the local equilibrium or through the direct simulation of the dynamics. These simulations produce a number of results: removing a plant species from the base of a simple food chain destroys the entire system (fig. 20.2, top left). But the loss of one of the several plant species utilized by a generalized herbivore in the more complex system would have much less ofan effect, because the herbivore is not so dependent on one species (fig. 20.2, middle). These effects are obvious and motivated MacArthur's (1955) argument that more complex communities would be more "stable." Less obvious are the effects of removing species from the top of food chains. Removing a predator from a monophagous herbivore probably leaves the plant at a lower density, but it is likely to survive (fig. 20.2, top right). Special conditions are required for a predator to eliminate its sole prey before it too becomes extinct. In the more complex system, the predator's absence may lead to the herbivore exterminating all but the one resistant plant species, which then regulates the herbivores' numbers (fig. 20.2, bottom).

These results suggest a number of effects relevant to the subject of species introductions. There seem to be three "don'ts" for species introductions. Impacts will likely be severe when

1. Species are introduced into places where predators are absent; this is equivalent to removing predators from communities. Examples would include the introduction of large herbivores to islands, or predators that feed high in a

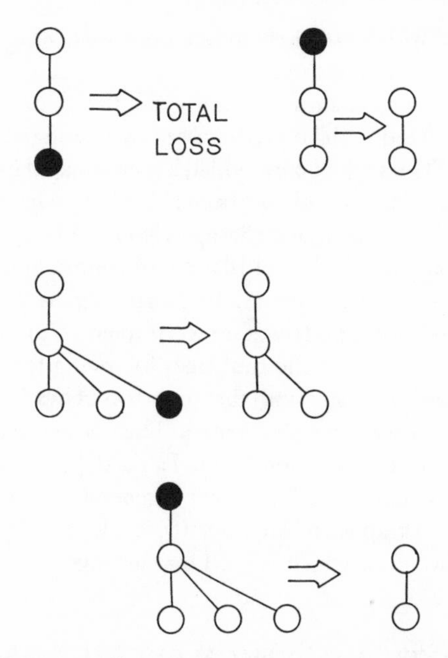

FIGURE 20.2 Effects of removing species from food webs. Circles represent species, lines represent feeding interactions between species. Removing the plant at the base of a simple food chain results in the total loss of all the species. Removing a predator from the top of a simple food chain probably leaves the remaining species extant, even if at different densities. Now consider a more complex system with a polyphagous herbivore. Removing a plant from the base of this system may cause no extinctions. Removing a predator from a polyphagous herbivore may be disastrous, as the herbivore eliminates all but one of the plant species. (From Pimm 1987.)

food chain and which themselves lack predators. In a similar way, we might expect introductions to have more severe impacts when competitors are also absent.

2. Introductions of highly polyphagous species are likely to be more severe, because, for example, polyphagous herbivores can eat a wider variety of plant species to extinction.

3. Species are introduced into relatively simple communities where the removal of a few plant species will cause the collapse of entire food chains.

There is empirical support for these theories. By searching the literature on over 800 introductions, Ebenhard (1988) has classified those species that have an effect in changing abundance, or species composition, or in causing extinctions. Herbivorous mammals introduced to continents were recorded as having effects in 26% of 86 cases. On islands close to continents, the figure is a comparable 28%

of 89 cases. Yet on oceanic islands, which are likely to lack predators, 50% of 363 introductions show effects.

Almost all introduced herbivorous mammals have generalized diets, but among five introductions of specialized herbivores, discussed by Ebenhard, none were recorded as having an effect. The snake *Boiga,* which has eliminated most of the bird species on the island of Guam, has probably been able to do so because it also feeds on the abundant lizards on that island (Savage 1987).

For the third effect, we can compare the fate of island and continental birds following the destruction of various plant species. In North America, there appear to be no extinctions of vertebrates and few of insects (Opler 1978) that are attributable to the loss of the once abundant chestnut trees following the introduction of chestnut blight. Contrast this result with that found on Hawaii, where about half of the surviving bird species are threatened. The species seem so vulnerable because of their dietary specialization. Thus Palila depends on the seeds of one species of tree, the Mamane, which does not regenerate when large herbivores are present, and the Akiapolaau, an insectivore, depends on the presence of large Koa trees, the wood of which is used for carvings.

SECONDARY EXTINCTIONS: WHICH SPECIES ARE IMPORTANT?

Some species are likely to be much more important than others and, given that we probably cannot save all the species threatened with extinction, identifying "important" species becomes essential. What makes a species "important"? Species may be important because, like giant pandas, cheetahs, and rhinos, they are charismatic. Yet other species may be important because they are economically valuable to us. However, species may also be important in preserving communities. Keystone species are those species whose loss is likely to trigger the greatest number of secondary extinctions. Clearly, special efforts must be made to identify and to preserve keystone species, because the fate of so many other species is tied to them. We have discussed in the previous section ecological theory that identifies some species that may be keystone species. The loss of predators of generalized herbivores, for example, might be predicted to have major consequences. The experience in the U.S. National Park system, with the eliminations of top predators and great subsequent increases in large herbivores, suggests the need to preserve predators for reasons other than their being charismatic (Chase 1986).

CONCLUSIONS

The human species needs desperately to find a new way to navigate its ship of technology and population. One problem is that we are drowning too many other species in our wake. A successful answer will involve components from all areas of human thought and inquiry. At the strategic level, philosophy, politics, and human psychology will certainly be the most important. But at the tactical and day-to-day level—the crisis level of conservation biology—our knowledge of ecologic, population, and genetic processes will be crucial.

In this chapter we have identified the four horsemen of the environmental apocalypse—habitat loss, introductions, overexploitation, and secondary effects. We have discussed the immediacy of the problems and the degree to which ecologic, population, and genetic theory can guide the corrective steps that must be taken. We do not pretend that our models will prove to be perfect; our science is too young, our study systems too multivariate. Nonetheless, our models are greatly superior to ignorance. And, in employing our models and theories against the challenges of the environmental crisis, we will have the best possible chance to test and to correct our theoretical understanding.

REFERENCES

Beddington, J. R., and R. M. May. 1977. Harvesting natural populations in a randomly fluctuating environment. *Science* 197:463–65.

Bonner, J. T. 1965. *Size and Cycle: An essay on the Structure of Biology.* Princeton University Press, Princeton, N.J.

Cawthorne, R. A., and J. H. Marchant. 1980. The effects of the 1978–1979 winter on British bird populations. *Bird Study* 27:163–72.

Chase, A. 1986. *Playing God in Yellowstone.* Atlantic Monthly Press, Boston.

Diamond, J. M. 1984. Historic extinctions: A Rosetta stone for understanding prehistoric extinctions. In P. S. Martin and R. F. Klein, eds., *Quaternary Extinctions: A Prehistoric Revolution.* University of Arizona Press, Tucson.

Diamond, J. M., and T. J. Case. 1986. *Community Ecology.* Harper and Row, New York.

Ebenhard, T. 1988. Introduced birds and mammals and their ecological effects. *Swedish Wildlife Research* 13, no. 4.

Fuerst, P. A., and T. Maruyama. 1986. Considerations on the conservation of alleles and of genic heterozygosity in small managed populations. *Zoo Biology* 5:171–79.

Gilpin, M. E. 1988. Extinction of metapopulations in correlated environments. In B. Shorrocks, ed., *Life in a Patchy Environment.* Oxford University Press, Oxford.

Gilpin, M. E., and J. M. Diamond. 1980. Subdivision of nature reserves and the maintenance of species diversity. *Nature* 285:567–68.

Gilpin, M. E., and M. E. Soulé. 1986. Minimum viable populations: The processes of species extinction. In M. E. Soulé, ed., *Conservation Biology: The Science of Scarcity and Diversity.* Sinauer, Sunderland, Mass.

Gilpin, M. E., and E. A. Thompson. In prep. The optimal management of pedigrees.

Goodman, D. 1987. The demography of chance extinction. In M. E. Soulé, ed., *Viable Populations for Conservation.* Cambridge University Press, Cambridge, England.

Higgs, A. J., and M. B. Usher. 1980. Nature reserve design: Could smallness be an aim? *Nature* 285:568–69.

Lande, R., and G. F. Barrowclough. 1987. Effective population size, genetic variation, and their use in population management. In M. E. Soulé, ed., *Viable Populations for Conservation.* Cambridge University Press, Cambridge, England.

Leigh, E. G. 1981. The average lifetime of a population in a varying environment. *J. Theor. Biol.* 90:213–39.

Long, J. 1981. *Introduced Birds of the World.* David and Charles, London.

Lovejoy, T. E., R. O. Bierregaard Jr., A. B. Rylands, J. R. Malcome, C. E. Quintela, L. H. Harper, K. S. Brown Jr., A. H. Powell, G.V.N. Powell, H.O.R. Schubart, and M. B. Hays. 1986. Edge and other effects of isolation on Amazon Forest fragments. In M. E. Soule, ed., *Conservation Biology: The Science of Scarcity and Diversity,* Sinauer, Sunderland, Mass.

MacArthur, R. H. 1955. Fluctuations of animal populations and a measure of community stability. *Ecology* 36:533–36.

MacClure, J. W., J. L. Vandeberg, B. Read, and O. A. Ryder. 1986. Pedigree analysis by computer simulation. *Zoo Biology* 5:147–60.

McLellan, C. H., A. P. Dobson, D. S. Wilcove, and J. M. Lynch. 1986. Effects of forest fragmentation on New and Old World bird communities: Empirical observations and theoretical implications. In J. Verner, M. Morrison, and C. J. Ralph, eds., *Modeling Habitat Relationships of Terrestrial Vertebrates.* University of Wisconsin Press, Madison.

May, R. M. 1984. *Exploitation of Marine Communities.* Springer-Verlag, Berlin.

May, R. M. 1986. The cautionary tale of the black-footed ferret. *Nature* 320:13–14.

May, R. M., and A. M. Lyles. 1987. Conservation biology: Problems in leaving the arc. *Nature* 326:245–46.

May, R. M., J. R. Beddington, C. W. Clark, S. J. Holt, and R. M. Laws. 1979. Management of multispecies fisheries. *Science* 205:267–77.

Mooney, H. A., and J. A. Drake. 1986. *Ecology of Biological Invasions of North America and Hawaii.* Springer-Verlag, Berlin.

Moulton, M. P., and S. L. Pimm. 1986. The extent of competition is shaping an introduced avifauna. In J. M. Diamond and T. J. Case, eds., *Community Ecology.* Harper and Row, New York.

O'Neill, R. V., and B. Rust. 1979. Aggregation error in ecological models. *Ecol. Model.* 7:91–105.

Opler, P. A. 1978. Insects of american chestnuts: Possible important conservation concerns. *Proc. Amer. Chestnut Symp.,* pp. 83–85.

Pimm, S. L. 1982. *Food Webs.* Chapman and Hall, New York.

Pimm, S. L. 1984. Food chains and return times. In D. R. Strong Jr., D. Simberloff, L. F. Abele, and A. B. Thistle, eds., *Ecological Communities: Conceptual Issues and the Evidence,* pp. 397–412. Princeton University Press, Princeton, N.J.

Pimm, S. L. 1987. Determining the effects of introduced species. *Trends in Ecol. & Evol.* 2:106–108.

Pimm, S. L., and A. J. Redfearn. 1988. The variability of population densities. *Nature* 334:613–14.

Pimm, S. L., L. Jones, and J. M. Diamond. 1988. On the risk of extinction. *Amer. Nat.* 132. In press.

Ralls, K., K. Brugger, and J. Ballou. 1979. Inbreeding and juvenile mortality in small populations of ungulates. *Science* 206:1101–1103.

Richter-Dyn, N., and N. S. Goel. 1972. On the extinction of a colonizing species. *Theor. Pop. Biol.* 3:406–23.

Savage, J. A. 1987. Extinction of an island forest avifauna by an introduced snake. *Ecology* 68:660–68.

Shaffer, M. L. 1981. Minimum population sizes for species conservation. *Bioscience* 31:131–34.

Soulé, M. E., M. E. Gilpin, W. Conway, and T. Foose. 1986. The Millennium Ark: How long a voyage, how many staterooms, how many passengers? *Zoo Biology* 5:101–13.

Sugihara, G. 1984. Ecosystem dynamics. In R. M. May, ed., *Exploitation of Marine Communities.* Springer-Verlag, Berlin.

Templeton, A. R., and B. Read. 1984. Factors eliminating inbreeding depression in a captive herd of Speke's gazelle. *Zoo Biology* 3:177–99.

Thompson, E. A. 1986. Ancestry of alleles and extinction of genes in populations with defined pedigrees. *Zoo Biology* 5:161–70.

Walters, C. 1986. *Adaptive Management of Renewable Resources.* Macmillan, New York.

Wilcove, D. S., C. H. McLellan, and A. P. Dobson. 1986. Habitat fragmentation in the temperate zone. In M. E. Soulé, ed., *Conservation Biology: The Science of Scarcity and Diversity.* Sinauer, Sunderland, Mass.

Chapter 21

Discussion: Ecology and Resource Management—Is Ecological Theory Any Good in Practice?

PAUL R. EHRLICH

One is tempted to answer the question posed in the title of this essay with a simple "yes"—ecological theory can help in solving many of the most important problems facing humanity. After all, the most important ecological theory was put forth 130 years ago by Charles Darwin, explaining, among other things, the crucial observation that antibiotics and pesticides become less effective with continued use, to say nothing of putting all of biomedical science into a coherent whole rather than leaving a mishmash of unconnected phenomena.

But Darwin's theory is so pervasive in its application that it is taken for granted by biologists (many of whom, outside of population biology, are, ironically, not very familiar with it). And to many people, evolution's very designation as a "theory" implies that its validity is suspect, making it vulnerable to challenge by the supporters of various cults and superstitions. Indeed, widespread misunderstanding of the scientific meaning of "theory," in my view, not only fuels public support for ignorant anti-evolutionism but also accounts for the negative attitudes toward theoretical ecology sometimes found among ecological empiricists.

In the first case, the public confuses two dictionary definitions: the scientific one of "theory" as a well-tested frame of reference for a field of inquiry (evolutionary theory) as opposed to the more common "theory" as an unsubstantiated conjecture (the theory that creationists have lower IQs than evolutionists). When using the second definition, empiricists tend to equate theory with mathematical theory, to undervalue the latter, and to misunderstand the purpose and applicability of both. One result is that the state of ecological knowledge is often misrepresented and the undervaluation of ecology by scientists in other disciplines and citizens in general is unwittingly encouraged.

As discussed elsewhere (Ehrlich 1986), there is a tendency to expect too much predictive power from ecological theory—to hold it to a standard higher than, say, theory in many areas of physics. Good theory abstracts the essential features of a system from the clutter of detail that occurs in the unhappily stochastic real world. It cannot, then, be expected to serve as a tool for flawlessly predicting features of that clutter. Theory allows the forest to be seen in spite of the trees; it would be preposterous, then, to complain that the theory is inadequate because it won't allow us to choose which individual trees to harvest. Very often ecological theory is like meteorological theory. Meterorologists can explain and predict seasonal changes very well, but they often fail at predicting the precise weather in a given location on a particular day. Ecologists can predict with great accuracy that destruction of 75 percent of a forest and fragmentation of the remainder will cause a decline in the number of species of forest birds present, but they are only beginning to be able to predict which ones will go extinct and in what order.

For instance, island biogeographic theory tells us that, everything else being equal, a large island will have a higher equilibrium number of species than a small island. Of course, everything else is never equal, and so small islands sometimes have more species than large ones. To say that this renders the theory useless is about as sensible as concluding that the result of repeatedly tossing a two-headed coin invalidates Bernoulli's theorem. In the absence of information on other factors, one would still be advised to make a reserve out of a large island rather than a small one, just as one would ordinarily bet tails on a coin toss if offered two-to-one odds.

ECOLOGICAL THEORY IN PRACTICE

What can be said in response to the charge of our group at this workshop: evaluating the applicability of theoretical ecology to questions of policy—especially those related to the "human predicament" in general and resource management in particular?

Let us start at a relatively simple level. A large body of theory in population dynamics and population genetics can provide guidance on such questions as, "What is the optimum yield that should be taken from a given fishery?" or "What is the smallest breeding population or organism *X* that will be relatively free from the deleterious effects of inbreeding?" Indeed, at that level, empirical studies to test the theory are lagging far behind. Particularly lacking are studies of natural populations in which demographic units are well defined (Brown and Ehrlich 1980; Ehrlich and Murphy 1987), changes in population size have been recorded over many generations, and the information is available for estimating N_e (Ehrlich, Launer, and Murphy 1984; Mueller et al. 1985). The difficulties of gathering such data are substantial, and in most cases the time, funds, and techniques are not available for the effort.

This absence of data can make the application of the most highly developed theory difficult—for example, fisheries biologists make do with very simplified "logistic-type" or "dynamic pool" models to estimate the impact of harvesting on fisheries stocks (Schaefer 1968). Unfortunately, ignorance of population structure (and also community relationships, the population genetic consequences of harvesting, and the impacts of naturally occurring changes in the physical environment) may be an important factor in failures of management strategies developed from such models. The Peruvian anchoveta fishery collapsed in 1972 and has never recovered, in part because of a failure to follow prescribed management strategies, and quite likely also because those strategies were not based on knowledge of population structure.

Theory, both mathematical and nonmathematical, can be very useful even where it can only provide general understanding of a system without allowing very accurate predictions to be made. In conservation biology this has proven to be the case in the notion of "minimum viable population" (e.g., Frankel and Soulé 1981; Gilpin and Soulé 1986; Soulé 1987). The mathematical concept is rooted in rather simple theory of population genetics and more complex and incomplete demographic theory (e.g., Richter-Dyn and Goel 1972; Strebel 1985; Goodman 1988). The theory does not make it possible to say exactly what size will provide, say, a 5% chance of a population persisting for one hundred generations (in part because it requires more thorough empirical data than are normally available). It does allow managers to be alert to situations in which there is danger of inbreeding depression or stochastic extinction. And the theory shows promise that with further development its predictive value can be greatly increased.

Overall, though, when planning to harvest or conserve a single resource, managers can lean on a well-established body of theory in ecology. In dealing with the properties of communities, on the other hand, theory is less well established. A great deal has been done in mathematical modeling of competition-structured communities; much less on those where predation or environmental stochasticity dominate. A lack of both agreement on models and of empirical tests has limited the application of mathematical community ecology to practical problems. For instance, it has become increasingly important to understand biological invasions because of the enhanced spread of exotic organisms and pathogens due to the increased mobility of humanity, because of the rapid rise of genetic engineering and because of the increasing need to establish exotic organisms as biological control agents. Most of the theory dealing with invasions is nonmathematical (e.g., Elton 1958; Ehrlich 1986; Moulton and Pimm 1986; and most other articles in Mooney and Drake 1986), but that is changing, as shown by the development of an invasions model based on coevolution and niche theory by Jonathan Roughgarden's group (e.g., Roughgarden, Heckel, and Fuentes 1983) and recent excursions into mathematical epidemiology by Bob May, Roy Anderson, and their colleagues (e.g., Dobson and May 1986; May and Anderson 1983, 1987).

The equilibrium theory of island biogeography is the theory that has attracted the most attention in the rapidly developing area of conservation biology, and it has stimulated a great deal of empirical research, especially in the study of the impacts of fragmentation on biotas. Its use has been quite controversial, however. The principal criticism of the theory (e.g., Simberloff and Abele 1982) is that it is too simplified a representation of the real world to be useful. For example, the theory predicts that a single large (SL) preserve will, on average, suffer less rapid attrition of its species diversity than several small (SS) preserves. But, it is claimed, there are many circumstances in which SS will suffer less rapid decay of biodiversity, and therefore the theory misleads (although there is no disagreement about the need for theory—indeed, much of the controversy centers on what theoretical species/area relationship should be used). The SL or SS (SLOSS) controversy has generated a substantial literature (e.g., Diamond 1976; Diamond and May 1976; Higgs and Usher 1980; Simberloff and Abele 1982; Soulé and Simberloff 1986; Wilcox and Murphy 1985; Simberloff 1986).

Similarly, while the utility of corridors between reserves has been viewed as a corollary to island biogeographic theory, corridors may present biological and economic costs that would make their establishment and maintenance counterproductive to the cause of conservation (Simberloff and Cox 1987; but see Noss 1987). The problems recognized by Simberloff and his colleagues are real, but they reflect more the failure of many ecologists to understand the uses of theory than an inadequacy of the theory itself. The applicability of the theory (and other biogeographic theories) clearly will vary from situation to situation (e.g., Case and Cody 1987). The main point, however, is that in the various controversies, the equilibrium theory of biogeography has served two of the main functions of theory—providing a coherent basis for discussing complex issues, and transforming the way an entire area of intellectual endeavor was approached.

An area of conservation biology where theory has not yet had sufficient impact is that of determining the diversity and distribution of Earth's biota and quantifying the rate of loss of diversity. It is clearly impossible to name, describe, and plot the distribution of *all* life forms. Some sort of sampling approach is required (Ehrlich 1964). At the moment we have no theory to guide us in that sampling, for instance to help to transform distributional data into estimates of genetic differentiation. I have found it difficult, for instance, to confirm the assertion (which I believe to be correct) that the average animal species in a tropical rain forest will consist of many fewer populations (however defined) than the average temperate zone species. The time is ripe for systematists to stop arguing over how many cladists can dance on the head of a pin, and start asking critical questions about their rapidly disappearing subjects of study.

A great deal of the effort of mathematical theorists in ecology has gone into the development of competition and niche theory (for summary, see Ehrlich and Roughgarden 1987). While this body of theory has formed a useful framework for understanding the structure of natural communities, it has not yet had much impact in applied problems. The potential is clearly there, however, as indicated by Roughgarden's work. It is often crucial to be able to make general predictions about the response of communities to perturbation.

Suppose there were a guild of four coexisting herbivores, A, B, C, and D. Would relaxation of predation pressure on a population of species A after a great reduction in its numbers result in a resurgence of A, a population explosion of one other species in the guild, say C, with little change in the size of populations of B and D, or some other combination? What kind of pattern would depression of the numbers of B produce in A, C, and D? What if two of the four are subject to heavy predation and then relaxation? Three of the four?

Answers to such questions are often of great practical importance, most obviously in fisheries management. It is still not clear, for instance, why the overfished northern populations of the California sardine failed to recover when fishing pressure was removed. One possibility is that a change of community structure occurred that prevents the restoration of sardine abundance. There is no fundamental reason why ecological theoreticians could not to determine the answer—here again a lack of data on guild structure and community relationships is probably more of a barrier than a failure of theory. Where that information *is* available, however, considerable insight into actions and reactions within a community has been gained for managers, based on nonmathematical food-web theory; for example, this theory allowed untangling the impact of rinderpest on the biota of the Serengeti (e.g., Sinclair 1979).

Development of mathematical theory to deal with the complexities of ecosystems is in its infancy, and will, as Si Levin points out, require a large variety of models. It will also require a great deal of empirical work. We have barely begun, for instance, to document and quantify the services supplied to humanity by natural ecosystems (Holdren and Ehrlich 1974; Ehrlich and Ehrlich 1981), let alone model the delivery of most. On the bright side, however, considerable progress has been made in developing a largely nonmathematical theory of reconstruction ecology (e.g., Cairns, Dickson, and Herricks 1977; Bradshaw and Chadwick 1980; Bradshaw 1983). In addition, I think that the much-discussed problems of scale in ecosystem (or other) ecology are overrated. Obviously, one must choose appropriate scales for any kind of ecological work, empirical or theoretical, and find ways to tie together work at different levels. There is clearly a need to incorporate different scales of time and space more explicitly in various models, and there seem to be no insurmountable barriers to doing so. The challenge of developing ecosystem theory is daunting, but progress is being made, and much more is possible if the incentives and funding are available. This is true even for the complicated problems of modeling the effects of various perturbations on the climate, a key need in any broad theoretical approach to ecosystemic components of the human predicament.

For example, in 1983 ecologists (broadly defined) finally turned their attention to the question of the ecological effects of a large-scale nuclear war (e.g., Turco et al. 1983; Ehrlich et al. 1983). The critical issue was how the atmosphere would behave if huge amounts of dust, soot, and smoke were injected into it by multiple nuclear explosions. The answer, first derived from a one-dimensional atmospheric model, was that the Northern Hemisphere would get extremely dark and cold. Atmospheric physicists (e.g., Thompson et al. 1984) quickly produced more refined estimates of postwar atmospheric conditions, based on three-dimensional global circulation models. Interestingly, the strictly ecological conclusions have remained robust and have conformed to the nonquantitative estimates of the environmental impacts of nuclear war made much earlier by ecologists (e.g., Ehrlich 1970). Thus, on critical issues where only qualitative assessments are needed by policy makers, the accumulated nonquantitative theory of ecology can be very useful.

Many of the most serious global problems require assessment and corrective action long before comprehensive models can be constructed to predict their outcomes in detail. Changes in climate resulting from the increase in greenhouse gases and (perhaps) depletion of ozone and from deforestation and other landscape changes will be extremely difficult to model, especially since some of the critical component processes are not well understood, still less their possible interactions. Developing theory to predict the disruption of natural and agricultural ecosystems caused by climatic changes, acid deposition, ozone depletion, and the interactions among them will also be exceedingly difficult. Most complex of all, the more advanced models must be able to take into account the costs and benefits of various actions that might be taken in attempts to ameliorate the problems and integrate them with at least ballpark estimates of risks (including zero-infinity dilemmas).

Things get much more complex when one leaves the purely biological sphere, even at levels of organization much simpler than that of an ecosystem. Although current ecological theory does speak to the SLOSS problem, it does not provide much help in predicting, say, the reproductive decisions of the inhabitants of the region surrounding the reserve, or the politico-economic decisions on the guarding of its boundaries. Here, as in virtually all aspects of the application of the theory of population biology for the benefit of humanity, we must either gain the cooperation of social scientists or invade their turf. The former is infinitely preferable, for obvious reasons.

Ecologists already have, for example, considerable experience in the area of bioeconomics, beautifully summarized in Colin Clark's paper in this volume. Theory here has already proven enormously useful both in comprehending major features of complex systems and in providing a basis for choosing among policy instruments for regulation. Combining ecological and economic theory produces insights that have yet to pervade either professional community. Many biologists still assume that the goal in exploiting a renewable resource is to obtain a maximum sustainable yield. Bioeconomic theory explains why this is ordinarily not the case in exploitation of renewable resource stocks with low *r*'s—as has been shown repeatedly in the real world. The Japanese whaling industry has deliberately overexploited its basic resource. A higher return on capital could be won by exterminating the whales and then investing the capital in exterminating another resource than in harvesting a sustainable yield of whales. As Dick Southwood put it in the discussion at the conference, if the natural biological interest rate is sufficiently high, people may be willing to harvest a resource on a sustainable-yield basis, "but if the interest rate is low, as it is in a tropical rain forest, then they will take the capital."

ECOLOGICAL THEORY AND THE HUMAN PREDICAMENT

Biologists are naive in their neglect of the economic dimension of the system; economists, businessmen, and politicians are naive in their neglect of the inevitable limits on the number of exterminatable resources—much of their behavior is based on the fallacy of infinite substitutability. Bioeconomic theory can help cure both groups of their naiveté; population dynamic parameters, discount rates, uncertainty, and what I would like to call "meta-resource depletion" could be worked into a comprehensive framework for regulating the relationship of the human population to both common property (Hardin and Baden 1977) and privately owned resources. Meta-resource depletion is the steady exhaustion of the resource of resources (the entire set or pool of resources). Economists do not seem to realize that humanity is sampling resources without replacement from a finite urn—a process that should be explicitly modeled by bioeconomic theoreticians.

The largest remaining theoretical problem here, in my view, lies in dealing with meta-resource depletion, which involves both living and inorganic resources. The classical formulation of the economic doctrine of infinite substitutability was given by Barnett and Morse (1963, p. 11):

> Advances in fundamental science have made it possible to take advantage of the uniformity of energy/matter—a uniformity that makes it feasible without preassignable limit, to escape the quantitative constraints imposed by the character of the earth's crust. . . . Nature imposes particular scarcities, not an inescapable general scarcity. Man is therefore able, and free, to choose among an indefinitely large number of

> alternatives. There is no reason to believe that these alternatives will eventually reduce to one that entails increasing cost—that it must sometime prove impossible to escape diminishing quantitative returns. Science, by making the resource base more homogeneous, erases the restrictions once thought to reside in the lack of homogeneity. In a neo-Ricardian world, it seems, the particular resources with which one starts increasingly become a matter of indifference.

Such gibbering nonsense should not be treated lightly by ecologists, as it clearly represents the fundamental theoretical framework accepted by the vast majority of economists today (see Daly 1977 for a refreshing exception), the framework that underlies much economic decision making. Even the manifest difficulties already encountered in making inorganic substitutions (e.g., nuclear power for fossil fuels, aluminum wire for copper wire) have failed to dim the enthusiasm of nonscientists for this fallacy. They are still less aware that there are severe problems in making organic substitutions (Ehrlich and Mooney 1983). This, of course, is due in no small part to the enormous apparent success of some substitutions, such as plastics for other structural materials and petroleum for coal. The success of the computer industry in reducing the materials and energy required in handling information is viewed as the ultimate proof that humanity can do anything it sets its collective mind to and improve its environment in the process.

The largely unquantified ultimate social and environmental costs of the "successful" substitutions eventually may provide an entirely different perspective on them, but at the moment there is no theory to guide us in making the calculations. To develop a theory for use in evaluating future substitutions will be a monumental task. The complexity of the problem quickly escalates from that usually encountered in dealing with the behavior of single populations (resources) to the scale encountered at the community-ecosystem (society) level. But, if we do not begin to shape such a theory, the future of technological *Homo sapiens* seems dim indeed. Theory, remember, is not just useful when it is providing specific predictions, but also when it supplies an overall framework that is useful in policy making. For example, evolutionary and epidemiological theory cannot now predict the course of the AIDS epidemic, but the former alerts us to the possibility of changing transmission characteristics of the virus, and the latter has already indicated the kinds of information that should be gathered to permit projection of the future course of the disease (May and Anderson 1987).

The first steps would probably involve retrospective model building. For one example, the social and environmental benefits and costs of producing handy portable fuels (and thus generating a society governed by freeways and airways) might be modeled, and then the model could be used to estimate the benefits and costs of taking a different course. Suppose that coal, petroleum, and natural gas were reserved primarily for use as feedstocks, for producing process heat and generating electricity (and supplemented in the latter areas by solar and nuclear energy), and for powering mechanical transport that was largely public and on rails. Some aspects of the problem would be relatively (*only* relatively) simple to model—such as the influence of the mass introduction of automobiles on patterns of human settlement. That submodel could then be used to estimate the consequences for society if bicycles and horse-drawn conveyances had remained the primary modes of personal transportation and of distribution of goods from railheads (an unlikely, but conceivable, "Amish model").

Things would rapidly grow more complex as one extended the scope of questions asked. What kind of social structure would have evolved, say, if suburbanization had been nipped in the bud (as I suspect it would have been with a different transportation system)? Would desegregation have occurred in the United States? What would have been the demographic consequences? How would national power and patterns of warfare have been influenced? How likely is it that nuclear weapons, computers, or high-tech medicine would have been developed by societies that eschewed personal transport powered by internal combustion engines? Would the environmental impact of such societies have increased to the point of threatening the integrity of the atmosphere-climate system? Would their epidemiological situation be likely to become one that favored the epidemic spread of diseases like AIDS?

Ecological theoreticians working with colleagues in the social sciences are unlikely ever to develop a comprehensive model that would definitively answer such questions, but models could be constructed that would be very useful tools in present-day policy making. It seems likely that patterns of automobile use will have to be drastically altered in nations like the United States in the next few decades. If this is to be done rationally, what substitutions would appear to be the most beneficial and least costly? Would the environmental and social costs of greatly heightening people's dependence on mass transit (including the costs of crime control to make it safe) be greater or less than those of a system of enforced or subsidized car pooling? How much of the effort should be put into population control and into further reducing the size and increasing the efficiency of individual vehicles?

Unfortunately, many of the critical questions at the ecology-society interface involve economics, and most economists are poorly equipped to deal with them. I am afraid that it is time for population biologists to start investigating issues that have been considered to fall within the purview of economists. For example, ecological theoreticians might start to modify input-output models of economies so they can be used to find the least socially disruptive ways of reducing economic throughput. Only a substantial reduction in throughput will give industrial societies a fighting chance of preserving the ecosystem services that support the economic system. The scale of the global human economy relative to what can be sustained by Earth's life-support systems must become a central focus of economics—even if ecologists must do the focusing.

The most difficult part of modeling the meta-resource aspect of the human predicament is familiar to all theoreticians—dealing with uncertainties. One of the most critical uncertainties here is the course of human innovation in the future. Once again, however, retrospective studies can help, and it should be possible to estimate reasonable limits on innovation in specific cases. For example, the injection of chlorofluorocarbons into the atmosphere probably can be eliminated through satisfactory technological substitutions much more easily than the injection of CO_2 by human societies can be reduced by half.

NOT UNDERESTIMATING THE PRESENT VALUE OF THEORY

Finally, let me expand on something I have said at the start. The inability of ecological theory to predict precisely future population sizes, the rate at which a fauna will collapse following insularization of its habitat, or the response of an ecosystem to a complex series of insults does not necessarily represent failure of the theory. Physicists, after all, cannot predict which of two identical radioactive nuclei will decay first or which of a series of nearly identical missiles launched from the same silo will come closest to the target. The systems modeled by ecological theorists are much more complex—composed of diverse, often little-known organisms interacting with each other and with diverse, complex, usually barely studied physical environments. The problem of balancing precision against generality is much more difficult for ecological theoreticians than it is for theoretical physicists.

Even though ecological theoreticians usually cannot generate probability distributions similar to those associated with nuclear decay, what they have managed to do is far from trivial. Indeed, starting with nonmathematical theory generated by people such as Charles Elton, H. G. Andrewartha, and Charles Birch, and moving on through the flowering of mathematical theory stimulated by G. E. Hutchinson and Robert MacArthur, theory has changed ecology from a purely descriptive discipline (reflected in its 1940s definition as "the science that called a spade a geotome") into one with a theoretical framework that is growing in quality and comprehensiveness. Theory has already transformed ecology and will become increasingly important to ecologists trying to help transform the relationship between *Homo sapiens* and its environment. *At present, neglect of the theory ecologists have already developed is costing many billions of dollars annually and possibly mortgaging the future of our species.*

ACKNOWLEDGMENTS

I thank Jared Diamond, Anne Ehrlich, Bob May, and Jon Roughgarden for helpful comments on the manuscript. Ideas from my colleagues at the meetings, especially Bob May and Dick Southwood, have been shamelessly appropriated without attribution.

REFERENCES

Barnett, H. J., and C. Morse. 1963. *Scarcity and Growth.* Johns Hopkins University Press, Baltimore, Md.

Bradshaw, A. D. 1983. The reconstruction of ecosystems. *J. Appl. Ecol.* 20:1–17.

Bradshaw, A. D., and M. J. Chadwick. 1980. *The Restoration of Land: The Ecology and Reclamation of Derelict and Degraded Land.* Blackwell, Oxford.

Brown, I. L., and P. R. Ehrlich. 1980. Population biology of the checkerspot butterfly *Euphydryas chalcedona:* Structure of the Jasper Ridge Colony. *Oecologia* 47:239–51.

Cairns, J., K. L. Dickson, and E. E. Herricks, eds., 1977. *Recovery and Restoration of Damaged Ecosystems.* University Press of Virginia, Charlottesville.

Case, T. J., and M. L. Cody. 1987. Testing theories of Island Biogeography. *Amer. Sci.* 75:402–11.

Daly, H. E. 1977. *Steady-State Economics.* Freeman, San Francisco.

Diamond, J. M. 1976. Island biogeography and conservation: Strategy and limitations. *Science* 193:1027–29.

Diamond, J. M., and R. M. May. 1976. Island biogeography and the design of natural reserves. In R. M. May, ed., *Theoretical Ecology: Principles and Applications,* pp. 163–86. W. B. Saunders, Philadelphia.

Dobson, A. P., and R. M. May. 1986. Patterns of invasions by pathogens and parasites. In Mooney and Drake (1986), pp. 58–76.

Ehrlich, P. R. 1964. Some axioms of taxonomy. *Syst. Zool.* 13:109–23.

Ehrlich, P. R. 1970. Population control or Hobson's choice. In L. R. Taylor, ed., *The Optimum Population for Britain,* pp. 151–62. Academic Press, London.

Ehrlich, P. R. 1986. Which animal will invade? In Mooney and Drake (1986), pp. 79–95.

Ehrlich, P. R., and A. H. Ehrlich. 1981. *Extinction: The Causes and Consequences of the Disappearance of Species.* Random House, New York.

Ehrlich, P. R., J. Harte, et al. 1983. Long-term biological consequences of nuclear war. *Science* 222:1293–1300.

Ehrlich, P. R., A. E. Launer, and D. D. Murphy. 1984. Can sex ratio be defined or determined? The case of a population of checkerspot butterflies. *Amer. Natur.* 124:527–39.

Ehrlich, P. R., and H. A. Mooney. 1983. Extinction, substitution, and ecosystem services. *BioScience* 33:248–54.

Ehrlich, P. R., and D. D. Murphy. 1987. Conservation lessons from long-term studies of checkerspot butterflies. *Conserv. Biol.* 1:122–31.

Ehrlich, P. R., and J. Roughgarden. 1987. *The Science of Ecology.* Macmillan, New York.

Elton, C. S. 1958. *The Ecology of Invasions by Animals and Plants.* Chapman and Hall, London.

Frankel, O. H., and M. E. Soulé. 1981. *Conservation and Evolution.* Cambridge University Press, Cambridge, England.

Gilpin, M. E., and M. E. Soulé. 1986. Minimum viable populations: processes of species extinction. In M. E. Soulé, *Conservation Biology: The Science of Scarcity and Diversity,* pp. 19–34. Sinauer, Sunderland, Mass.

Goodman, D. 1988. Considerations of stochastic demography in the design and management of nature reserves. *Nat. Res. Model.* 1:205–34.

Hardin, G., and J. Baden, eds. 1977. *Managing the Commons.* Freeman, San Francisco.

Higgs, A. J., and M. B. Usher. 1980. Should nature reserves be large or small? *Nature* 285:568–69.

Holdren, J. P., and P. R. Ehrlich. 1974. Human population and the global environment. *Amer. Sci.* 62:282–92.

May, R. M., and R. M. Anderson. 1983. Parasite-host coevolution. In D. J. Futuyma and M. Slatkin, eds., *Coevolution,* pp. 186–206. Sinauer, Sunderland, Mass.

May, R. M., and R. M. Anderson. 1987. Transmission dynamics of HIV infection. *Nature* 326:137–42.

Mooney, H. A., and J. A. Drake, eds. 1986. *Ecology of Biological Invasions of North America and Hawaii.* Springer-Verlag, New York.

Moulton, M. P., and S. L. Pimm. 1986. Species introductions to Hawaii. In Mooney and Drake (1986), pp. 231–49.

Mueller, L. E., B. A. Wilcox, P. R. Ehrlich, and D. D. Murphy. 1985. A direct assessment of the role of genetic drift in determining allele frequency variation in populations of *Euphydryas editha. Genetics* 110:495–511.

Noss, R. F. 1987. Corridors in real landscapes: A reply to Simberloff and Cox. *Conserv. Biol.* 1:159–64.

Richter-Dyn, N., and R. S. Goel. 1972. On the extinction of a colonizing species. *Theor. Pop. Biol.* 3:406–33.

Roughgarden, J., D. Heckel, and E. Fuentes. 1983. Coevolutionary theory and the biogeography and community structure of *Anolis*. In R. Huey, E. Pianka, and T.

Schoener, eds., *Lizard Ecology: Studies on a Model Organism.* Harvard University Press, Cambridge, Mass.

Schaefer, M. B. 1968. Methods of estimating effects of fishing on fish populations. *Trans. Amer. Fish. Soc.* 97:231–41.

Simberloff, D. 1986. Design of nature reserves. In M. B. Usher, ed., *Wildlife Conservation Evaluation,* pp. 315–37. Chapman and Hall, London.

Simberloff, D., and L. G. Abele. 1982. Refuge design and island biogeographic theory: Effects of fragmentation. *Amer. Natur.* 120:41–50.

Simberloff, D., and J. Cox. 1987. Consequences and costs of conservation corridors. *Conserv. Biol.* 1:63–71.

Sinclair, A.R.E. 1979. Dynamics of the Serengeti ecosystem: Process and pattern. In A.R.E. Sinclair and M. Norton-Griffiths, eds., *Serengeti: Dynamics of an Ecosystem,* pp. 1–30. University of Chicago Press, Chicago.

Soulé, M. E., ed. 1987. *Viable Populations for Conservation.* Cambridge University Press, Cambridge, England.

Soulé, M. E., and D. Simberloff. 1986. What do genetics and ecology tell us about the design of nature reserves? *Biol. Conserv.* 35:19–40.

Strebel, D. E. 1985. Environmental fluctuations and extinction—single species. *Theor. Pop. Biol.* 27:1–26.

Thompson, S. L., V. V. Aleksandrov, G. L. Gtenchikov, S. H. Schneider, C. Covey, and R. M. Chervin. 1984. Global consequences of nuclear war: Simulations with three-dimensional models. *Ambio* 13:236–43.

Turco, R. P., O. B. Toon, T. P. Ackerman, J. B. Pollack, and C. Sagan. 1983. Nuclear winter: Global consequences of multiple nuclear explosions. *Science* 222:1283–1300.

Wilcox, B. A., and D. D. Murphy. 1985. Conservation strategy: The effects of fragmentation on extinction. *Amer. Natur.* 125:879–87.

ECOLOGY OF PESTS AND PATHOGENS VIII

Chapter 22

The Population Biology of Host-Parasite and Host-Parasitoid Associations

MICHAEL P. HASSELL AND ROBERT M. MAY

In the study of any population of plants or animals, two central questions are: What is the basic reproductive rate, R_0 (the number of surviving female offspring that each female could on average produce in the absence of external constraints), and what particular mixture of density-independent and density-dependent effects acts to keep the long-term average of the effective reproductive rate, R, around unity for most populations? The first question is vividly illustrated by the Smithsonian Museum of Natural History's exhibit on evolution, in which a kitchen swarms with all the cockroaches one female would produce if all survived. Our ignorance about the second question is borne home by our current lack of understanding of what held human population densities roughly constant for several hundred thousand years before the Agricultural Revolution, despite our manifest capacity for rapid population growth.

For most natural populations, estimates of R_0 are difficult because it is hard to study populations in the absence of factors that obscure their intrinsic capacity for population growth. By the same token, it is often difficult (and sometimes unethical) to design field experiments on sufficiently large spatial and temporal scales to unravel the density-dependent factors that ultimately prevent average population densities from increasing or decreasing indefinitely. It is, however, obviously important to understand how populations are regulated—either to roughly steady values, or in pronounced but regular oscillations, or in irregular fluctuations. Such understanding bears on fundamental questions about the number and relative abundance of species, and on applied questions having to do with sustainable harvesting, control of agricultural pests and vectors of disease, design of vaccination and other programs aimed at reducing the abundance of infectious disease agents, preservation of endangered species of plants and animals, and in general with the response of natural populations to various kinds of disturbances.

Many associations between insect hosts and their hymenopteran or dipteran parasitoids or between hosts and infectious disease agents have features that make it somewhat easier to investigate the dynamics, by combining field and laboratory experiments with mathematical models, than is the case for associations among larger and longer-lived species. (In what follows, we use the term "parasite" to embrace a wide spectrum of agents of infectious diseases, including viruses, bacteria, protozoans, and fungi, along with the more conventionally defined helminth and arthropod parasites.) For one thing, the relatively small physical size and relatively short generation time of most parasites and parasitoids make it possible to bring them into the laboratory, and run replicated and controlled experiments over many generations, in a way that simply is impossible for lions or hornbills. For another thing, the numerical responses—which encapsulate the biological processes whereby one generation of parasites or parasitoids produces the next generation—are, for all their complexity, arguably simpler and more amenable to mathematical modeling than are the numerical responses of lions to their prey: for directly transmitted parasites. new infections (which in a sense represent parasite reproduction) appear at a rate determined by contacts between various categories of infected and susceptible hosts; eggs, larvae, or pupae produced by adult hosts in one generation either produce a host (if unparasitized) or a parasitoid (if parasitized) in the next generation, in proportions that depend mainly on how parasitoids search. In these situations, mathematical models can be built up from components (essentially representing the numerical and functional responses of parasites or parasitoids) that are based on specific field or laboratory studies. The dynamical behavior of the nonlinear systems, thus constructed, can then be explored, and predictions tested against different field or laboratory studies. This kind of synthetic approach can be used, on the one hand, as a guide to tentative generalizations (shedding light on the fundamental questions posed in the opening paragraph), and, on the other hand, as a tool for building a detailed and predictive understanding of particular systems (which have direct relevance to pest control and public-health planning).

In what follows, we first outline the basic ideas that have guided work on the dynamics of host-parasitoid and host-parasite associations, drawing a distinction between microparasites and macroparasites as we do so. We then sketch some experimental background and some complications, after which we discuss some of the things we think have been learned from this work. We conclude with a "view to the future," giving a list of problems that we believe will repay further study.

BASIC IDEAS

Host-Parasitoid

More than 10% of all metazoan species are insect parasitoids, a term first applied by Reuter (1913) to describe insects that develop as larvae on or in the tissues of other arthropods, which they kill as a developmental necessity. The population dynamics of such associations were first explored systematically by Nicholson and Bailey (1935), using equations that can be generalized to read

$$N_{t+1} = FN_t f(N_t, P_t), \tag{22.1a}$$

$$P_{t+1} = cN_t[1 - f(N_t, P_t)]. \tag{22.1b}$$

Here N and P are the host and parasitoid populations, respectively, in successive generations, t and $t + 1$; F is the per capita reproductive rate of the host population (F will in general depend on the host density, $F(N_t)$, but many studies treat it as a constant); and c is the average number of adult female parasitoids emerging from each parasitized host (c therefore includes the average number of eggs layed per host parasitized, the survival of these progeny, and their sex ratio). The functional response, f, defines the fraction of hosts escaping parasitism, which in general will depend on both host and parasitoid population density. Nicholson and Bailey assumed parasitoids to search independently randomly in a homogeneous environment, where f is the zero*th* term in a Poisson distribution ($f = \exp(-aP)$, with a the "searching efficiency" of the parasitoid); such a system always exhibits oscillations of diverging amplitude.

Host-Macroparasite

Anderson and May (1979) have found it useful to distinguish between "microparasites" and "macroparasites"; this distinction is made on the basis of population biology, rather than conventional taxonomy. Broadly speaking, microparasites are those having direct reproduction, usually at very high rates, within the host. As exemplified by most viral and bacterial and many protozoan infections, microparasites tend to be small in size and to have short generation times. Although there are many exceptions, the duration of infection is typically short relative to the average life span of the host, and hosts that recover usually possess immunity against reinfection, often for life. Microparasitic infections are thus characteristically of a transient nature. In contrast, macroparasites typically have no direct reproduction within the host, and are larger and have much longer generation times than microparasites. This category embraces essentially all parasitic helminths and arthropods. When an immune response is elicited, it usually depends on the number of parasites present in the host, and tends to be of relatively short duration. Thus macroparasitic infections are typically of a chronic or persistent nature, with hosts being continually reinfected.

It follows that for microparasitic infection processes, the host population can usefully be divided into several distinct categories or compartments (e.g., susceptible, infected, recovered-and-immune), with the mathematical models taking the form of differential equations describing the flows among compartments. For macroparasites, on the other hand, the various factors characterizing the interaction—egg output per parasite, pathogenic effects upon the host, evocation of an immune response in the host, parasite death rates, and so on—all tend to depend on the number of parasites present in a given host. This requires more complicated mathematical models that account for the distribution of parasites among hosts.

The parasitologist Crofton (1971a,b) was the first to model the dynamics of a host-macroparasite association. He assumed a host population with discrete, nonoverlapping generations (as in eq. (22.1)), and he showed by numerical simulations that the macroparasites could regulate their host population provided they were distributed among hosts in a sufficiently clumped or aggregated manner; with insufficient clumping, he found diverging oscillations. Crofton emphasized the importance, for the dynamics, of the observed clumping of helminth parasites among their populations of human or nonhuman hosts, where often 10% of the hosts harbor 90% of the macroparasites.

Anderson and May (1978; May and Anderson 1978) have presented a systematic framework for more general studies of the dynamics of host-macroparasite systems, based on differential equations (for continuous changes in host and macroparasite populations). Their basic model has the form:

$$dN/dt = (a - b)N - \alpha P, \tag{22.2a}$$

$$dP/dt = \beta WN - (\mu + \alpha + b)P - \alpha[(k+1)/k]P^2/N, \tag{22.2b}$$

$$dW/dt = \lambda P - (d + \beta N)W. \tag{22.2c}$$

Here N, P, and W are the population densities of hosts, adult macroparasites (established inside hosts), and free-living transmission stages of the parasites, respectively. The per capita host birth and death rates (assumed to be density independent) are a and b; the parasite-induced host death rate (or, equivalently, depression of the birth rate) is taken to be linearly proportional to the parasite burden, i, in a given host, at a rate α per parasite; μ is the natural mortality rate of adult parasites; λ is the rate of production of infective stages by mature adult parasites; d is the death rate of these infective stages; and β is the transmission rate, such that hosts acquire parasites at a rate βWN. The parasites are assumed to be distributed among hosts according to a negative binomial distribution, with mean P/N and "clumping parameter" k; the smaller k, the more clumped the distribution (more specifically, the mean, m, and variance, σ^2, of the distribution are related by $\sigma^2 = m + m^2/k$). This distribution of macroparasites, or something very like it, is widely observed in nature, though why it should be so remains a puzzle that is discussed below.

The basic system of eq. (22.2) can exhibit a stable equilibrium or stable cycles. Alternatively, if the macroparasite's reproductive rate is less than that of the host, discounted by a factor $(k + 1)/k$ to allow for the clumped distribution of parasites, the host population will grow exponentially, although at a rate below the disease-free rate $(a - b)$.

Host-Microparasite

For host-microparasite associations, a basic model has the form (Anderson and May 1979):

$$dX/dt = aN - bX - \beta XY, \tag{22.3a}$$

$$dY/dt = \beta XY - (b + \alpha + v)Y, \tag{22.3b}$$

$$dZ/dt = vY - bZ. \tag{22.3c}$$

Here X, Y, Z are the densities of susceptible, infectious, and recovered-and-immune hosts, respectively; the total host density is $N = X + Y + Z$. The per capita host birth and death rates, taken to be density independent, are a and b; v is the recovery rate; β is the transmission rate (such that new infections are produced at the rate βXY); and immunity has been assumed lifelong. Mortality produced by the microparasites, which can have important ecological implications, is represented by the disease-induced death rate, α. The overall population density then obeys

$$dN/dt = (a-b)N - \alpha Y. \tag{22.3d}$$

The system settles—albeit via very weakly damped oscillations in some cases—to a stable equilibrium, provided the microparasite is sufficiently pathogenic, $\alpha > (a-b)[1 + (v/b)]$. For insufficiently pathogenic parasites, the host population will continue to grow, again at an exponential rate below the disease-free rate $(a - b)$.

Human Hosts and Microparasites or Macroparasites

An important subclass of host-parasite models deals with human or other hosts whose population densities change very slowly in relation to all other time scales in the system. It is then conventional to assume the host population, N, is constant and that disease-induced deaths are negligible, $a = b$ and $\alpha = 0$. The focus is now on asking what fraction of the population experiences infection, what average worm burdens are, what fraction needs to vaccinated in order to eradicate infection, how best to deploy antihelminthic drugs, and so on. Although simpler than eqs. (22.2) and (22.3) in that N = constant, the basic models usually have the compensating complication of dealing with the age structure within the host population (so that variables like N are functions of time t and host age a, replacing eq. (22.2) and (22.3) with partial differential equations).

EXPERIMENTAL TESTS AND COMPLICATIONS

We now sketch the kinds of empirical evidence on which such models are based and against which they are tested. Such confrontations with real examples require various refinements to the basic models, which differ from case to case as outlined below.

Host-Parasitoid

As reviewed by Hassell (1978), field and laboratory studies show that parasitoids do not search randomly. Many parasitoids tend to take shorter steps and turn more frequently for a period of time after a successful encounter with a host, lengthening step and turning less as time goes by; the result of this individual behavior is a tendency to aggregate in patches of high host density (Hassell and May 1974). Conversely, other parasitoids have egg-laying behavior that results in their attack rates per host being lower in patches of high host density. Both of these different patterns of searching behavior can help to stabilize host-parasitoid associations, provided hosts are distributed patchily and parasitoids move freely among patches (Hassell 1985).

More generally, patterns of dispersal of hosts and of parasitoids can result in both hosts and parasitoids having clumped distributions, entirely independent of each other. Such nonuniformities in spatial distributions can, in some circumstances, promote the long-term persistence of the overall host-parasitoid association (Chesson and Murdoch 1986).

Whether arising from foraging behavior or from more general aspects of dispersal behavior, such population-level consequences of individual behavior can be studied by applying eq. (22.1) to each of many patches, with differing host and parasitoid densities in individual patches, in each generation, and then summing over all patches and redistributing some or all hosts and parasites according to observed dispersal patterns. More simply but more crudely, the outcome of nonrandom search or dispersal can be described phenomenologically by taking

$f(P)$ to be a negative binomial (with clumping parameter k constant or, more realistically, depending on host density) or other distribution; for a more full discussion, see Chesson (1986).

Parasitoids take a finite time to capture, oviposit in, or otherwise "handle" prey, which tends to saturate attack capacities at high host densities; this is a destabilizing feature (Holling 1959; Hassell 1978). Host rates of increase in general exhibit other density-dependent effects, which may act at various stages in the life cycle. Such density dependencies can affect the dynamics in different ways, depending on their strength and timing in the life cycle (Wang and Gutierrez 1980; May et al. 1981). If strong enough, such effects can keep hosts at densities too low for parasitoids to maintain themselves (that is, below "threshold" for the parasite; May and Hassell 1987). The presence of other parasitoid species, of hyperparasitoids, or of other generalist predators with switching behavior can further complicate the dynamics of host-parasitoid systems, sometimes in surprising ways (May and Hassell 1987).

Figure 22.1 shows data for a laboratory study of the effects of a pteromalid parasitoid, *Anisopteromalus calandrae,* on the bruchid bettle, *Callosobruchus chinensis.* Figure 22.2 shows the observed densities of winter-moth larvae in Nova Scotia, before and after control by the parasitoid *Cyzenis albicans,* as observed and

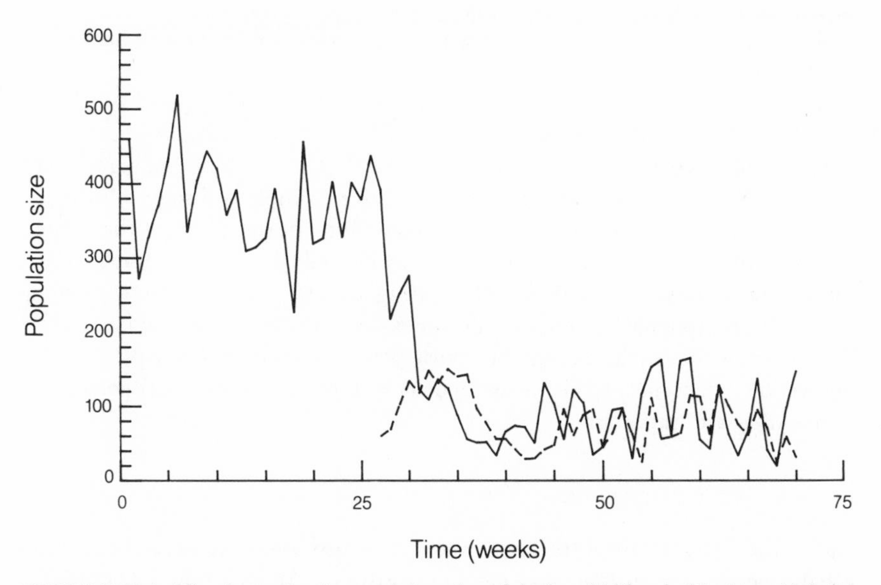

FIGURE 22.1. The depression in population size of the bruchid beetle (*C. chinensis*) (solid line) following the introduction of the pteromalid parasitoid, *A. calandrae* (dashed line) in week 26 of the interaction; the populations were maintained in the laboratory at 30°C and 70% RH with black-eyed beans as the host resource.

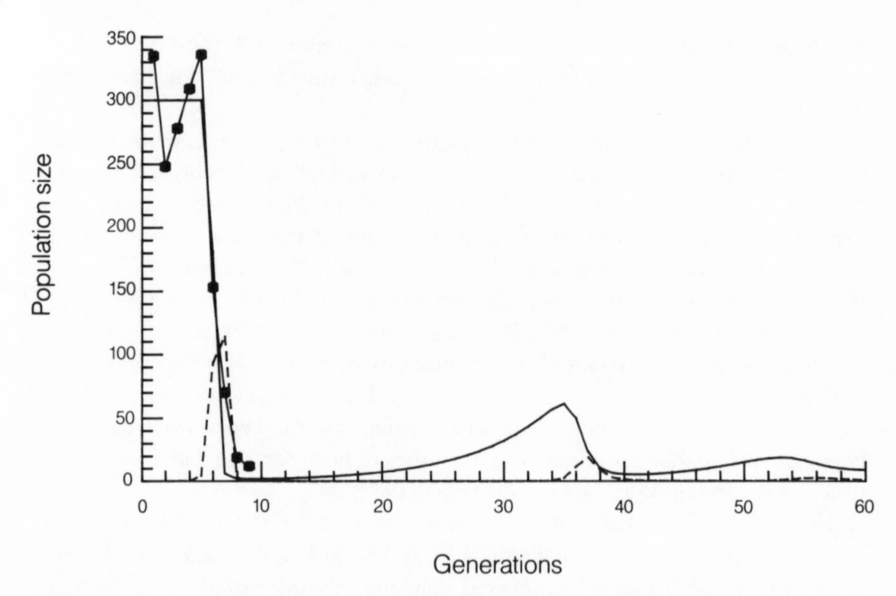

FIGURE 22.2. Biological control of the winter moth in Nova Scotia. Solid and broken lines give the host and parasitoid populations per square meter canopy area, respectively, predicted by eq. (22.1) with $f = (1 + aP/k(N))^{-k(N)}$. The parameter values, determined from field studies, are $F = 1.2$, $c = 0.65$, $a = 0.14\ \mathrm{m}^2$, and $k = 0.28 + 0.006N_t$. Solid points give the observed populations of winter-moth larvae (Embree 1966). For further details, see Hassell (1980).

as predicted by eq. (22.1) (again with $f = (1 + aP/k)^{-k}$ and with the parameters estimated from field and laboratory studies); this example is discussed in more detail, and contrasted quantitatively with the different dynamics of the winter moth/*C. albicans* system in Wytham Wood, by Hassell (1980). These examples are representative of what can be done by teasing such systems apart and studying individual components of the overall interaction usually within the laboratory. The worry is that such "deconstructionist" studies may overlook aspects of the interaction—especially spatial heterogeneity and intergenerational dispersal patterns—that are important.

Host-Macroparasite

Again, the components of the basic model described by eq. (22.2) can be investigated, one by one, in field and laboratory studies. Anderson and May (1978, 1985a) review such studies for nonhuman and human populations, respectively.

The basic model can be modified to take account of a wide variety of biological complications specific to particular host-macroparasite interactions. Such refinements are summarized in table 22.1, along with the way they tend to affect the dynamics of the association. Of particular interest is the effect that the nutritional state of the host can have on the interaction. Malnourished hosts tend to have lowered immunological competence, which, broadly speaking, results in the effective pathogenicity of the parasite tending to increase as host density rises to a level where competition for available food is severe; that is, α increases with N. A consequence can be the existence of two alternative stable states, with the system switching discontinuously from low to high levels of infection following a disturbance severe enough to cross some "breakpoint." Possible examples are provided by the observed association between populations of the Red Grouse, *Lagopus scoticus,* and the directly transmitted nematode, *Trichostrongylus tenuis,* in Britain (Lack 1954; Hudson and Dobson 1988) and outbreaks of fowl cholera, caused by the bacterium *Pasteurella multocida* among populations of wild ducks in North America (Petrides and Bryant 1951).

As has been reviewed in detail elsewhere (Anderson and May 1978), the clumped distribution of macroparasites among hosts can almost always be char-

TABLE 22.1
Some complications to the basic host-macroparasite model of eq. (22.2) and their dynamical consequences (for a full discussion, see Anderson and May 1978, and May and Anderson 1978).

Effects that tend to facilitate persistence of hosts and macroparasites, and/or to enhance the regulatory potential of the parasites	*Processes that have a destabilizing effect, possibly producing oscillations in host-macroparasite systems*
Aggregated distributions of parasite numbers per host	Parasite-induced reduction in the host birth rate (this effect can, however, enhance the regulatory capacity of the parasite)
Density-dependent parasite mortality or reproduction	Direct growth of parasite populations within individual hosts
Rates of parasite-induced host mortality that increase faster than linearly with parasite burden	Time delays in parasite developmental processes
Parasite-induced host mortality rates that are more severe at high host densities ("stress")	Random or underdispersed (regular) distributions of parasite numbers per host
Host immune responses (possibly slowly evoked and rapidly fading) that rise with the burden of established parasites, or with the number of challenging infections	

acterized by a negative binomial. Although this fact is abundantly documented, the underlying mechanisms remain unclear. We return to this later.

Applying this work to parasitic infections of humans, figures 22.3a,b show data and theory for the age-specific prevalence (fraction harboring worms) and mean worm burden, respectively, of *Ascaris* in human populations in Iran (from Croll et al. 1982). The theoretical curves come from an age-specific version of eq. (22.2), with parameters determined from field studies (in particular, the clumping parameter $k = 0.57$). With all the parameters of the model thus fixed, figures 22.3c and d show theoretical predictions and observed data for the rise in prevalence and mean worm burden, respectively, following the removal of worms by

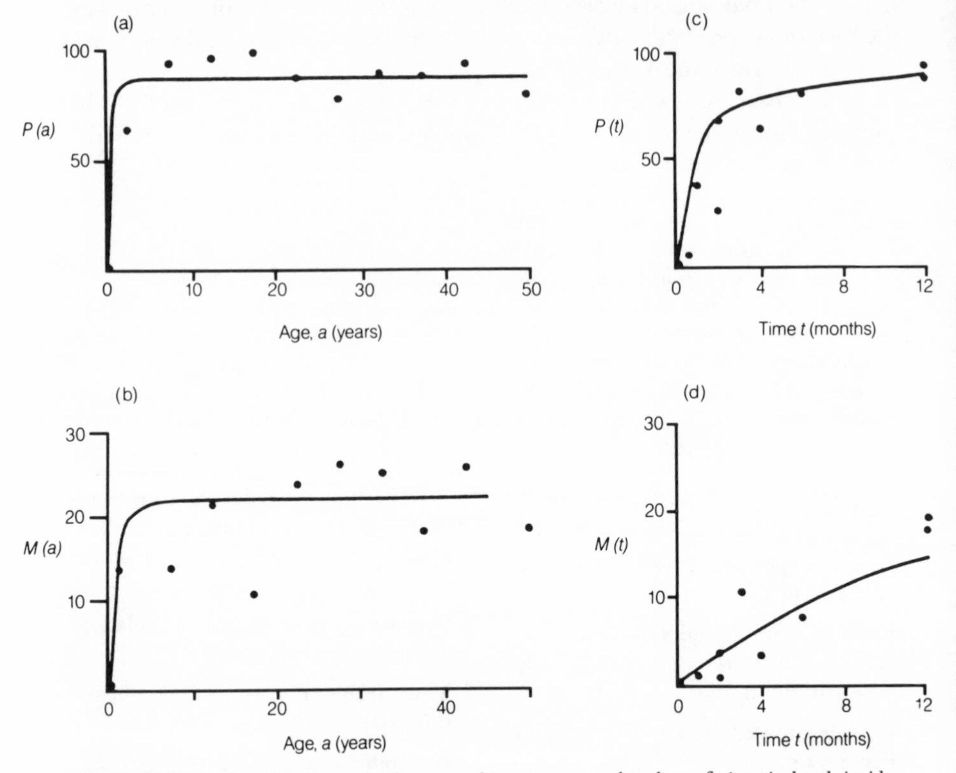

FIGURE 22.3. Data for the prevalence and mean worm burden of *Ascaris lumbricoides* among humans in Iran are compared with theoretical results: (a) prevalence and (b) mean worm burden, as functions of age, before intervention; (c) rise in prevalence and (d) mean worm burden, with time, following elimination of worms by chemotherapy at $t = 0$. Parameters characterizing the worm distribution and abundance of free-living transmission stages were deduced from the data in figures 22.3a and b, so that there are no free parameters in the fit between data and theory in figures 22.3c and d. (From Croll et al. 1982 and Anderson and May 1982.)

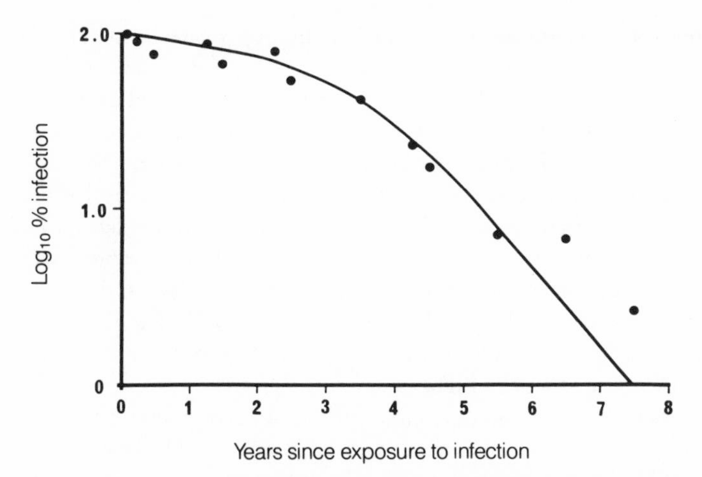

FIGURE 22.4. The decay in prevalence of infection with *Wuchereria bancrofti* within a sample of people who are no longer exposed to infection. The solid circles are observed values (Webber 1975) and the solid line is the theoretical prediction, essentially from the basic model of eq. (22.2) (Anderson and May 1985a). The parameter values, estimated by a nonlinear least-squared technique, are $k = 0.61$, $\mu = 1.1 \text{ yr}^{-1}$, and the initial mean worm burden is $P/N = 39.2$.

antihelminthic drugs; the agreement between theory and data encourages the hope that such models may be of use in planning control programs. As another example, figure 22.4 shows the observed decline in prevalence of *Wuchereria bancrofti* infection among a group of people who moved from Tahiti to France, thus removing themselves from new infections. The theoretical curve (again based on an age-dependent version of eq. (22.2)) explains these results, with an initial phase in which the decline depends on the way worms are distributed among hosts and a later phase where most hosts have either one or no worms (whence the decline in prevalence depends only on the death rate of adult worms). These examples are representative of a wide variety of studies which combine mathematical models with field data, as surveyed by Anderson and May (1985a).

Host-Microparasite

The study of microparasitic infections in human populations, where they are assumed not to affect the overall population dynamics, has a long history going back to the work of Ross (1911) and Macdonald (1957) on malaria, Bartlett (1957) on measles, and others (see Bailey 1975). The assumption that transmission is proportional to the rate of encounter between susceptible and infected hosts accords reasonably well with data on epidemics among human (Bailey 1975) and invertebrate (Anderson and May 1981a) host populations.

Studies along the lines of eq. (22.3), where microparasites are assumed to affect the host population dynamics, are more recent. Anderson and May (1979) give detailed theoretical fits to the laboratory data of Greenwood et al. (1936) and Fenner (1948, 1949) for the effects of *Ectromelia* and *Pasteurella muris* on the dynamics of laboratory mice populations; these studies have the simplifying feature that new mice were introduced into the arena at controlled rates, thus avoiding the complications attendant upon replenishment by natural birth processes. Detailed theoretical studies of the observed effects of rabies virus on the dynamics of fox populations in Europe (Anderson et al. 1981), and of tuberculosis on badger populations in England (Anderson and Trewhella 1985), have the extra difficulty of dealing with the density-dependent factors that influence the host populations in the absence of disease.

Invertebrate hosts can mount cellular or humoral responses to microparasitic infections, but they do not manifest acquired immunity. The basic model is thus somewhat simpler, having only susceptible and infected categories. On the other hand, one or more of a variety of other complications are present in particular interactions between invertebrate hosts and microparasites. Some of these complications, and their effects upon the overall dynamics, are summarized in table 22.2; for a more detailed discussion, with specific examples, see Anderson and May (1981a).

A major complication arises when the free-living transmission stages of the pathogen are long-lived. Such free-living infective stages include the spores of many bacteria, protozoans and fungi, and the capsules, polyhedra, or free particles of viruses (Tinsley 1979). In particular, baculoviruses of univoltine insects of temperate forests (especially of lepidopteran, hymenopteran, and dipteran species) tend to have such long-lived infective stages, partly because the soil environment of temperate forests affords relative protection from the ultraviolet components of sunlight (Jaques 1977). In this event, we need to acknowledge the dynamics of the population of free-living infective stages, $W(t)$, and to treat transmission as proportional to the rate of encounters between susceptible hosts and infective stages of the parasite, βWX (cf. eq. (22.2c)). Provided infected hosts produce transmission stages of the parasite at a sufficiently fast rate, the pathogen can again regulate its host population so long as $\alpha > (a - b)$. The regulated state may, however, be a stable point or it may be a stable cycle. Cyclic solutions tend to arise for infections with high pathogenicity that produce large numbers of long-lived infective stages; in effect, the "seedbank" of long-lived transmission stages can induce oscillations by virtue of the time lags introduced into the system. Many microspordian protozoan and baculovirus infections of insects appear to possess the combination of relatively large α and long life span that produces cyclic changes in host abundance. Anderson and May (1981a) review data for several forest insect pests and associated pathogens, which may explain observed cycles with periods in the general range of 3–20 years for these pests.

TABLE 22.2
This table summarizes the way in which some realistic complications can affect the ability of a microparasite to regulate an invertebrate population, and to persist within such a population of hosts that either is of low density or fluctuates widely in abundance (condensed from Anderson and May 1981a).

Complicating factor	*Effect of regulatory capacity of pathogen (relative to basic model's criterion $\alpha > a - b$).*	*Effect on threshold host density to maintain pathogen (relative to basic model)*
Diminution in reproductive capacity of infected hosts	Regulation is easier; regulated host population density is lower	No effect on threshold criterion
Vertical transmission	No effect on regulatory condition; regulated host population density is lower	Threshold host density lower (can be arbitrarily small)
Latent period of infection	Regulation is harder; regulated host population density is higher	Threshold host density is higher
Pathogenicity is stress-related	Can always regulate if pathogenicity is sufficiently strongly enhanced by stress	Pathogen can persist provided transmission efficiency is high enough
Other density-dependent constraints on host population growth	Disease-related depression of host population density maximized for an intermediate degree of pathogenicity	Pathogen cannot persist if threshold host density is too high (if $N_T > K$)
Free-living infective stages of pathogen	Regulatory criterion as for basic model; regulated state may be a stable point or cyclic oscillations	Easier to maintain pathogen, especially if infective stages are long-lived

The most abundant data are, of course, for viral, bacterial, and protozoan infections of human populations. Here the host population is treated as constant on time scales of epidemiological interest, but realistic models are made more complicated by the need to include age structure, variability in rates of acquiring new sexual partners (for sexually transmitted diseases), and other relevant heterogeneities; for reviews, see Bailey (1975), Dietz (1975), Anderson and May (1983,

1985b), and Hethcote and Yorke (1984). In order to use such models as reliable tools for planning vaccination programs or other strategies of intervention, it is usually necessary to take account of age-specific variabilities in transmission probabilities. That is, the probability, $\lambda(a,t)$, that a susceptible of given age, a, will acquire infection may be taken as $\lambda(a,t) = \int \beta(a,a')Y(a',t)da'$, where $\beta(a,a')$ represents the probability that a susceptible of age a will acquire infection from an infected of age a', per unit time, and $Y(a',t)$ is the population density of infectious individuals of age a', at time t. There are substantial difficulties in obtaining reliable estimates of $\beta(a,a')$ from available data, mainly because one requires a two-dimensional array of information from serological data that give only its one-dimensional shadow, $\lambda(a)$ (Dietz and Schenzle 1985; Anderson and May 1985b).

Figure 22.5 illustrates the outcome of a representative calculation of this kind. The proportion seropositive to measles in Britain is shown, by age class (from 0 to

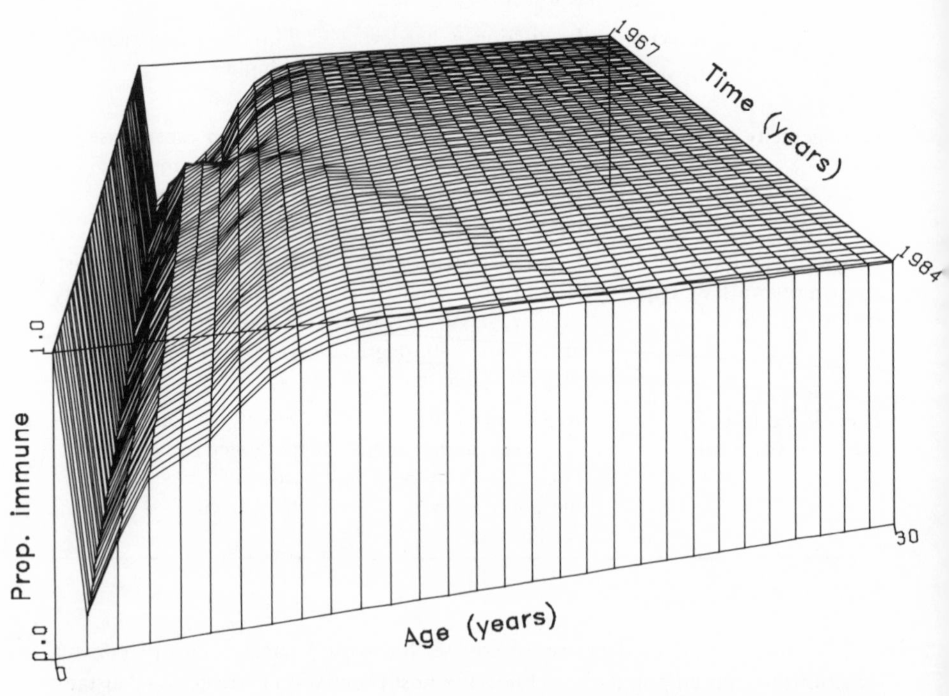

FIGURE 22.5. As discussed in the text, the proportion seropositive to measles in Britain is shown as a function of age, each year from 1967 (when vaccination began) to 1984. Age-dependent transmission parameters are deduced from the 1967 data, and reported vaccination rates used in conjunction with partial differential equations for changes with age and time, to predict changes in serological profiles. For details, see Anderson and May (1985c).

30 years of age) each year from 1967—when vaccination began—to 1984. The 1967 data are used to assess age-dependent transmission coefficients, and a set of partial differential equations is used to describe the rates of change with age and time in the proportions protected by maternal antibodies, susceptible, infected but latent, infectious, and recovered/immune, as a result of the reported age-specific vaccination schedules each year. The age-specific serological profile in 1984 is then essentially a theoretical prediction, based on the structure of the model, the 1967 data, and the observed vaccination rates; figure 22.6a compares this 1984 prediction with the actual serological data. Figure 22.6b shows a similar comparison between predicted and observed age-specific seropositivity for rubella in Britain in 1984 (roughly fourteen years after vaccination of 10–15-year-old girls was begun).

SOME THINGS THAT HAVE BEEN LEARNED

Nonlinear systems can exhibit stable equilibria, or stable cycles, or apparently random "chaotic" fluctuations. The astonishing array of dynamical behavior that can be manifested by simple, natural, and purely deterministic population models has only recently been appreciated (Li and Yorke 1975; May and Oster 1976). One of the earliest of these studies was motivated by ideas about how a fungal pathogen may regulate populations of periodical cicadas (Lloyd, pers. comm.; May 1974). These results (as surveyed by May 1976) have had a wide and growing impact on many areas outside population biology, and the important insights they give us into the dynamical behavior possible in ecological models is now taken for granted (sometimes by people who paradoxically seem to believe that simple and general models are useless).

Host-parasite and host-parasitoid associations have an intrinsic propensity to oscillate, with periods that often depend in a simple way on the characteristic lifetimes of host and parasite individuals. The Lotka-Volterra metaphor for prey-predator interactions has purely neutral stability, but more realistic models often nave stable limit cycles whose amplitude and period are uniquely determined by the biological parameters. Figure 22.7 shows, for example, the Fourier spectrum for the time series generated by weekly case notifications of measles in England and Wales for 1948–68; the minor peak corresponds to annual variations in incidence (associated with seasonality and/or school openings), while the more pronounced peak corresponds to two-year cycles. Similar "inter-epidemic" cycles are seen, though less unambiguously, for pertussis, rubella, mumps, chickenpox, and other childhood infections in developed countries. The basic host-microparasite models exhibit weakly damped oscillations at around the correct periods in each case, although there remains debate about the mechanisms that may "pump" the system to sustain the oscillations (candidates are demographic stochasticity, sea-

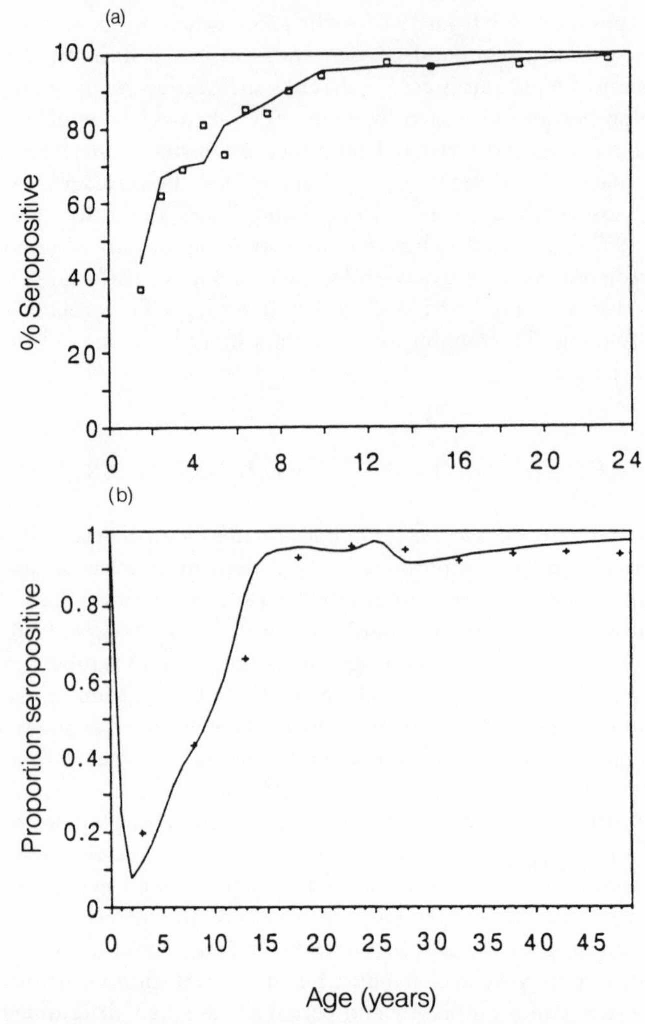

FIGURE 22.6. (a) A comparison between the predictions of the model in figure 22.5 for age-specific seroprevalence of measles antibodies in Britain in 1979–82 with reported age-seroprevalence data 1979–82 (note that the theoretical curve uses only the 1967, prevaccination, serological data and the reported age-specific vaccination schedules; the seroprevalence data in this figure play no part in computing the theoretical curve). (b) A similar comparison between theory and data for the age-specific serological profile for rubella antibodies in females in England and Wales in 1984; predictions are based on an average age at infection of 10 years before control, and the age-specific vaccination rates in the United Kingdom (vaccination of teenage girls between the ages of 10 and 15 years over the period 1970–84. For details, see Anderson and May (1985b,c).

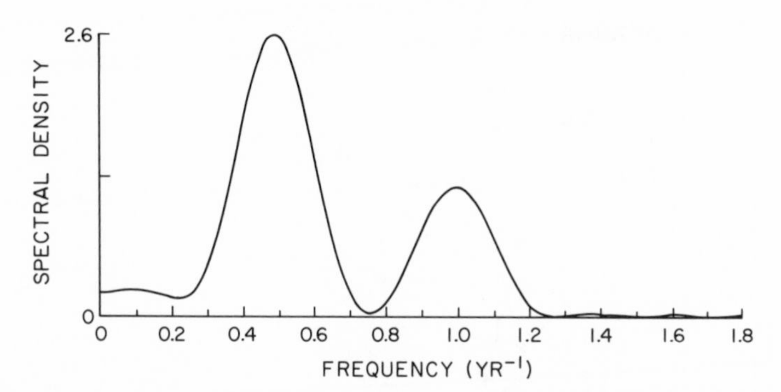

FIGURE 22.7. Frequency spectrum constructed from the weekly measles reports for England and Wales, 1948–68. For details of the analysis underlying this figure, see Anderson, Grenfell, and May (1984).

sonal variations in transmission efficiency, age-structure effects, or combinations of these and other processes; for reviews, see Anderson, Grenfell, and May 1984, and Schaffer and Kot 1985, 1986).

The existence of such oscillatory propensities can have practical implications. When, for instance, a vaccination program is begun, we would not necessarily expect the incidence of infection to decline monotonically, but rather to exhibit oscillations en route to some lower level or eradication. Such oscillations appear to be observed, for example, in the incidence of congenital rubella syndrome (associated with rubella infections in pregnant women) in Britain and in the United States, following the introduction of different vaccination programs in the two countries; see figure 22.8. Such oscillations may appear alarming, and their public health implications may be worrisome, if the dynamics of the system are not properly appreciated (Knox 1980; Anderson and May 1983, 1985b).

There are threshold densities of hosts, below which parasites or parasitoids cannot maintain themselves. The threshold concept has been most fully developed for human host-microparasite associations, where it dates back to Kermack and McKendrick (1927) and Bartlett's (1957) explicit studies of threshold population densities for maintenance of measles. It follows that immunization programs do not have to reach everyone, but rather must have sufficient coverage so that the unimmunized population is effectively below threshold density (Arita, Wickett, and Fenner 1986; Anderson and May 1981b); the eradication criteria essentially derive from the requirement that the parasite population cannot maintain itself unless it has a basic reproductive rate, R_0, in excess of unity. Macdonald's (1957) extension of Ross's analysis of malaria uses these ideas in arriving at the influential conclusion that control measures targeted against adult mosquitos would be more effective than those targeted against larvae (basically because of the factor

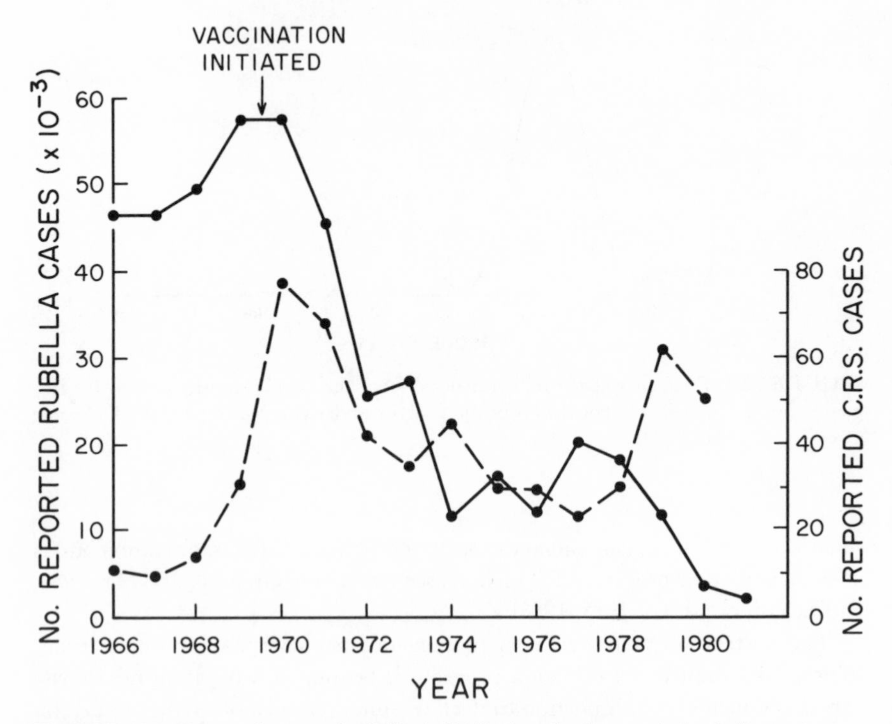

FIGURE 22.8. Reported cases of rubella (solid line) and of congenital rubella syndrome, CRS (dashed line), in the United States between 1966 and 1981. Note the resurgence in cases of CRS about 10 years after initiation of the immunization program, which is what the theoretical models predict. (From Anderson and May 1983, where the reasons for the lower numbers before 1970 are also discussed.)

$\exp(-\mu T)$ associated with adult mortality at a rate μ over the latent period T for infected mosquitos to become infectious).

These concepts are also important in host-parasitoid and host-macroparasite systems. As discussed by May and Hassell (1987), appropriate modifications of these basic ideas can be used to assess the densities at which parasitoids or pathogens must be released in control programs that aim to eradicate a pest species.

In situations where spatial patchiness or other heterogeneities are significant, one must first study the dynamics and then average; first averaging and then studying the dynamics can mislead. This is an important caveat for the kinds of strongly nonlinear systems that often arise by virtue of prey-predator interactions. Cautionary examples abound. The carefully designed project to interrupt malarial transmission in West Africa essentially failed owing to heterogeneities in mosquito behavior (with exophilic mosquitos playing a disproportionate role in maintaining transmission in the presence of residual insecticides; Molineaux and Gramiccia 1980). Hethcote and Yorke (1984) showed that the transmission dynamics of endemic gonorrhea in the United States cannot be understood using average rates of acquisition of new sexual partners, but that core groups of highly active individuals must be distinguished; similar considerations apply to current HIV/AIDS epidemics (May and Anderson 1987).

A number of other interesting topics may be more usefully discussed in terms of potential future developments. This we now do.

A VIEW TO THE FUTURE

In this concluding section, we list some aspects of host-parasite and host-parasitoid associations which we think are interesting foci for current and future research.

1. *Nonlinear Dynamics.* As reviewed by Lauwerier (1986) and May (1987), pairs of coupled first-order difference equations for particular host-parasite or parasitiod associations (in which both species have discrete, nonoverlapping generations) can exhibit extremely rich bifurcation phenomena and very chaotic dynamics. Even if environmental noise and spatial heterogeneity could legitimately be stripped away, there remain interesting problems in elucidating the panoply of nonlinear behavior that such systems can display.

One new approach to the analysis of data for host-microparasite and prey-predator associations uses methods developed for situations where only one variable can be measured in a system that possesses many independent variables (Packard et al. 1980; Takens 1981). If indeed some multidimensional attractor underlies the observed time series, it may be reconstructed (without any understanding of the fundamental mechanism that generates it) by choosing some fixed time lag, T, and plotting values of the variables $x(t)$, $x(t + T)$, $x(t + 2T)$, . . . , $x(t + [m - 1]T)$ in m-dimensional space; the value chosen for T is not critical. The value of m is selected so that increasing its value by unity does not apparently result in any additional structure. These methods have been applied to the recorded numbers of cases of chickenpox, mumps, and measles per month in New York and Baltimore before mass vaccination (Schaffer and Kot 1985, 1986) and to the Canadian data on apparent cycles in lynx abundance (Schaffer 1984). In all cases, Poincaré cross sections of the phase plots suggest that the flows are indeed confined to a nearly two-dimensional conical surface, corresponding to some nearly one-dimensional map. This phenomenological approach is different in spirit from conventional approaches which seek to understand dynamical behavior in terms of specific models based on underlying biological mechanisms.

This approach holds the promise of providing a method for distinguishing between population fluctuations that are caused by random environmental variability and apparently random population fluctuations that arise from deterministic dynamical mechanisms (where nonlinear interactions among populations produce chaotic dynamics). One problem is that these Packard-Takens "embedding" techniques require longer time series of data than are commonly available for biological populations.

2. *Stochastic Effects.* Environmental and demographic stochasticity can interact with nonlinear regulatory processes in complicated ways. The result is that environmental noise can help or hinder the persistence of host-parasitoid or parasite associations, depending on the frequency spectrum of the noise and on how it interacts with density-dependent mechanisms. Likewise, depression of host populations due to "harvesting" by parasites or parasitoids can amplify or damp environmentally driven fluctuations, depending on the details. Chesson (1986) and others have made a start in codifying these effects, but much remains to be done.

3. *Spatial and Other Heterogeneities.* As mentioned earlier, Hassell (1978) and others have observed that searching behavior which results in parasitoid aggregation in response to nonuniform distributions of hosts can promote the persistence of host-parasitoid associations. Murdoch et al. (1984) and others have, however, emphasized that in some well-studied examples of persistent biological control of pests, the parasitoids fail to exhibit any such aggregative behavior. A possible explanation is that Hassell's work deals largely with interactions where there is complete redistribution of host and parasitoid individuals within the habitat in each generation, while in Murdoch and others' work (on scale insects and their natural enemies, or prey-predator interactions involving mites) the hosts and parasitoids tend to interact within a patch over several generations, with only a certain amount of mixing and recolonization of patches in each generation. In the latter case, local extinction of hosts and parasitoids within patches can easily occur, and the persistence of the system as a whole is dominated by the degree of asynchrony between the states of the different patches rather than by the spatial distribution of parasitism as such (Sabelis and van der Meer 1986). There is thus a continuum from cases of fairly complete mixing of the subpopulations each generation (as for winter moth) where heterogeneity in the distribution of parasitoid attack is likely to be important for regulation, through to cases in which large fractions of the host and parasitoid populations do not move from their patches (as for many scale insects and mites) where persistence is largely governed by the temporal-refuge effect of host patches being colonized at different times by the natural enemies. These tentative ideas may provide a constructive resolution of recent controversies, although further theoretical and empirical studies are clearly needed (Waage and Greathead 1988).

The fact that macroparasites typically are distributed in a highly aggregated way among hosts has important implications for the dynamics of the association. There has, however, been relatively little progress toward explaining the mechanisms that produce these distributions. One possibility is that the distributions are produced by environmental factors having to do with the way hosts acquire worms in a heterogeneous setting. Alternatively—and there is considerable controversy on this point (see, e.g., Cheever 1986, versus Anderson, Crombie, and May 1986)—it may be that the worm burden in any particular host is dependent to a significant degree on mechanisms within that host which regulate its eventual worm burden (by regulating worm establishment or longevity in a density-dependent way), and that genetic heterogeneity within the host population results in the regulated worm level being different in different hosts. This second type of explanation, which is supported by a certain amount of experimental data on laboratory mice and on humans, implies there are "wormy hosts," who in a controlled setting will end up with higher average parasite burdens following exposure to some constant level of infection, or, alternatively, who will return to their previously high level following chemotherapeutic clearing of worms followed by reinfection. Whether the explanation lies along one of the above lines or otherwise, the pervasive patterns of macroparasite clumping within host populations cry out for closer experimental and theoretical analysis.

For host-microparasite interactions, we have already discussed the important problems associated with age-dependent transmission coefficients. Other heterogeneities that must be included to bring mathematical models closer to real applications to biological systems include: genetic and other heterogeneities in intrinsic susceptibility to infection; geographical heterogeneity (big cities versus small villages); social factors (individuals in disadvantaged classes may be malnourished and more crowded); and others.

In particular, host-microparasite concepts are being applied to modeling the transmission dynamics of HIV/AIDS infections in developed and developing countries. In the models, it is important to take account of the substantial variability in rates of acquiring new heterosexual or homosexual partners that are found in different risk groups. Such mathematical models for the transmission dynamics of HIV/AIDS help us both to define clearly the kinds of data we need and to understand the dynamical implications of different kinds of behavioral changes (May and Anderson 1987; Hyman and Stanley 1988; Dietz 1988).

4. *Elucidating Density-Dependent Mechanisms.* For most populations of hosts and parasites or parasitoids, all three of the above factors roil together in nature, with the populations distributed patchily in a heterogeneous environment, and with density dependence and environmental stochasticity acting differently on subpopulations at different densities in individual patches. Mathematical models of such situations show that the overall dynamics can be very complicated. In particular, for populations with discrete generations, it will often be that average population densities in any one generation are determined mainly by stochastic and density-dependent processes acting among patches of varying density *within* a generation, with overall average densities between generations being relatively

unimportant. But many empirical studies, particularly of the population dynamics of insects, tend to deal with average population densities in successive generations (this is essentially the basis of conventional "K-factor analysis" of field data); under the circumstances just described, such analyses may be too crude to uncover the density-dependent mechanisms that in fact regulate the population.

These complications have been pursued in numerical simulations of host-parasitoid systems (Hassell 1986) and in the analysis of data for populations of the whitefly, *Aleurotrachelus jelinekii,* on viburnum bushes at Silwood Park (Hassell, Southwood, and Reader 1987). There is a serious possibility that the large body of existing data on life tables is not a reliable source for investigating the causes of regulation of many natural populations; within-generation heterogeneity can be a powerful promoter of population stability, and in the presence of environmental noise such regulatory effects can be missed in analyses that use only data on mean population densities each generation. All this suggests a need to rethink the design and analysis of some field studies aimed at determining density-dependent mechanisms.

5. *Complex Life Cycles.* Most work on host-parasitoid associations has focused on first-order difference equations, representing synchronized interactions between populations with discrete, nonoverlapping generations. This makes sense for many temperate-zone situations, where seasonality tends to synchronize the populations and to provide a natural interval between the appearance of successive generations. In the tropics, however, life cycles of hosts and of parasitoids are often of significantly different lengths, and generations also overlap continuously. Following earlier studies (Auslander, Oster, and Huffaker 1974; Nisbet and Gurney 1982), Godfray and Hassell (1987) have shown that these more general host-parasitoid systems tend to exhibit steady populations of hosts and parasitoids if the duration of the parasitoid life cycle is around an integer multiple of that of the host, and to exhibit pronounced cycles (with periods close to the duration of the host life cycle) if parasitoid life cycles are one-half or three-halves that of the host; see figure 22.9. Godfray and Hassell (1987) observe that some laboratory and field studies of tropical host-parasitoid systems lend support to these ideas, with a tendency for discrete generations to show up when the ratio of the lengths of host and parasitoid life cycles is roughly 2:1. These complicating features are of fundamental interest, as well as being relevant for biological control in the tropics.

6. *Population Genetics and Population Dynamics.* Another area deserving more attention is the interplay between the genetics and the dynamics of host-parasite or parasitoid associations.

For host-parasitoid associations, the work of Tabashnik and Croft (1982; Tabashnik 1985) suggests that the dynamics of such interactions explains why orchard pests tend to evolve resistance to pesticides faster than do their natural enemies. When both host and parasitoid populations are depressed by pesticide application, the host population tends to bounce back or "overcompensate" be-

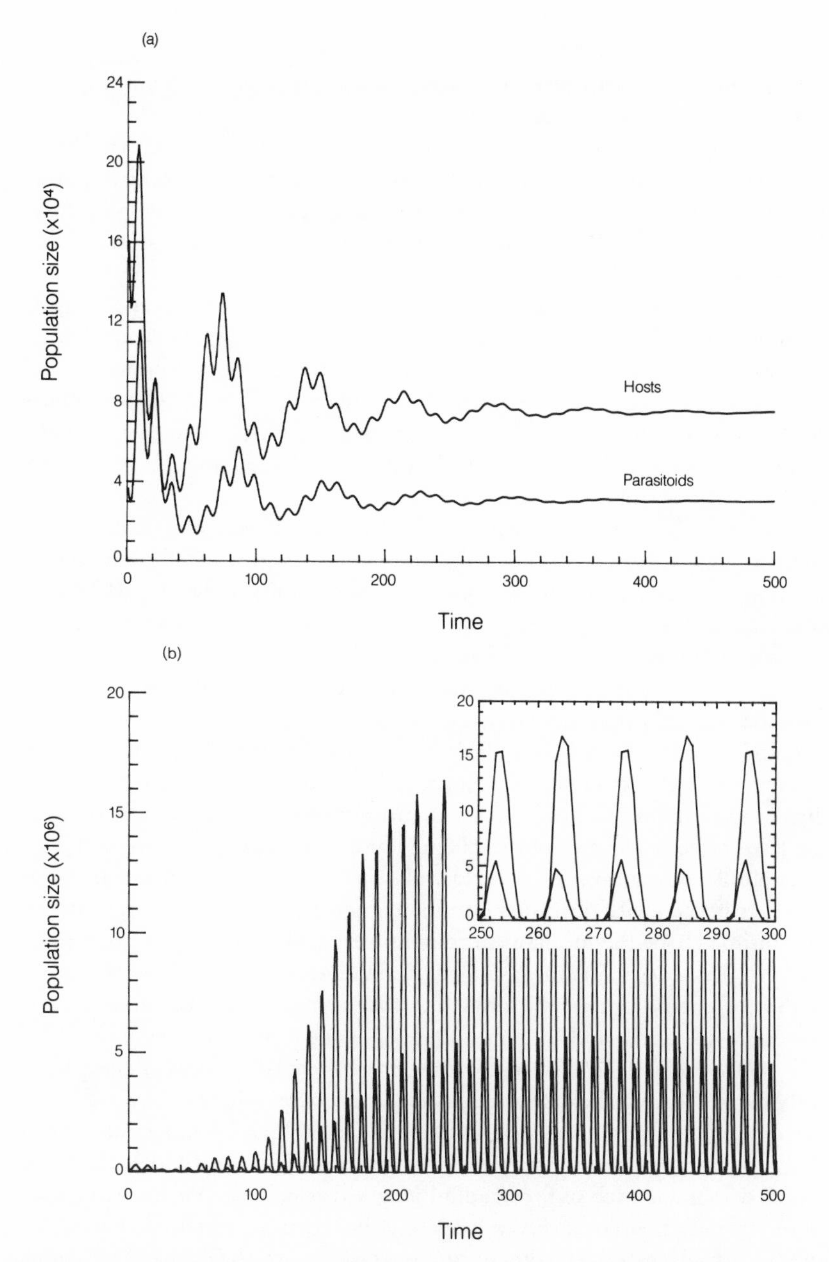

FIGURE 22.9. Simulations for a tropical host-parasitoid association with continuously overlapping generations. (From Godfray and Hassell 1987.) (a) Host and parasitoid have average life cycles of equal duration, and the system settles to stable values with all life stages simultaneously present. (b) The parasitoid life cycle is half that of the host, and there are stable cycles of host and parasitoid abundance with periods roughly equal to the duration of the host life cycle (i.e., the system effectively exhibits discrete, nonoverlapping generations).

cause enemies are relatively scarce (thus making the resistance-retarding immigration of other susceptible hosts from untreated regions less important), whereas the parasitoid population tends to recover slowly or "undercompensate" because hosts are scarce (thus making resistance-retarding immigration more important). This and other work combining population genetics with dynamics may help in the design of strategies of pesticide application that minimize the rate of evolution of resistance.

For host-macroparasite associations, we have already mentioned how genetic heterogeneity in host immune responses might combine with the dynamics of the interaction to explain observed patterns of pronounced worm clumping. As for host-parasitoid associations, empirical and theoretical studies of the interplay between population genetics and dynamics could help us find strategies for administering antihelminthic drugs to humans and other animals in such a way as to retard the evolution of drug resistance.

For host-microparasite associations, Gillespie (1975) has combined genetic and epidemiological considerations to suggest how pathogens could maintain polymorphic populations of resistant and susceptible hosts; Hamilton (1980, 1982) and May and Anderson (1983) have extended the analysis to show how gene frequencies in such polymorphisms are likely to vary cyclically or chaotically. Going beyond the familiar observation that coevolution of hosts and microparasites may be responsible for much of the genetic diversity found in natural populations, Hamilton (1980, 1982) has even suggested that—provided polymorphisms are fluctuating very chaotically—the greater inertia possessed by diploid systems over haploid ones (with rare alleles smuggled into heterozygotes) can lead to geometric mean fitness being more than twice as high in diploid than in haploid populations. That is, Hamilton suggests the frequency-dependent effects associated with transmission and maintenance of microparasitic infections may ultimately be the mechanism responsible for maintaining sexual reproduction. All these ideas warrant systematic development, along with more empirical studies of coevolutionary histories like that of myxoma virus and rabbits in Australia (see Levin and Pimentel 1981; May and Anderson 1983).

7. *Understanding Population Dynamics in Terms of Individual Behavior.* Hassell and May (1974, 1985) have provided detailed examples where the overall population dynamics of specific host-parasitoid associations may be understood in terms of the behavior of the constituent individuals. Kareiva and Odell (1987) have supplied a similar analysis for an aphid-ladybird association. We hope to see more empirical and theoretical research relating the dynamics of populations of hosts and parasitoids, macroparasites or microparasites to the behavior of individuals.

More generally, we would like to see more behavioral ecologists working up toward the population consequences of behavior, and more population biologists working down toward the individual behavior that determines population parameters. In this way, we can link population and community properties, via the behavior of individuals, to the evolutionary forces that have shaped them.

REFERENCES

Anderson, R. M., and R. M. May. 1978. Regulation and stability of host-parasite population interactions: I, Regulatory processes. *J. Anim. Ecol.,* 47:219–47.

Anderson, R. M., and R. M. May. 1979. Population biology of infectious diseases: Part I. *Nature* 280:361–67.

Anderson, R. M., and R. M. May. 1981a. The population dynamics of microparasites and their invertebrate hosts. *Phil. Trans. Roy. Soc.* (B) 291:451–524.

Anderson, R. M., and R. M. May. 1981b. Directly transmitted infectious diseases: Control by vaccination. *Science* 215:1053–60.

Anderson, R. M., and R. M. May. 1982. The population dynamics and control of helminth infections. *Nature* 297:557–63.

Anderson, R. M., and R. M. May. 1983. Vaccination against rubella and measles: Quantitative investigations of different policies. *J. Hyg.* 90:259–325.

Anderson, R. M., and R. M. May. 1985a. Helminth infections of humans: Mathematical models, population dynamics and control. *Adv. Parasitol.* 24:1–101.

Anderson, R. M., and R. M. May. 1985b. Age-related changes in the rate of disease transmission: Implications for the design of vaccination programmes. *J. Hyg.* 94:365–436.

Anderson, R. M., and R. M. May. 1985c. Vaccination and herd immunity to infectious diseases. *Nature* 318:323–29.

Anderson, R. M., and R. M. Trewhella. 1985. Population dynamics of the badger (*Meles meles*) and the epidemiology of bovine tuberculosis (*Mycobacterium bovis*). *Phil. Trans. Roy Soc.* (B) 310:327–81.

Anderson, R. M., J. A. Crombie, and R. M. May. 1986. Predisposition to helminth infections in man. *Nature* 320:195–96.

Anderson, R. M., R. T. Grenfell, and R. M. May. 1984. Oscillatory fluctuations in the incidence of infectious disease and the impact of vaccination: Time series analysis. *J. Hyg.* 93:587–608.

Anderson, R. M., H. Jackson, R. M. May, and T. Smith. 1981. The population dynamics of fox rabies in Europe. *Nature* 289:765–71.

Arita, I., J. Wickett, and F. Fenner. 1986. Impact of population density on immunization programs. *J. Hyg.* 96:459–66.

Auslander, D., G. F. Oster, and C. Huffaker. 1974. Dynamics of interacting populations. *J. Franklin Inst.* 297:345–75.

Bailey, N.J.T. 1975. *The Mathematical Theory of Infectious Diseases,* 2d ed. Macmillan, New York.

Bartlett, M. S. 1957. Measles periodicity and community size. *J. Roy. Stat. Soc.* (Ser. A) 120:48–70.

Cheever, A. W. 1986. Predisposition to helminth infection in man. *Nature* 320:195.

Chesson, P. L. 1986. Environmental variation and the coexistence of species. In J. Diamond and T. J. Case, eds., *Community Ecology,* pp. 240–56. Harper and Row, New York.

Chesson, P. L., and W. W. Murdoch. 1986. Aggregation of risk: Relationships among host-parasitoid models. *Amer. Natur.* 127:696–715.

Crofton, H. D. 1971a. A quantitative approach to parasitism. *Parasitology* 63:179–93.

Crofton, H. D. 1971b. A model of host-parasite relationships. *Parasitology* 63:343–64.

Croll, N. A., R. M. Anderson, T. W. Gyorkos, and E. Ghadirian. 1982. The population biology and control of *Ascaris lumbricoides* in rural communities in Iran. *Trans. Roy. Soc. Trop. Med. Hyg.* 76:187–97.

Dietz, K. 1975. Transmission and control of arbovirus diseases. In D. Ludwig and K. L. Cooke, eds., *Epidemiology,* pp. 104–21. Society for Industrial and Applied Mathematics, Philadelphia.

Dietz, K. 1988. On the transmission dynamics of HIV. *Math. Biosci.* In Press.

Dietz, K., and D. Schenzle. 1985. Mathematical models for infectious disease statistics. In A. C. Atkinson and S. E. Fienberg, eds., *A Celebration of Statistics,* pp. 167–204. Springer-Verlag, New York.

Embree, D. G. 1965. The population dynamics of the winter moth in Nova Scotia, 1954–1962. *Mem. Ent. Soc. Can.* 46:1–57.

Fenner, F. 1948. The epizootic behavior of mouse pox (infectious ectromelia). *Brit. J. Exp. Path.* 29:69–91.

Fenner, F. 1949. Mouse pox (infectious ectromelia of mice): A review. *J. Immunol.* 63:341–73.

Gillespie, J. H. 1975. Natural selection for resistance to epidemics. *Ecology* 56:493–95.

Godfray, H.C.J., and M. P. Hassell. 1987. Natural enemies can cause discrete generations in tropical insects. *Nature* 327:144–47.

Greenwood, M., A. Bradford Hill, W.W.C. Topley, and J. Wilson. 1936. *Experimental Epidemiology.* Special Report Series, No. 209, Medical Research Council. HMSO, London.

Hamilton, W. D. 1980. Sex versus non-sex versus parasite. *Oikos* 35:282–90.

Hamilton, W. D. 1982. Pathogens as causes of genetic diversity in their host populations. In R. M. Anderson and R. M. May, eds., *Population Biology of Infectious Disease Agents,* pp. 269–96. Springer-Verlag, New York.

Hassell, M. P. 1978. *The Dynamics of Arthropod Predator-Prey Associations.* Princeton University Press, Princeton, N.J.

Hassell, M. P. 1980. Foraging strategies, population models and biological control: A case study. *J. Anim. Ecol.* 49:603–28.

Hassell, M. P. 1985. Parasitism in patchy environments: Inverse density dependence can be stabilizing. *IMA J. Math. Appl. Med. Biol.* 1:123–33.

Hassell, M. P. 1986. Insect natural enemies as regulating factors. *J. Anim. Ecol.* 54:323–34.

Hassell, M. P., and R. M. May. 1974. Aggregation of predators and insect parasites and its effect on stability. *J. Anim. Ecol.* 43:567–94.

Hassell, M. P., and R. M. May. 1985. From individual behaviour to population dynamics. In R. Sibly and R. Smith, eds., *Behavioural Ecology,* pp. 3–32. Blackwell, Oxford.

Hassell, M. P., T.R.E. Southwood, and P. M. Reader. 1987. The dynamics of the viburnum whitefly (*Aleurotrachelus jelinekii*): A case study on population regulation. *J. Anim. Ecol.* 75:283–300.

Hethcote, H. W., and J. A. Yorke. 1984. *Gonorrhea Transmission Dynamics and Control.* Springer Lecture Notes in Biomathematics, No. 56. Springer-Verlag, New York.

Holling, C. S. 1959. The components of predation as revealed by a study of small mammal predation of the European Pine Sawfly. *Can. Entomol.* 91:293–320.

Hudson, P. J., and A. P. Dobson. 1988. The population dynamics of *Trichostronglylus tenuis* in red grouse. *Parasitology.* In press.

Hyman, J. M., and E. A. Stanley. 1988. A risk-based model for the spread of the AIDS virus. *Math. Biosci.* In press.

Jaques, R. P. 1977. Stability of entomopathogenic viruses. *Misc. Pub. Entomol. Soc. Amer.* 10:99–116.

Kareiva, P., and G. M. Odell. 1987. Swarms of predators exhibit "preytaxis" if individual predators use area restricted search. *Amer. Natur.* 130:233–70.

Kermack, W. O., and A. G. McKendrick. 1927. A contribution to the mathematical theory of epidemics. *Proc. Roy. Soc. Lond.* (A) 15:700–21.

Knox, E. G. 1980. Strategy for rubella vaccination. *Inst. J. Epidemiology* 9:13–23.

Lack, D. L. 1954. *The Natural Regulation of Animal Numbers.* Clarendon Press, Oxford.

Lauwerier, H. A. 1986. Two-dimensional iterative maps. In A. V. Holden, ed., *Chaos,* pp. 58–95. Princeton University Press, Princeton, N.J.

Levin, S. A., and D. Pimentel. 1981. Selection for intermediate rates of increase in parasite-host species. *Amer. Natur.* 117:308–15.

Li, T-Y, and J. A. Yorke. 1975. Period three implies chaos. *Amer. Math. Monthly* 82:985–92.

Macdonald, G. 1957. *Epidemiology and Control of Malaria.* Oxford University Press, Oxford.

May, R. M. 1974. Biological populations with nonoverlapping generations: Stable points, stable cycles, and chaos. *Science* 186:645–47.

May, R. M. 1976. Simple mathematical models with very complicated dynamics. *Nature* 261:459–67.

May, R. M. 1987. Chaos and the dynamics of biological populations. *Proc. Roy. Soc.* A413:27–44.

May, R. M., and R. M. Anderson. 1978. Regulation and stability of host-parasite population interactions: II, Destabilizing processes. *J. Anim. Ecol.* 47:249–67.

May, R. M., and R. M. Anderson. 1983. Epidemiology and genetics in the coevolution of parasites and hosts. *Proc. Roy. Soc. Lond.* 219:281–313.

May, R. M., and R. M. Anderson. 1987. Transmission dynamics of HIV infection. *Nature* 326:137–42.

May, R. M., M. P. Hassell, R. M. Anderson, and D. W. Tonkyn. 1981. Density dependence in host-parasitoid models. *J. Anim. Ecol.* 50:855–65.

May, R. M., and M. P. Hassell. 1988. Population dynamics and biological control. *Phil. Trans. Roy. Soc.* B318:129–69.

May, R. M., and G. F. Oster. 1976. Bifurcations and dynamic complexity in simple ecological models. *Amer. Natur.* 110:573–99.

Molineaux, L., and G. Gramiccia. 1980. *The Garki Project: Research on the Epidemiology and Control of Malaria in the Sudan Savanna of West Africa.* World Health Organization, Geneva.

Murdoch, W. W., J. D. Reeve, C. B. Huffaker, and C. E. Kennett. 1984. Biological control of olive scale and its relevance to ecological theory. *Amer. Natur.* 123:371–92.

Nicholson, A. J., and V. A. Bailey. 1935. The balance of animal populations, Part I. *Proc. Zool. Soc. Lond.* 1:551–98.

Nisbet, R. M., and W.S.C. Gurney. 1982. *Modelling Fluctuating Populations.* Wiley, New York.

Packard, N. H., J. P. Crutchfield, J. D. Farmer, and R. S. Shaw. 1980. Geometry from a time series. *Phys. Rev. Lett.* 45:712–16.

Petrides, G. A., and C. R. Bryant. 1951. An analysis of the 1949–50 fowl cholera epizootic in Texas panhandle waterfowl. *Trans. N. Amer. Wildlife Conf.* 16:193–216.

Reuter, O. M. 1913. *Lebensgewohnheiten und Instinkte der Inseckten bis zum Erwachen der Sozialen Instinkte.* Friedlander, Berlin.

Ross, R. 1911. *The Prevention of Malaria.* John Murray, London.

Sabelis, M. W., and J. van der Meer. 1986. Local dynamics of the interaction between predatory mites and two-spotted spider mites. In J.A.J. Metz and O. Diekmann, eds., *Dynamics of Physiologically Structured Populations,* pp. 322–44. Springer-Verlag, Berlin.

Schaffer, W. M. 1984. Stretching and folding in lynx fur returns: Evidence for a strange attractor in nature? *Amer. Natur.* 124:798–820.

Schaffer, W. M., and M. Kot. 1985. Nearly one-dimensional dynamics in an epidemic. *J. Theor. Biol.* 112:403–27.

Schaffer, W. M., and M. Kot. 1986. Differential systems in ecology and epidemiology. In A. V. Holden, ed., *Chaos,* pp. 158–178. Princeton University Press, Princeton, N.J.

Tabashnik, B. E. 1985. Computer simulation as a tool for pesticide resistance management. In *Pesticide Resistance: Strategies and Tactics for Management,* pp. 194–206. National Academy Press, Washington, D.C.

Tabashnik, B. E., and B. A. Croft. 1982. Managing pesticide resistance in crop-arthropod complexes: Interactions between biological and operational factors. *Environ. Entomol.* 11:1137–44.

Takens, F. 1981. Detecting strange attractors in turbulence. *Lecture Notes in Math.* 898:366–81.

Tinsley, T. W. 1979. The potential of insect pathogenic viruses as pesticidal agents. *Ann. Rev. Entomol.* 24:63–87.

Waage, J., and D. J. Greathead. 1988. Biological control: Concepts and opportunities. *Phil. Trans. Roy. Soc.* B318:111–28.

Wang, Y. H., and A. P. Gutierrez. 1980. An assessment of the use of stability analyses in population ecology. *J. Anim. Ecol.* 49:435–52.

Webber, R. H. 1975. Theoretical considerations in the vector control of filariasis. *Southeast Asian J. Trop. Med. Pub. Health* 4:544–48.

Chapter 23

Discussion: Ecology of Pests and Pathogens

ROY M. ANDERSON

This chapter records the discussion that took place at the conference on the general theme of the ecology of pests and pathogens. The aim is to survey current developments and, more importantly, interesting directions for future work both on fundamental aspects of the population biology of associations between hosts and pathogens, parasites or predators, and on applications of this work to the control of pests and diseases. I also make an attempt, where appropriate, to indicate areas of research that might benefit from adopting methods or ideas that have proved their worth in other contexts.

The chapter begins with a discussion of the population dynamics of pathogenic agents in human populations, and continues with a more general account of diseases among other animals and among plants. A brief discussion of the "epidemiology" of toxicants in the environment follows. The population biology and population genetics of pest populations are considered next, and this discussion leads to some broader thoughts about the dynamics and coevolutionary properties of "resource-consumer" systems (plant-herbivore, detritivores). The chapter concludes with a survey of some methodological issues and some general morals that might be drawn.

DISEASES IN HUMAN POPULATIONS

Current developments in the modeling of microparasitic infections of human populations are briefly surveyed by Hassell and May in chapter 22 (along with a definition of the term "microparasite"; see also Anderson and May 1985). To become reliable tools in the design of immunization programs or other public health measures, these models must take realistic account of age-dependent effects in transmission probabilities (a task which has begun), and of other heterogeneities—in geographical location, nutritional or social status, genetic makeup, and so on—that affect transmission and pathogenicity (a task which has received very little attention so far). That is, we need to tailor the epidemiological models to specific infections in specific countries or communities, building detail and realism upon a sound foundation of the understanding of transmission dynamics conferred by simpler models.

All this is easier said than done. Estimates of the vaccination coverage required to eradicate an infection such as measles can be sensitive to the assumptions made about the age-dependence in transmission rates. At present, the best way to estimate these rates is from serological studies of the proportion that is seropositive, age class by age class (Anderson and May 1985). But such data are not always available. There is need for closer coordination between data gatherers and modelers to ensure that epidemiological and other information is collected in such a way that it can be used to deduce the values of key epidemiological parameters, and thence to deduce the degree of control likely to be achieved by a given coverage level. In short, epidemiological models must be more closely engineered to the specifics of particular infections, but to accomplish this we must have appropriate data available.

Although an increasing number of epidemiological models deal with age-dependent effects, spatial complications, and other such things, most of the existing models assume all hosts are genetically identical. The interactions between individual hosts and pathogens are characterized by simple phenomenological rates of recovery or disease-associated mortality. But in fact these interactions between individual hosts and pathogenic organisms have a rich population biology of their own, as populations of viruses, bacteria, or protozoans contend with defending populations of cells (e.g., lymphocytes) that form part of the host's immune system. The dynamical behavior of these nonlinear interactive systems may vary from host to host, depending on genetic constitution and other factors. Many aspects of the overall epidemiology, including the acquisition of short-term or long-lasting immunity against subsequent infection, will benefit from a better understanding of the dynamics of the immune system. Some such work has been done on the way malaria may evoke an immune response, which fades in the absence of reinfections by mosquito bites; this work may explain age-dependent patterns in the prevalence of malaria in endemic regions and has important implications for control programs (Aron 1983). But there is need for more work both on fundamental aspects of the immune system as a dynamic collection of interacting cell and antibody populations, and on application of the emerging understanding to the design of public health programs in communities and to methods for the treatment of disease and the control of infection in individuals.

One conspicuous application of ecologically based epidemiological models is to the current HIV/AIDS epidemic (Anderson et al. 1986; May and Anderson 1987; Hyman and Stanley 1988; Dietz 1988). Building on earlier models for gonorrhea by Hethcote and Yorke (1984), which have been used by the Centers for Disease Control in designing control programs in the United States, such models are useful in many ways. First, they help define the data needed to make medium- or long-term predictions. Whether or not an epidemic can spread among a particular category of people depends on the ecologist's "basic reproductive rate," R_0, of the infection (which for microparasitic infections is essentially the average number of secondary infections produced by each infection in the early stages of the epidemic, before "density dependent" effects reduced R_0 below its intrinsic value). Thus, for example, the probability of a self-sustaining epidemic of HIV/AIDS by purely heterosexual chains of transmission depends on the value of R_0 for the infection among this population, which in turn requires specific information about transmission probabilities, duration of infectiousness, and the statistical distribution in rates of acquiring new sexual partners; mathematical models are playing an important part in giving focus to programs aimed at acquiring the appropriate data (NAS, 1988). Second, the models can, in some circumstances, be used to make indirect estimates of epidemiological parameters that are hard to determine directly. For example, transmission probabilities (among homosexual males, or among needle-sharing IV-drug users, etc.) are not easy to assess directly, largely because HIV infections are usually asymptomatic. But these transmission probabilities determine secular trends in seropositivity, which can be determined directly (or, less reliably, indirectly by deconvolution of secular trends in AIDS cases against the estimated distribution of incubation times); see May and Anderson 1987. In this way, individual-level parameters (transmission probabilities) may be estimated from population-level data (longitudinal serological studies).

Other lessons can be learned from ecologically based epidemiological models for HIV/AIDS. One might think that many epidemiological parameters could be estimated, along the lines just indicated, by constructing multiparameter epidemiological models of beguiling complexity, and then choosing that constellation of parameters that gave the best statistical fit to the available data (on seroprevalence, number of AIDS cases, and so on, over time). But a deeper understanding suggests that many of these secular trends depend (at least to the accuracy available in present data) on combinations of epidemiological parameters, rather than on the parameters one by one. For instance, new infections appear at a rate determined essentially by the probability of infecting any one partner times the average rate at which new partners are acquired, and so it is difficult to deduce more than the product of these two quantities from incidence data (if, indeed, one varied the two parameters separately and chose the pair that gave the best fit to the data as indicated by some statistical package, one would be likely to be fitting the noise in the data rather than to be seeing any real trend).

In much of Africa, HIV/AIDS appears to be spread by heterosexual transmission. There seems little doubt that the disease will have a significant impact on age profiles and on overall demographic patterns in sub-Saharan Africa. By combining ecological/demographic considerations with the above epidemiological considerations, the impact of HIV/AIDS—which is transmitted both horizontally and vertically to the offspring of infected mothers—can be estimated under various assumptions. In particular, preliminary studies suggest that HIV/AIDS may lead to little change in the fraction of the population below the age of 15 years, because the severe effects of AIDS causing adult mortality is largely counterbalanced by the change toward a less "pyramidal" age profile resulting from lowered rates of overall population growth (Anderson, May, and McLean 1988; Bongaarts 1988).

In brief, mathematical models for HIV/AIDS are proving valuable in helping to define the kinds of data that are needed to make projections, in estimating the likely effects of different kinds of behavioral changes, in providing indirect estimates of individual-level parameters from population-level data, and in helping to predict likely social changes and social needs (numbers of hospital beds; changes in "dependency ratios" in developing countries; and so on). As understanding grows, more complicated and more realistic models can be built on the foundation provided by the current generation of relatively simple ones.

For macroparasites (as defined in chapter 22 by Hassell and May; see also May and Anderson 1979), two avenues of research seem to be indicated, in order both to gain a better ecological understanding of host-macroparasite associations and to help design effective programs of control.

As summarized by Hassell and May, the distribution of most helminth parasite populations among human and other animal hosts are markedly nonuniform. As emphasized by Crofton, a "practical parasitologist" who made fundamental theoretical contributions, these aggregated or "clumped" distributions of macroparasites have implications both for the population biology of the association and for the design of effective strategies of intervention (Crofton 1971). But the fundamental causes of these clumped patterns remain poorly understood. Controversy remains as to whether the clumped patterns derive from environmental factors (affecting different hosts differently) or from individual hosts having different susceptibility (which in turn may derive from intrinsic genetic factors, or from environmental factors affecting exposure and nutrition, or from some combination of the two).

A second set of research problems concerns the extent to which individual hosts mount effective (short-term) immune responses to protozoan and helminth infections, along with the possibility that such immune responses may vary systematically from person to person owing to genetic, nutritional, or other sources of heterogeneity.

These two categories of research problems may well be entwined. If there are indeed intrinsically "wormy" people, who for immunological reasons (having to do with their genetic makeup, behavior, or geographical location) are predisposed to heavy infection, then clumped parasite distributions can be explained. Elucidation of these questions will require better understanding of the population dynamics of the immune system along the lines discussed above, laboratory studies of systems of "animal models" (usually mice) in the laboratory, and, ideally, also in the field, as well as careful studies of patterns of infection among humans in natural settings (in which ecologists and epidemiologists are accompanied by anthropologists who are sensitive to the subtleties of human behavior in relation to environmental factors).

Many of the research topics described above depend, directly or indirectly, on a better understanding of the spatial spread of infection.

Most existing studies of the spatial spread of microparasitic or macroparasitic infections among human populations use diffusion models (for large-scale phenomena) or chain binomial models (for small outbreaks). The propagation of infections of HIV/AIDS, for example, is not likely to be accurately characterized in so simple a fashion. It is possible that new approaches could come by borrowing from other areas of biology or from physics, for example, cellular automata (Wolfram 1984) or percolation theory (Stauffer 1985); whatever the source, some new approaches seem called for.

The application of ideas from epidemiology in conjunction with conventional demographic concepts was noted above, in relation to HIV/AIDS. More generally, infectious diseases continue to take a severe toll, particularly among children, in most developing countries, as can be seen from the dramatic difference between age-specific survivorship curves in typical developed and developing countries. Yet most epidemiological studies treat human populations as being constant, and focus on the dynamics of infection. While this may make sense in the short run, the ecological/demographic impact of infectious diseases in developing countries is considerable, and insofar as programs of immunization or vector control may currently be reducing mortality and morbidity, there will be concurrent demographic, and thus social and economic, changes. There is need for more studies that combine epidemiology with demography, in pursuit of a better understanding of the likely impacts of specific programs of intervention (McLean and Anderson 1988).

DISEASE IN NONHUMAN POPULATIONS

Several authors have recently reviewed evidence in support of the suggestion that viral, bacterial, protozoan, and helminth infections may affect the numerical magnitude or geographical distribution of many animal populations (Anderson and May 1978, 1979,; Toft 1986; Scott 1988; Allison et al. 1982). In particular,

Scott (1988) has reviewed field and laboratory studies showing the influence that infectious diseases can have on population dynamics. There is need, however, for more work that identifies and estimates the epidemiological and demographic parameters characterizing particular host-parasite associations; in many cases, knowledge of these parameters bears directly on management problems having to do with control of insect pests or vectors of disease or with the conservation of endangered species.

For example, Anderson and May (1981) have shown that if viral, bacterial, or protozoan pathogens are released at rates that exceed some threshold value, they may be capable of eradicating a target insect population; these authors have also suggested ways in which the threshold release rate may be estimated from studies of the natural history of the host-pathogen association. This theoretical work further indicates that the most lethal pathogens are not necessarily the best for sustained control, because such agents may often require high host densities for effective transmission and/or maintenance. Infectious agents of intermediate pathogenicity may often be more effective in controlling or eradicating target species (Anderson 1979). Interesting work has been done on the use of pathogens as control agents, for example, *Bacillis thuringensis* against gypsy moth. But most such use of pathogens as control agents has been by trial and error, with little or no attempt made to assess the epidemiological and demographic parameters that ultimately determine the dynamical behavior of the association. It would be interesting to see more collaborations between theorists and empiricists in this area, with the ultimate aim of selecting infectious agents for their overall ecological characteristics in relation to the target population, rather than trying to select the most pathogenic agent.

The problems of disease among small, and often unnaturally crowded, populations in zoos and reserves are familiar, arising from new diseases or from old diseases in new settings. Tourism, or even the presence of researchers, can be a factor in introducing or spreading such infections. These problems of disease in small populations are often exacerbated by the associated genetic impoverishment of the population. These problems underline the extent to which infectious diseases are agents of selection in natural populations, and they raise many fundamental questions about the ecological significance of genetic diversity and about the effects of population "bottlenecks" in reducing such diversity. These questions are likely to receive increasing attention in conservation contexts (Ralls and Ballou 1988).

In other conservation contexts, diseases can lead to situations where the biological issues are relatively clear, but where the social or political issues are complex. Thus in parts of southwestern Britain badgers are reservoirs for bovine tuberculosis, which affects the dairy cattle on whose pastures badgers forage at night. Anderson and Trewhella (1985) have elucidated the population biology of this host-parasite situation, and it seems that the existing policy is sound from a narrowly epidemiological standpoint, though other methods of controlling bovine TB are possible. Rabies in fox populations in Europe provides another example where epidemiological models, in conjunction with appropriate information about the natural history and demography of the host population, can give insights into the likely effects of different control strategies (Anderson et al. 1981). Studies of this kind can clarify policy options, provided the relevant epidemiological and demographic parameters can be assessed.

Other recent work has focused on the coevolution of hosts and their pathogens (Hamilton 1980, 1982; May and Anderson 1983; Bremmerman 1980; Allison et al. 1982; Ewald 1983). This work ranges from questions about the extent to which parasites evolve toward diminished virulence (and the possible implications for control programs, or for the future of newly invading pathogens), to fundamental questions about the role of pathogens in the evolution of sex (Hamilton 1980, 1982; Allison et al. 1982). This research is relatively recent, and the preliminary conclusions are that no simple generalizations can be made about the coevolutionary trajectories of host-pathogen associations. In principle, pathogens may evolve toward diminished virulence, toward intermediate virulence, or toward increasing virulence, depending on the way their life histories determine tradeoffs among virulence, transmissibility, and host resistence. If this is indeed the case, then important practical conclusions about the evolution of pathogens will require that we understand how their "natural history" affects these tradeoffs in each specific instance.

The population biology and ecology of plasmids are relevant to the issues just discussed, as well as being of interest in their own right (Levin et al. 1982; Levin and Lenski 1983). As shown by Levin and co-workers, the ecology and genetics of plasmid interactions can be illuminated by theoretical models in combination with laboratory experiments. Systematic pursuit of such inquiries seems likely to give insights that relate to a range of practical problems, including appropriate protocols for the release of genetically engineered microorganisms and the more efficient disposal of sewage.

TOXICOLOGY

The flow of nonliving toxicants of various kinds into the environment, and their effects on biota, shares some features with microparasites and macroparasites. The environmental effects of many such toxicants exhibit, for example, threshold phenomena. There may be scope for borrowing other ideas that have proved useful in epidemiological studies.

Ecological theory could shed light on problems having to do with efficient monitoring of toxicant flows (possibly by indicator species or keystone species).

Ecosystem concepts seem likely to find increasing application in helping to understand the effects of toxicants upon populations and community structure.

INSECT PESTS

The growing capacity and accessibility of computers, together with a growing understanding of the population dynamics of insect and other pests, hold the promise of better programs of integrated control. But it is important that this go hand in hand with wider appreciation of evolutionary concepts, with their implication that pest populations evolve in response to control strategies of all kinds.

First, as surveyed by Hassell and May in chapter 22, field and laboratory studies of natural enemies of insect pests (parasitoids in particular) are giving us a clearer idea of factors that govern their persistence and their efficacy in controlling target populations. But much remains to be done before integrated control becomes a reliably predictable science. Among other things, we need more theoretical and empirical work on the way the dynamical behavior of populations of insect predators are influenced by sex-ratio effects (themselves derived from considerations of local mate competition, or other behavioral ecological factors), by dispersal behavior and foraging patterns (which themselves may depend on spatial patterns of prey distribution), and by the details of complex life histories (particularly in tropical environments); for recent reviews, see Chesson and Murdoch (1986), May and Hassell (1988). The discussion in chapter 22 shows that there currently are many unresolved questions about how patterns of host distribution and aggregation relate to the dispersion and the attack patterns of their natural enemies; Hassell and May's chapter and the associated refeences sketch various future directions.

Multispecies complications have provoked much speculation over the years, with some ecologists arguing that the best control is likely to come from release of a mixture of natural enemies (who are left to sort out for themselves which is the best species or combination of species), while others have argued that the optimal course is systematically to determine the best natural enemy (in some defined sense) and then release only it. Although they are no substitute for manipulative experiments in field and laboratory, theoretical models are helpful in suggesting there is no single answer, and that different combinations of life history characteristics among the contending natural enemies can lead to different answers to the original question (Hassell and May 1985). One generalization does seem to emerge from recent theoretical and experimental work: generalist predators have dynamical behavior that tends often to be messier than that of specialists (Southwood, pers. comm.). This is, of course, partly because generalists by definition tend to be embedded in multispecies webs, which, for example, makes threshold phenomena less predictable.

The evolution of pesticide resistance is a growing problem for programs aimed at control of crop pests and disease vectors. While elucidation of the molecular mechanisms ultimately responsible for resistance is one piece of the puzzle, other pieces involve better understanding of the way population structure and genetics affect the evolution of resistance in heterogeneous habitats. Some directions for future research are sketched in a recent NAS (1986) study, where Via shows how current models for the population genetics of quantitative inheritence have implications for the evolution of pesticide resistance, Uyenoyama discusses how pleiotropic effects may affect resistance, and Tabashnik (using computer simulations) and May and Dobson (using analytic methods) explore how the population dynamics and movement patterns of insect pests and their natural enemies may affect rates of evolution of resistance to pesticides.

Harking back to the opening chapters of this book, much of the discussion upon which this chapter is based reiterated the need to relate population-level parameters to the physiology and/or behavior of individual organisms.

The small size and short life span of most insects mean that manipulative experiments can be performed and replicated in a way that is difficult, if not impossible, for many large species of vertebrates. Entomological chauvinists embrace this "comparative advantage" with understandable enthusiasm. As emphasized elsewhere in this book, however, we must be wary of uncritical generalizations; insights or techniques that are appropriate for some categories of organisms may not hold for others (particularly if their population biology is on different spatial and temporal scales).

OTHER RESOURCE-CONSUMER SYSTEMS

For all their complexity, the dynamics of host-pathogen, host-parasite, and host-parasitoid systems are in many ways simpler than those of other "resource-consumer" systems. Infected individuals produce new infections, and lepidopteran larvae produce either a new host or a new parasitoid, in ways that are more accurately captured in simple mathematical models than is the case, say, for zebras producing lions. But many of the basic ideas about nonlinear threshold phenomena, and about the coevolution of prey and predator, have or can be applied to other resource-consumer systems.

It was suggested that the population biology of bacterial, fungal, and other detritivores (which has received relatively little attention from most ecologists) shares some features with the better-studied kinds of systems discussed above. Plant-herbivore associations are, of course, the subject of much attention from behavioral and population ecologists, although even here systematic comparisons with parallel subdisciplines might be of interest. Such systematic comparisons of similarities and differences among a range of resource-consumer systems could lead to a better understanding of coevolution.

SOME METHODOLOGICAL ISSUES

There are a variety of new ideas and technological innovations that could conceivably have applications in the control of pests and pathogens.

As also mentioned in chapter 18, satellite data (for example, "landsat" images) can provide information about environmental conditions on broad scales, which can be used to assess likely distributions of insect and other populations. Such information, in conjunction with improving climatological models, can be a powerful tool for monitoring trends and making predictions. Rogers and Randolph (1986), for example, have shown that such satellite data lead to conclusions about the distribution of tsetse flies in Central Africa that accord with more conventional distributional studies, and they outline some possible future developments.

At a less grand level, Rainey (pers. comm.) has observed that recent developments in airborne systems for radar monitoring offer the hope of controlling outbreaks of locusts in Africa. The well-established system of international reporting of locusts, along with the known characteristics of locust behavior which make them amenable to such radar monitoring, means that it may be possible to keep track of their population densities and distributions, generation by generation, in a manner not previously possible. Such techniques, moreover, hold similar possibilities for other insect pest populations.

As time goes by, an understanding of the molecular basis of host-pathogen interactions deepens, and it seems likely that infectious agents may be found—or even designed—to provide "magic bullets" for deliberately eradicating species that are unwanted invaders. Deliberate or accidental introductions of rats, cats, goats, and other animals, particularly to previously isolated islands, have caused much destruction of native flora and fauna, and are responsible for the endangered status of many species. Possible collaborations with molecular biologists on the production of such viral, bacterial, protozoan, or helminth agents for use against target species must, of course, be accompanied by many precautions against unwanted effects on other organisms (Dobson 1988).

There is increasing recognition that we must understand the clumped or other patterns of spatial distribution of species if we are to understand their population dynamics; that is, the long-term dynamical behavior of many populations simply cannot be understood by dealing just with overall average densities. As we give more attention to such distribution patterns, there is an understandable tendency to look for some second parameter—perhaps the variance, perhaps a clumping parameter—that helps characterize the distribution. But many areas of science are being profoundly altered by recognition of the fractal geometry of much of nature (Mandelbrot 1977). One implication is that many actual distributions are not well characterized by intuitive ideas about smooth distributions (derived essentially from Gaussian and like distributions, which are well characterized by means and variances), but rather many real distributions are, as it were, jagged on every scale. As the implications of these ideas for the analysis of "fractal distributions" become better developed, they may well affect the way ecologists think about (and collect data on) the spatial distribution of populations.

As in many other situations, management decisions about the control of pests or pathogens must often be made in advance of a good understanding of the system that is to be managed. When this happens, we should recognize that management is itself usually an experiment. By deliberately using different management strategies in different subregions of the overall area being managed, we can learn from comparisons among different management regimes, and also hedge our bets (by not putting all the eggs in one management basket). In brief, such deliberate use of management as experiment offers many advantages and opportunities, which are too frequently neglected, in actual programs of intervention against crop pests, insect vectors, and pathogens.

SOME GENERAL CONCLUSIONS

As Hassell and May emphasize in chapter 22, for the markedly nonlinear circumstances that often characterize the dynamics of natural populations we must be wary of characterizing behavior by overall averages. For instance, the observed changes in seropositivity to HIV over time among groups of homosexual men in large cities are qualitatively different from those suggested by simple models that assume new homosexual partners are acquired at some average rate; to describe the observed trends, we must study the dynamics of transmission under the wide distribution in rates of acquiring new partners that are reported among homosexual men, and deal with overall average levels of seroprevalence only as the final step (May 1987). Molineaux and Gramiccia (1980) and Cohen (1979) have shown how the Garki project, aimed at eradicating malaria in a part of Central Africa, failed partly because of unforeseen heterogeneities in the resting behavior of mosquito vectors. When nonlinearities are important, the average value of some characteristic variable in a dynamical system can be very different from the dynamical behavior of the "average system." Such problems do not arise for linear dynamical systems, and it is unfortunate that so much of our intuition is formed by studies of such linear systems.

In general, an ecologist's perspective on infectious diseases and on insect pests provides an emphasis on population phenomena (thresholds and breakpoints, critical levels of vaccination coverage or release of natural enemies) and on evolutionary effects (evolution of pesticide resistance or resistance to antibiotics) that are lacking in some empirical programs of pest control or public health. On the other hand, ecological theory that is not tied to the natural history of the system under consideration is of doubtful value. Current work shows that mathematical models, in combination with appropriately collected data, give useful and testable insights; there would seem to be many opportunities for applying proven methods in new areas.

REFERENCES

Allison, A. C., et al. 1982. Coevolution between hosts and infectious disease agents, and its effects on virulence. In R. M. Anderson and R. M. May, eds., *Population Biology of Infectious Diseases,* pp. 245–68. Springer-Verlag, New York.

Anderson, R. M. 1979. Parasite pathogenicity and the depression of host population equilibria. *Nature* 279:150–52.

Anderson, R. M., and R. M. May. 1978. Regulation and stability of host-parasite population interactions: I, Regulatory processes. *J. Anim. Ecol.* 47:219–47.

Anderson, R. M., and R. M. May. 1979. Population biology of infectious diseases: Part I. *Nature* 280:361–67.

Anderson, R. M., and R. M. May. 1981. The population dynamics of microparasites and their invertebrate hosts. *Phil. Trans. Roy. Soc.* (B) 291:451–524.

Anderson, R. M., and R. L. May. 1985. Vaccination and herd immunity to infectious diseases. *Nature* 318:323–29.

Anderson, R. M., and R. M. Trewhella. 1985. Population dynamics of the badger (*Meles meles*) and the epidemiology of bovine tuberculosis (*Mycobacterium bovis*). *Phil. Trans. Roy. Soc.* (B) 310:327–81.

Anderson, R. M., R. M. May, and A. R. McLean. 1988. Possible demographic consequences of AIDS in developing countries. *Nature* 332:228–234.

Anderson, R. M., H. Jackson, R. M. May, and T. Smith. 1981. The population dynamics of fox rabies in Europe. *Nature* 289:765–71.

Anderson, R. M., G. F. Medley, R. M. May, and A. J. Johnson. 1986. A preliminary study of the transmission dynamics of the human immunodeficiency virus (HIV), the causative agent of AIDS. *IMA J. Math. Appl. Med. Biol.* 3:229–63.

Aron, J. L. 1983. Dynamics of acquired immunity boosted by exposure to infection. *Math. Biosci.* 64:249–59.

Bongaarts, J. 1988. Modeling the demographic impact of AIDS in Africa. In *AIDS 1988: AAAS Symposium Papers.* In press.

Bremermann, H. J. 1980. Sex and polymorphism as strategies in host-pathogen interactions. *J. Theor. Biol.* 87:671–702.

Chesson, P. L., and W. W. Murdoch. 1986. Aggregation of risk: Relationships among host-parasitoid models. *Amer. Natur.* 127:696–715.

Cohen, J. E. 1979. Longitudinal studies of malaria. In S. A. Levin, ed., *Some Mathematical Questions in Biology,* vol. 10. American Mathematical Society, Providence.

Crofton, H. D. 1971. A quantitative approach to parasitism. *Parasitology* 63:179–93.

Dietz, K. 1988. On the transmission dynamics of HIV. *Math Biosci.* In press.

Dobson, A. P. 1988. Restoring island ecosystems: Potential of parasites to control introduced mammals. *Conserv. Biol.* 2:31–39.

Ewald, P. W. 1983. Host-parasite relations, vectors, and the evolution of disease severity. *Ann. Rev. Ecol. Syst.* 14:465–85.

Hamilton, W. D. 1980. Sex versus non-sex versus parasite. *Oikos* 35:282–90.

Hamilton, W. D. 1982. Pathogens as causes of genetic diversity in their host populations. In R. M. Anderson and R. M. May, eds., *Population Biology of Infectious Disease Agents,* pp. 269–96. Springer-Verlag, New York.

Hassell, M. P., and R. M. May. 1985. From individual behaviour to population dynamics. In R. Sibly and R. Smith, eds., *Behavioural Ecology,* pp. 3–32. Blackwell, Oxford.

Hethcote, H. W., and J. A. Yorke. 1984. *Gonorrhea Transmission Dynamics and Control.* Springer Lecture Notes in Biomathematics, No. 56. Springer-Verlag, New York.

Hyman, J. M., and E. A. Stanley. 1988. A risk-based model for the spread of the AIDS virus. *Math. Biosci.* In press.

Levin, B. R., and R. E. Lenski. 1983. Coevolution in bacteria and their viruses and plasmids. In D. J. Futuyma and M. Slatkin, eds., *Coevolution,* pp. 99–127. Sinauer, Sunderland, Mass.

Levin, B. R., *Population Biology of Infectious Diseases,* et al. 1982. Evolution of parasites and hosts (group report). In R. M. Anderson and R. M. May, eds., pp. 212–43. Springer-Verlag, New York.

McLean, A. R., and R. M. Anderson. 1988. Measles in developing countries: i, Epidemiological parameters and patterns. *Epidem. Inf.* 100:111–33.

Mandelbrot, B. B. 1977. *The Fractal Geometry of Nature.* Freeman, New York.

May, R. M. 1987. Nonlinearities and complex behavior in simple ecological and epidemiological models. *Ann. N.Y. Acad. Sci.* 504:1–15.

May, R. M., and R. M. Anderson. 1979. Population biology of infectious diseases: ii. *Nature* 280:455–61.

May, R. M., and R. M. Anderson. 1983. Epidemiology and genetics in the coevolution of parasites and hosts. *Proc. Roy. Soc. Lond.* (B) 219:281–313.

May, R. M., and R. M. Anderson. 1987. Transmission dynamics of HIV infection. *Nature* 326:137–42.

May, R. M., and M. P. Hassell. 1988. Population dynamics and biological control. *Phil. Trans. Roy. Soc.* B318:129–69.

Molineaux, L., and G. Gramiccia. 1980. *The Garki Project: Research on the Epidemiology and Control of Malaria in the Sudan Savanna of West Africa.* World Health Organization, Geneva.

NAS. 1986. *Pesticide Resistance: Strategies and Tactics for Management.* National Academy of Sciences, Washington, D.C.

NAS. 1988. Report of the workshop on mathematical modeling of the spread of HIV and the demographic impact of AIDS. National Academy of Sciences, Washington, D.C.

Ralls, K., and J. Ballou. 1988. Estimates of lethal equivalents and the cost of inbreeding in mammals. *Conserv. Biol.* In press.

Rogers, D. J., and S. E. Randolph. 1986. Distribution and abundance of tsetse flies. *J. Anim. Ecol.* 55:1007–25.

Scott, M. E. 1988. The impact of infection and disease on animal populations: Implications for conservation biology. *Conserv. Biol.* 2:40–56.

Stauffer, D. 1985. *Introduction to Percolation Theory.* Taylor and Francis, London.

Toft, C. A. 1986. Communities of species with parasitic life-styles. In J. Diamond and T. J. Case, eds., *Community Ecology,* pp. 445–63. Harper and Row, New York.

Wolfram, S. 1984. Cellular automata as models of complexity. *Nature* 311:419–28.

Appendix

Ecological Theory: A List of Books

This Appendix gives a list of books that deal with one or other aspect of ecological theory. The list does not include works that are intended primarily as graduate or undergraduate texts, and it is undoubtedly less encyclopedic than it might be in other respects as well. We nevertheless hope that such a catalogue will prove useful, if only to demonstrate the existence of a canonical literature in ecological theory.

Anderson, R. M., ed. 1982. *Population Dynamics of Infectious Diseases: Theory and Applications.* Chapman & Hall, London and New York.

Anderson, R. M., and R. M. May, eds. 1982. *Population Biology of Infectious Diseases.* Dahlem Conferences. Life Sciences Research Report 25. Springer-Verlag, Heidelberg.

Bailey, N.J.T. 1975. *The Mathematical Theory of Infectious Diseases,* 2d ed. Macmillan, New York.

Bartlett, M. S. 1960. *Stochastic Population Models in Ecology and Epidemiology.* Methuen, London.

Bartlett, M. S., and R. W. Hiorns, eds. 1973. *The Mathematical Theory of the Dynamics of Biological Populations.* Academic Press, New York.

Beverton, R.J.H., and S. J. Holt. 1957. *On the Dynamics of Exploited Fish Populations.* Ministry of Agriculture; Fisheries and Food. London Fish. Invest. SER 2(19).

Charlesworth, B. 1980. *Evolution in Age-Structured Populations.* Cambridge Studies in Mathematical Biology. Cambridge University Press, Cambridge, England.

Charnov, E. L. 1982. *The Theory of Sex Allocation.* Monographs in Population Biology 18. Princeton University Press, Princeton, N.J.

Childress, S. 1981. *Mechanics of Swimming and Flying.* Cambridge Studies in Mathematical Biology 2. Cambridge University Press, New York.

Clark, C. W. 1976. *Mathematical Bioeconomics: The Optimal Management of Renewable Resources.* Wiley, New York.

Clark, C. W. 1985. *Bioeconomic Modelling and Fisheries Management.* Wiley Interscience, New York.

Cohen, J. E. 1978. *Food Webs and Niche Space.* Monographs in Population Biology 11. Princeton University Press, Princeton, N.J.

Crow, J. F., and M. Kimura. 1970. *An Introduction to Population Genetics Theory.* Harper and Row, New York.

Diamond, J., and T. J. Case, eds. 1986. *Community Ecology.* Harper and Row, New York.

Diamond, J. M., and M. L. Cody, eds. 1975. *Ecology and Evolution of Communities.* Harvard University Press, Cambridge, Mass.

Ewens, W. 1969. *Population Genetics.* Methuen, London.

Freedman, H. I. 1980. *Deterministic Mathematical Models in Population Ecology.* Marcel Dekker, New York.

Gates, D. M. 1980. *Biophysical Ecology.* Springer-Verlag, New York.

Gause, G. F. 1934. *The Struggle for Existence.* Williams and Wilkins, Baltimore (reprinted 1971 by Dover, New York).

Goel, N. S., and N. Richter-Dyn. 1974. *Stochastic Models in Biology.* Academic Press, New York.

Goel, N. S., S. C. Maitra, and E. W. Montroll. 1971. *On the Volterra and Other Nonlinear Models of Interacting Populations.* Academic Press, New York.

Hallam, T. G., and S. A. Levin, eds. 1986. *Mathematical Ecology, An Introduction.* Biomathematics Vol. 17. Springer-Verlag, Berlin.

Hassell, M. P. 1978. *The Dynamics of Arthropod Predator-Prey Systems.* Monographs in Population Biology 13. Princeton University Press, Princeton, N.J.

Hethcote, H. W., and J. A. Yorke. 1984. *Gonorrhea Transmission Dynamics and Control.* Lecture Notes in Biomathematics 56. Springer-Verlag, Heidelberg.

Holling, C. S., ed. 1978. *Adaptive Environmental Assessment and Management.* IIASA International Series on Applied Systems Analysis. Wiley, New York.

Hoppensteadt, F. 1975. *Mathematical Theories of Populations: Demographics, Genetics, and Epidemics.* SIAM Reg. Conf. Series 20. Society for Industrial and Applied Mathematics, Philadelphia.

Hutchinson, G. E. 1978. *An Introduction to Population Ecology.* Yale University Press, New Haven.

Keyfitz, N. 1968. *Introduction to the Mathematics of Population.* (Revised 1977.) Addison-Wesley, Reading, Mass.

Kingsland, S. E. 1985. *Modeling Nature.* University of Chicago Press, Chicago.

Kostitzin, V. A. 1937. *Biologie Mathématique.* Librarie Armand, Colin, Paris.

Levin, S. A. (ed). 1978. *Studies in Mathematical Biology.* Part II: Populations and

Communities. Studies in Mathematics, vol. 1. Mathematical Association of America, Providence.

Levin, S. A., T. G. Hallam, and L. J. Gross, eds. 1988. *Applied Mathematical Ecology.* Biomathematics 19. Springer-Verlag, Heidelberg. In press.

Lewis, E. 1977. *Network Models in Population Biology.* Biomathematics 7. Springer-Verlag, Heidelberg.

Lighthill, Sir J. 1975. Mathematical Biofluiddynamics. Regional Conf. Series in Applied Mathematics 17, SIAM, Soc. for Industrial & Applied Mathematics, Phiadelphia.

Lotka, A. J. 1956. *Elements of Mathematical Biology.* Dover, New York (originally published 1925 as *Elements of Physical Biology,* Williams and Wilkins, Baltimore).

Ludwig, D. 1974. *Stochastic Population Theories.* Lecture Notes in Biomathematics, vol. 3. Springer-Verlag, Heidelberg.

MacArthur, R. H. 1972. *Geographical Ecology.* Harper and Row, New York. (Princeton University Press reprint, 1984.)

MacArthur, R. H., and E. O. Wilson. 1967. *The Theory of Island Biogeography.* Monographs in Population Biology I. Princeton University Press, Princeton, N.J.

Mangel, M. 1985. *Decision and Control in Uncertain Resource Systems.* Academic Press, New York.

May, R. M., ed. 1984. *Exploitation of Marine Communities.* Dahlem Conferences. Life Sciences Research Report 32. Springer-Verlag, Heidelberg.

May, R. M. 1974. *Stability and Complexity in Model Ecosystems,* 2d. ed. Monographs in Population Biology 6. Princeton University Press, Princeton, N.J.

May, R. M. (ed). 1976. *Theoretical Ecology: Principles and Applications.* Blackwell Scientific, Oxford.

Maynard Smith, J. 1968. *Mathematical Ideas in Biology.* Cambridge University Press, Cambridge, England.

Maynard Smith, J. 1982. *Evolution and the Theory of Games.* Cambridge University Press, Cambridge, England.

Metz, J.A.J. and O. Diekmann. 1986. *The Dynamics of Physiologically Structured Populations.* Lecture Notes in Biomathematics. Springer-Verlag, Heidelberg.

Michod, R. E., and B. R. Levin, eds. 1988. *The Evolution of Sex.* Sinauer, Sunderland, Mass.

Murray, J. D. 1988. *Mathematical Ecology and Biology.* Biomathematics 18. Springer-Verlag, Heidelberg.

Nisbet, R. M., and W.S.C. Gurney. 1982. *Modelling Fluctuating Populations.* Wiley, Chichester, England.

Nobel, P. 1983. *Biophysical Plant Physiology and Ecology.* Freeman, San Francisco.

Okubo, A. 1980. *Diffusion and Ecological Problems: Mathematical Models.* Biomathematics 10. Springer-Verlag, Heidelberg.

O'Neill, R. V., D. L. DeAngelis, J. B. Waide, and T.F.H. Allen. *A Hierarchical Concept of Ecosystems*. Monographs in Population Biology 23. Princeton University Press, Princeton, N.J.

Oster, G. F., and E. O. Wilson. 1978. *Caste and Ecology in the Social Insects*. Monographs in Population Biology 12. Princeton University Press, Princeton, N.J.

Pickett, S.T.A., and P. S. White. 1985. *The Ecology of Natural Disturbance and Patch Dynamics*. Academic Press, Orlando, Fla.

Pielou, E. C. 1977. *Mathematical Ecology*, 2d ed. Wiley, New York.

Pimm, S. L. 1982. *Food Webs*. Population and Community Biology Series. Chapman and Hall, New York.

Provine, W. 1971. *The Origins of Theoretical Population Genetics*. University of Chicago Press, Chicago.

Roughgarden, J. 1979. *Theory of Population Genetics and Ecology: An Introduction*. Macmillan, New York.

Scudo, F., and J. Ziegler. 1978. *The Golden Age of Theoretical Ecology*. Lecture Notes in Biomathematics. Springer-Verlag, Heidelberg.

Smith, D. P., and N. Keyfitz. 1978. *Mathematical Demography*. Biomathematics 6. Springer-Verlag, Heidelberg.

Steele, J. H., ed. 1978. *Spatial Pattern in Plankton Communities*. NATO Conference Series IV: Marine Sciences, vol. 3. Plenum Press, New York.

Steele, J. H. 1975. *The Structure of Marine Ecosystems*. Harvard University Press, Cambridge, Mass.

Thornley, J.H.M. 1976. *Mathematical Models in Plant Physiology*. Academic Press, New York.

Tilman, D. 1982. *Resource Competition and Community Structure*. Monographs in Population Biology 17. Princeton University Press, Princeton, N.J.

Vogel, S. 1983. *Life in Moving Fluids—The Physical Biology of Flow*. Princeton University Press, Princeton, N.J.

Volterra, V. 1931. *Leçons sur la théorie mathématique de la lutte pour la vie*. Gauthier-Villars, Paris.

Whittaker, R. H. 1970. *Communities and Ecosystems*. (2d. ed., 1975). Macmillan, New York.

Whittaker, R. H., and S. A. Levin. 1975. *Niche: Theory and Application*. Benchmark Papers in Ecology 3. Dowden, Hutchinson, and Ross, Stroudsburg, Penn.

Williams, G. C. 1975. *Sex and Evolution*. Monographs in Population Biology 8. Princeton University Press, Princeton, N.J.

Wilson, D. S. 1978. *The Natural Selection of Populations and Communities*. Benjamin/Cummings, Menlo Park, Calif.

Yodzis, P. 1976. *The Effects of Harvesting on Competitive Systems*. Lecture Notes in Biomathematics 56. Springer-Verlag, Heidelberg.

List of Participants

Anderson, Roy M., Imperial College of Science & Technology, Department of Pure and Applied Biology, Prince Consort Road, London SW7 2BB, UK.

Averner, Maurice, Biospherics Program, Code ECR, NASA Headquarters, Washington, D.C. 20546.

Caswell, Hal, Biology Department, Woods Hole Oceanographic Institution, Woods Hole, MA 01543.

Cherfas, Jeremy, *New Scientist,* Commonwealth House, 1-19 New Oxford St., London WC1A 1NG, UK.

Chesson, Peter, Department of Zoology, Ohio State University, 1735 Neil Ave., Columbus, OH 43210.

Clark, Colin, W. Department of Mathematics, University of British Columbia, 121-1984 Mathematics Road, Vancouver, B.C., V6T 1Y4, Canada.

Cody, Martin L., Department of Biology, University of California–Los Angeles, Los Angeles, CA 90024.

Cohen, Joel E., Department of Population Studies, Rockefeller University, Box 20, 1230 York Ave., New York, NY 10021.

Dorigan, Janet, Ecological Research Division, ER-75, Office of Health & Environmental Research, Office of Energy Research, Department of Energy, Washington, D.C. 20545.

Ehrlich, Paul R., Department of Biological Sciences, Stanford University, Stanford, CA 94305.

Falco, James, Office of Research & Development, RD 682, Environmental Protection Agency, Washington, D.C. 20460.

Feldman, Marcus W., Department of Biological Sciences, Stanford University, Stanford, CA 94305.

Flanagan, Patrick, General Ecology Program, National Science Foundation, 1800 G. Street N.W., Washington, D.C. 20550.

Forman, Richard T. T., Graduate School of Design, Harvard University, Cambridge, MA 02138.

Gilpin, Michael E., Department of Biology, University of California–San Diego, La Jolla, CA 92037.

Gross, Louis, Department of Mathematics, University of Tennessee, 121 Ayres Hall, Knoxville, TN 37916.

Harris, Frank, Associate Director, Biotic Systems and Resources, National Science Foundation, 1800 G. Street N.W., Washington, D.C. 20550.

Hassell, Michael P., Imperial College Field Station, Silwood Park, Department of Pure and Applied Biology, Ascot, Berks S15 7PY, UK.

Hubbell, Stephen P., Department of Biology, Princeton University, Princeton, NJ 08544

Kareiva, Peter, Department of Biology, NJ 15, University of Washington, Seattle, WA 98195.

Koehl, M.A.R., Department of Zoology, University of California–Berkeley, Berkeley, CA 94720.

Levin, Simon A., Section of Ecology and Systematics, 347 Corson Hall, Cornell University, Ithaca, NY 14853.

Lewin, Roger, Deputy Editor, Research News, *Science,* 1333 H Street N.W., Washington, D.C. 20005.

May, Robert M., Department of Biology, Princeton University, Princeton, NJ 08544.

McElroy, Robert, CELSS Program, NASA-AIMS, Moffat Field, California.

Mooney, Harold A., Department of Biological Sciences, Stanford University, Stanford, CA 94305.

Mueller, Laurence D., Department of Zoology, Washington State University, Pullman WA, 99164.

O'Neill, Robert V., Environmental Sciences Division, Oak Ridge National Laboratory Bldg. 1505, Oak Ridge, TN 37830.

Pacala, Stephen W., Department of Ecology and Evolutionary Biology, University of Connecticut, Storrs, CT 06268.

Paine, Robert T., Department of Zoology, University of Washington, Seattle, WA 98195.

Pimm, Stuart L., Department of Zoology, University of Tennessee, Knoxville, TN 37916.

Powell, Thomas M., Division of Environmental Studies, University of California–Davis, Davis, CA 95616.

Pulliam, H. Ronald, Department of Zoology, University of Georgia, Athens, GA 30602.

Real, Leslie, Department of Zoology, North Carolina State University, Raleigh, NC 27650.

Rosenzweig, Michael, Department of Ecology and Evolutionary Biology, University of Arizona, Tucson, AZ 85721.

Roughgarden, Jonathan, Department of Biological Sciences, Stanford University, Stanford, CA 94305.

Schlesinger, William H., Department of Botany, Duke University, Durham, N.C. 27706.

Simberloff, Daniel, Department of Biological Sciences, Florida State University, Tallahassee, FL 32306.

Sousa, Wayne, Department of Zoology, University of California–Berkeley, Berkeley, CA 94720.

Southwood, Richard, Department of Zoology, Oxford University, South Parks Rd., Oxford OX1 3PS, UK.

Stanley, Steven M., Department of Earth and Planetary Science, The Johns Hopkins University, Baltimore, MD 21218.

Steele, John H. Woods Hole Oceanographic Institution, Woods Hole, MA 02543.

Sugden, Andrew, *Trends in Ecology and Evolution,* Elsevier Publications, 68 Hills Road, Cambridge CB2 1LA, UK.

Sugihara, George, Scripps Institute of Oceanography, University of California–San Diego, La Jolla, CA 92093.

Taylor, Phillip, Biological Oceanography Program, Ocean Sciences Division, National Science Foundation, 1800 G Street N.W., Washington, D. C. 20550.

Tilman, David, Department of Ecology and Behavioral Biology, University of Minnesota, 318 Church St. S.E., Minneapolis, MN 55455.

Travis, Joseph, Department of Biological Sciences, Florida State University, Tallahassee, FL 32306.

Urban, Dean L., Department of Environmental Sciences, Clark Hall, University of Virginia, Charlottesville, VA 22903.

Wiegert, Richard, Department of Zoology, University of Georgia, Athens, GA 30602.

Wilson, Edward O., Museum of Comparative Zoology, Harvard University, Cambridge, MA 02138.

Author Index

Boldface numbers are pages on which references appear.

Subject Index